ORGANIC CHEMISTRY

The Made Simple series
has been created
primarily for self-education
but can equally well
be used as
an aid to group study.
However complex the subject,
the reader is taken
step by step,
clearly and methodically,
through the course. Each volume
has been prepared by
experts,
using throughout the
Made Simple technique of teaching.
Consequently the gaining
of knowledge now becomes
an experience to be enjoyed.

Accounting	Export
Acting and Stagecraft	Financial Management
Additional Mathematics	French
Advertising	Geology
Anthropology	German
Applied Economics	Human Anatomy
Applied Mathematics	Italian
Applied Mechanics	Journalism
Art Appreciation	Latin
Art of Speaking	Law
Art of Writing	Management
Biology	Marketing
Book-keeping	Mathematics
British Constitution	Modern Electronics
Business and Administrative	Modern European History
Organisation	New Mathematics
Calculus	Office Practice
Chemistry	Organic Chemistry
Childcare	Philosophy
Commerce	Photography
Company Administration	Physical Geography
Company Law	Physics
Computer Programming	Pottery
Cookery	Psychology
Cost and Management	Rapid Reading
Accounting	Retailing
Data Processing	Russian
Dressmaking	Salesmanship
Economic History	Secretarial Practice
Economic and Social	Social Services
Geography	Soft Furnishing
Economics	Spanish
Electricity	Statistics
Electronic Computers	Transport and Distribution
Electronics	Typing
English	Woodwork
English Literature	

ORGANIC CHEMISTRY
Made Simple

S. K. Murthy, M.A.
S. S. Nathan, Ph.D.

Advisory editor
John B. Holden
B.Sc., Ph.D.

Made Simple Books
W. H. ALLEN London
A Howard & Wyndham Company

© 1968 edition by W. H. Allen & Company Ltd.

Made and printed in Great Britain
by Richard Clay (The Chaucer Press), Ltd., Bungay, Suffolk
for the publishers W. H. Allen & Company
44 Hill Street, London, W1X 8LB

First edition April, 1968
Reprinted February, 1972
Reprinted April, 1974
Reprinted September, 1975
Reprinted March, 1979

ISBN 0 491 01900 9 Paperback

Foreword

Organic chemistry is the study of the compounds of carbon excluding the carbonates, metallic carbides and carbonyls.

Our bodies, the food we eat, the clothes we wear and many of the materials we use are composed of or are combined with compounds of carbon. Before the middle of the last century, organic chemical compounds were thought to originate exclusively from plants and animals. Living systems were thought to invest a 'vital force' in such compounds making them impossible to synthesise in chemical laboratories. When it was found that compounds identical to those from living systems could be made from inanimate materials it no longer made sense to classify compounds according to their origin. As nearly all the chemical compounds produced by plants and animals contained carbon the term 'organic' was reserved for carbon-containing compounds.

This book has been written for persons with some knowledge of general or inorganic chemistry who wish to obtain an understanding of organic chemistry. Whilst efforts have been made to make the book self-sufficient, reference to the companion volume *Chemistry Made Simple* will give additional information, particularly concerning the general principles of chemical bonding and the properties of inorganic compounds and radicals.

Although written for the enthusiastic amateur working on his own, this book more than covers the syllabus for the G.C.E. Advanced Level Chemistry course. It could serve as an organic chemistry textbook or companion reader for students studying for a Teacher's Certificate, Higher National Certificate or Advanced Chemical Technician's Certificate.

A systematic approach to the subject is adopted. The compounds are classified according to the functional groups present. A non-mathematical approach is applied to the modern theories of chemical structure and bonding.

Each chapter contains a summary and most conclude with a set of problems which the reader is urged to solve before moving on to the next chapter.

Organic chemistry is a practical subject. Unfortunately a complete practical course requires a well equipped and supervised laboratory. Many organic compounds are flammable or explosive so must be handled with care. However, some experiments for the student working on his own have been included.

JOHN B. HOLDEN

Table of Contents

Organic Chemistry Made Simple

*PART IV—HETEROCYCLIC COMPOUNDS, PHYSIOLOGICALLY
ACTIVE COMPOUNDS, AND POLYMERS*

PART I

INTRODUCTORY MATERIAL

CHAPTER ONE

INTRODUCTION

SCOPE OF THE SUBJECT

For a long time it has been known that substances obtained from the animate world—the plant and animal kingdoms—like sugar, alcohol, oils, etc., have certain characteristic properties that distinguish them from the substances obtained from the inanimate or the mineral world. From a historical point of view the domain of organic chemistry was a study of the chemistry of products obtained from plants and animals having organized cells. It was believed that for the elaboration of organic substances a vital force, which was presumed to be present in plants and animals, was necessary, and that man could not prepare them in the laboratory. This belief in the vitalistic theory was shaken by the German chemist WOHLER, who, in 1828, prepared **urea**, a typical organic substance excreted by mammals and birds, from **ammonium cyanate**, a typical inorganic substance. Wohler accomplished this revolutionary transformation by a remarkably simple method—evaporating an aqueous solution of **ammonium cyanate**—and certainly no special power was invoked. Wohler's experiment was repeated several times before it was accepted. With Wohler's synthesis of urea, the original distinction between organic chemistry and inorganic chemistry vanished. With a growing number of preparations of other organic substances in the laboratory, it became increasingly evident that organic substances are subject to the same chemical laws as inorganic substances, and the original connotation of organic chemistry lost significance.

Nevertheless, for the sake of convenience, we still use the term organic chemistry to designate the study of **carbon compounds**. There are about a hundred known elements, and the question naturally arises as to why the element carbon should be assigned such a unique place of honour—arrogating to itself a separate branch of the subject, while all the other elements and their compounds are put together to constitute the other branch called **inorganic chemistry**. Is it necessary to have one division of chemistry for each element? The mere thought would be appalling to the student. If not, why should carbon alone be singled out for this unique honour? There are justifications for the preferential treatment accorded to carbon. Carbon forms an extremely large number of compounds, a phenomenon unparalleled by any other element. About 700,000 carbon compounds are currently known, and this figure is ever increasing, about 30,000 new compounds being made every year. The reason for the astoundingly prolific nature of carbon is that the carbon atom possesses the ability, to a marked degree, to form

3

carbon chains of varying lengths and rings of different sizes. Further, the compounds of carbon, notwithstanding their large number, possess certain distinctive features as a class.

CHARACTERISTICS OF ORGANIC SUBSTANCES

Most organic substances are extremely sensitive to heat and melt or decompose below 300° C.—a temperature which leaves an inorganic substance like **sodium chloride** unaffected. Generally speaking, organic substances are not soluble in water, but are soluble in organic solvents like **ether, alcohol, chloroform**, etc. One other property of organic compounds is their **non-ionic** nature. This property arises from the position of carbon in the periodic table (see Chapter Two). It is half-way between the **electropositive alkali metals** and the **electronegative halogens**. Because of this neutral character of carbon, its compounds display a marked reluctance to ionize, that is, to split into electrically charged particles in solution. Organic compounds are **covalent**; they seldom ionize in solution, and organic reactions are as a rule slow. We shall go into the exact significance and meaning of the above statement in the next chapter.

ORGANIC CHEMISTRY AND MODERN LIVING

Having defined the scope of the subject, we shall try to assess the nature and magnitude of the impact of the development of organic chemistry on modern civilization and, in particular, its role in raising the standard of living. Such slogans as 'Better things for better living through chemistry' (du Pont) and 'Chemistry—the key to better living' (American Chemical Society) certainly can be taken to refer to organic chemistry and spotlight the practical importance of the subject. If we prepared a catalogue of substances used by a civilized nation, we would be impressed by the overwhelming preponderance of organic substances on the list. If we think of the prime necessities of man—food, clothing, and shelter—we encounter such organic substances as fats, proteins, starch, sugar, vitamins, cellulose, synthetic fibres, wood, etc. If we think of luxuries or other accessories we are intimately bound up with perfumes, cosmetics, etc. In sickness, sulpha drugs, aspirin, quinine, and so on come to our mind. So we find that in health and in sickness, from absolute necessities through all the gradations to absolute luxuries, we utilize organic substances. If we followed a man through his daily life from morning to night we would be surprised to note how often he uses organic substances or a mixture of them. It has been estimated that organic compounds are involved in 90% of the value of all the products of man's labour. Then it is no wonder that the standard of living parallels the development of organic chemistry.

If we take a simple item such as petrol, we can visualize what a revolution it has made in our way of life. It has enabled a person to live, work, and enjoy himself anywhere. Synthetic chemicals have been manufactured to meet the ever-increasing demands for specific purposes. Such products as drugs, foods, vitamins, plastics, and synthetic textiles

—Nylon, Orlon, and Dacron—come to mind. The synthetic organic chemist tries to emulate and improve on nature. The story of synthetic drugs, dyes, plastics, synthetic fibres, perfumes, cosmetics, fertilizers, etc., is an illuminating saga of human endeavour—to make life better, more comfortable, and perhaps luxurious. In the great adventure of life, organic chemistry does indeed play a vital role. One may hazard the opinion that the development of organic chemistry in a nation is an index of its progress.

Every student of history knows about the three revolutions that mark the eighteenth century: the French Revolution, the American Revolution, and the Industrial Revolution. Contemporaneously with them, we may trace a chemical revolution. About the time when Marie Antoinette was being guillotined, while the Declaration of Independence was being drafted, and steam power was first applied to weaving, LAVOISIER was bringing about the lesser-known chemical revolution. He, as a pioneer, emphasized the necessity for quantitative work in chemistry and the importance of understanding the composition of substances by means of elementary analysis. The dim beginnings of organic chemistry are discernible here. With the advent of DALTON's Atomic Theory and AVOGADRO's Hypothesis, a correct idea of the atomic weights of elements came into use, and the infant science of organic chemistry was ready to embark on a phase of vigorous growth. The subject is just about 160 years old, and in this comparatively short period it has developed amazingly. The growth has been phenomenal, especially during periods of world tensions. For example, during the First World War, when there was a shortage of camphor, the organic chemist stepped in and produced synthetic camphor. Again in the Second World War, when the supply of rubber was cut off, the national requirement was supplied by several varieties of synthetic rubber. Supplies of aviation spirit and other special types of fuels were maintained by the application of organic chemistry.

PRACTICAL WORK

Like other branches of chemistry, organic chemistry is a practical subject: all the theories are based on, and must be consistent with, experimental observations. Ideally the student of organic chemistry should pursue a systematic practical course simultaneously with his reading, to acquire appropriate skills and familiarity with the subject. But, because many organic substances are volatile, highly inflammable liquids, many more have toxic properties, and a fully representative set of apparatus and chemicals is expensive, such a course should only be taken under adequate supervision, in a specially equipped laboratory with several types of fire extinguisher and a full first-aid kit. Some laboratory manuals suitable for such a course are listed in Appendix C.

However, to ensure that the isolated reader gains some experience of practical techniques and becomes familiar with the properties and reactions of some classes of organic compounds, we have described in Appendix A a few experiments which the reader can carry out for himself.

One who is about to begin the study of organic chemistry is at the threshold of a fascinating and colourful new world. There will not be, and in fact should not be, a dull moment in the study of the subject. It is really 'organic' and the developments are dramatic and spectacular.

SUMMARY

The scope of organic chemistry; its definition as the chemistry of carbon compounds. The formation of an extraordinarily large number of carbon compounds is essentially due to the capacity of the carbon atoms to combine with carbon to form chains and rings. Distinctive properties of organic compounds. The impact of organic chemistry on modern life and its role in improving the standard of living. Its usefulness in times of peace and war.

CHAPTER TWO

THE ARCHITECTURE OF ATOMS AND MOLECULES

The chemistry of organic compounds is intimately linked with their structures. For an understanding of these structures, it is necessary to become familiar with such terms as **atoms, molecules, ions, valences,** etc.

CONCEPT OF ATOMS AND MOLECULES

The word **element** is quite well known. We use it to describe a simple substance that is incapable of being split further by chemical means. Thus **hydrogen, oxygen,** and **carbon** are all elements, since they cannot be split up into anything simpler. Elements combine to form substances called **compounds** which have certain features distinguishing them from pure elements, on the one hand, and mixtures, on the other. Firstly, all pure samples of the same chemical compound contain the same elements in the same proportions by weight. Secondly, when elements combine to form a compound the process is accompanied by an energy change which usually shows up as a temperature change: to split the compound up into its constituents again the energy change must be reversed. Thirdly, the properties of a compound are not those of the elements from which it was made. For example, carbon dioxide and water are both compounds: their compositions never vary; they can be formed by burning charcoal in oxygen and exploding hydrogen with oxygen, and in both cases a lot of energy is set free as heat, light, and sound—energy which must be replaced to split the compounds up into their constituent elements again. The properties of the colourless gas carbon dioxide are not at all like those of oxygen and the black solid charcoal, nor are those of the liquid water like gaseous hydrogen and oxygen. Carbon dioxide can be broken down by burning magnesium in it; to split up water electrical energy is used (electrolysis).

The smallest particle of an element which can take part in a chemical reaction is called an **atom**. In a similar way we can picture a compound subdivided so that we are left with the smallest possible piece which can have the properties of the compound. This will be a small group of atoms with at least one from each constituent element present, and it is called a **molecule**. A molecule of carbon dioxide contains two atoms of oxygen and one of carbon; a molecule of water contains two atoms of hydrogen and one of oxygen. In either case further subdivision is impossible without breaking up the compound.

The question immediately arises as to how the several atoms constituting a particular molecule are arranged and what force holds the atoms together and makes a particular molecule preserve its identity. It is not as though a molecule were a random mixture of atoms thrown

7

together in a haphazard manner without any rhyme or reason. Every molecule has a well-defined architecture, and an understanding of the structure of molecules is fundamental to organic chemistry. The organic chemist is an engineer who studies the buildings which he calls molecules, tries to grasp the structural pattern of different molecules, and with this knowledge constructs molecules made to order—the synthetic compounds.

Only a poor engineer would begin the construction of a building without fully understanding the nature of the raw materials he uses and how the several materials interact to give the finished product. Similarly, no organic chemist can make any headway unless he thoroughly understands the raw materials he uses in the construction of molecules—namely the atoms of carbon, hydrogen, oxygen, nitrogen, etc. He must have an intimate knowledge of the behaviour of his raw materials under different conditions, the rules governing their combination, and the particular preferences or antipathies of one atom to another. In short, he must know them intimately.

STRUCTURE OF ATOMS

DALTON regarded the atom as the smallest indivisible particle of an element. The Daltonian concept of the atom has undergone a revision in the light of modern research. The study of the discharge of electricity through rarefied gases has led to the recognition of the **electron** as a fundamental constituent of all atoms. The electron has negligible mass (about $\frac{1}{1850}$ of the hydrogen atom) and carries a unit negative charge. Since matter as a whole is electrically neutral, it follows that in order to compensate the negative charge due to the electron, the atom must also contain some positively charged particles. RUTHERFORD'S classical experiments on the bombardment of atoms by projectiles (like α-particles and neutrons) led to the discovery of the **proton** as another constituent of all atoms. The proton has unit mass and carries a single positive electric charge. Later experiments revealed the presence of another particle in the atom called the **neutron**, which has no charge and has a mass of one. Electrons, protons, and neutrons are the bricks with which the several kinds of atoms are built. The atom has now come to be regarded as mostly empty space. Most of the mass of an atom is concentrated in a minute nucleus which is positively charged. The nucleus is made of a tight packing of neutrons and protons. A sufficient number of electrons necessary to compensate the positive charges on the nucleus occupy the space round the nucleus in some way.

It is realized nowadays that electrons do not behave entirely like particles; in their motion about atomic nuclei they show wavelike properties, too. One consequence is that we cannot know exact details of the position and motion of an electron: instead we assess the probability of finding an electron with a particular amount of energy at various places round the nucleus. Another is that electrons moving round a nucleus can only possess certain fixed amounts of energy. These ideas will be extended in Chapter Eleven, and for the present it is sufficient to know that the more energy an electron has, the bigger is the volume in which it can be found 90% of the time. Consequently an atom can be pictured as being a positive nucleus surrounded by electrons occupying concentric shells, the larger the electron's energy, the larger the shell. Each shell has a definite capacity for electrons. The first shell can accommodate two electrons, the second shell eight electrons, the third shell eighteen electrons, and the fourth shell thirty-two electrons.

With the above picture in mind, we can visualize the structures of atoms of some well-known elements. We can represent the nucleus by a small circle in which the protons (P) and the neutrons (N) are packed. The number of electrons in each shell is denoted by a number placed on the shell. Thus we have the following structures:

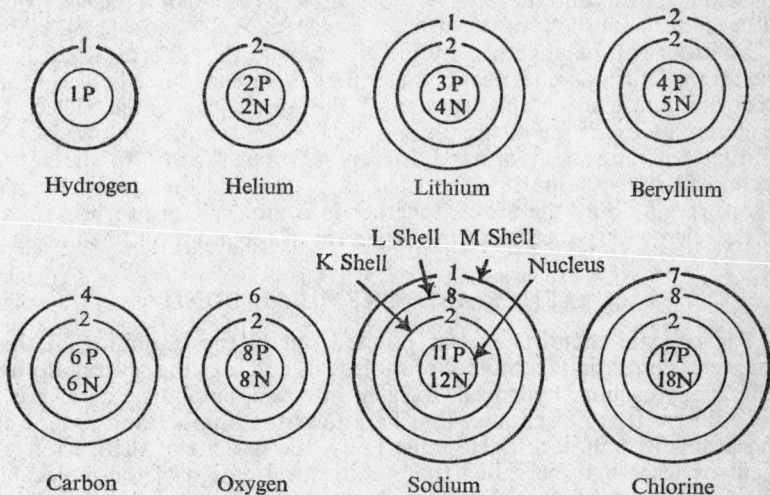

Hydrogen   Helium   Lithium   Beryllium

Carbon   Oxygen   Sodium   Chlorine

Fig. 1

The lighest element, **hydrogen**, contains one proton in the nucleus and one extranuclear electron. Since the electron has negligible mass, the weight of the hydrogen atom is mainly due to the single proton in the nucleus. The proton has unit mass, and hence the approximate mass of the hydrogen atom, or the mass number of hydrogen, is one. The second lightest element, **helium**, contains two protons and two neutrons in the nucleus, which gives it a mass of four units. There are two electrons in the extra-nuclear part to make the atom electrically neutral. The first shell can accommodate only two electrons, and saturation of the first shell is reached. Hence the next element, **lithium**, consisting of three protons and four neutrons in the nucleus, has three planetary electrons, two in the first shell called the **K shell** and one in the second shell called the **L shell**. The atoms of other elements are progressively built up. Thus, in the case of **carbon** there are six protons and six neutrons in the nucleus, two electrons in the K shell, and four electrons in the L shell. **Sodium** containing eleven protons and twelve neutrons in the nucleus has two electrons in the K shell, eight in the L shell, and one in the **M shell**. The M shell begins with sodium, since both the K and L shells have their saturation limits of two and eight electrons respectively, and the eleventh electron has to be accommodated in a new shell, M. **Chlorine** has seventeen protons and eighteen or twenty neutrons in the nucleus, making mass numbers 35 and 37 and an atomic weight of 35·5, and there are two electrons in the K shell, eight in the L shell, and seven in the M shell.

Now, when two atoms come together, only the peripheries of the atoms are likely to come into contact or have some sort of mutual interplay. The nuclei of the two atoms concerned have no chance to come together at all. Hence in chemistry we are not concerned with the nuclei at all, but are concerned only with the interplay of the extra-nuclear electrons and that too, in most cases, is confined to the electrons of the outermost shells of the atoms concerned. The chemical behaviour of an atom will be governed by the electrons in the outermost shell of the atom. During a chemical change there is an encounter between the electrons constituting the frontier of the atoms concerned. There is probably an exchange of fire among the frontier guards, followed by a mutual rearrangement or readjustment of these guards to the satisfaction of both the parties and then all is quiet on the atomic fronts. The forces holding the atoms together in a molecule are to be sought in this elementary picture of the structure of the atom outlined above.

VALENCE AND CHEMICAL BOND

The organic chemist, as was pointed out in the beginning of this chapter, is essentially a builder of molecules and uses the several atoms, principally carbon, hydrogen, oxygen, nitrogen, sulphur, etc., to erect his edifices. It is not enough that he knows the number and nature of the atoms in a molecule. He must know the pattern of their arrangement or what may be called the architectural design of the molecule. The fundamental factor that governs the architecture of the molecules is the valence of the elements making up the molecule. What is this concept of valence? Varying definitions of the term valence are given, changing with the level at which a student is studying. A school student will have a definite and simple idea of valence. The undergraduate and, more so, the graduate student are not so sure of the concept. Again a physicist may have a view which differs greatly from those of the students mentioned above. Ideas of valence have evolved considerably over the last 160 years, and in their higher stages of development they have an abstract beauty which is difficult to describe in terms of a simple physical picture. However, valence theory is a tool for handling and interpreting structural data and reaction mechanisms: we will choose the tool appropriate for our needs. Fortunately, for elementary organic chemistry a simple one will suffice. Later, when we need them, we will describe and use more sophisticated ones.

The word valence is used to describe the combining power of an element or what may be called, in common parlance, its mating power. In putting down the picture of a compound (the chemist calls it a structural formula) straight lines are used to represent the union between two atoms, and these lines are referred to as **valence bonds**.

Hydrogen is the simplest atom; let us look at the picture of some of the simple compounds it forms:

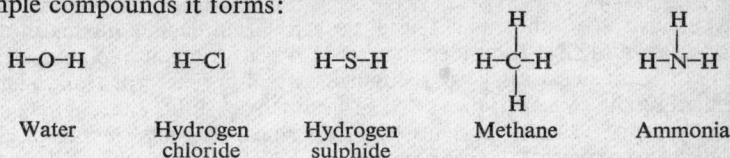

H—O—H   H—Cl   H—S—H   H—C—H   H—N—H
                                                                    |
                                                                    H (above C)
                                                                    |
                                                                    H (below C)

Water   Hydrogen   Hydrogen   Methane   Ammonia
               chloride   sulphide

The valence of hydrogen is 1 and is adopted as the standard. **The valence of an element** may now be defined as **the number of hydrogen atoms which will combine with one atom of the element in question.** From the formulae given above, oxygen has a valence of 2, nitrogen has a valence of 3, and carbon has a valence of 4. This is a very simple picture. But the moment we try to get a more definite picture of the nature of the valence bond and its physical significance we seem to be getting out of our depth. The valence bond is one of the most useful symbols ever invented, notwithstanding our lack of knowledge of its precise physical significance or nature. When we write down the symbols for two atoms and join them together, as when we write H—Cl, we imply that the two atoms are held together and leave the nature of the bond of linkage or the force that holds the atoms together conveniently vague. What are these forces like? Are they mechanical like grappling hooks? Are they magnetic or are they a case of some other type of attraction? We are not very sure of the answer to these questions, though modern advances like the **quantum theory** and **wave mechanics** do offer an intellectually satisfying picture. Notwithstanding this lack of precision, the valence bond is a very useful symbol. We shall try to understand a little more about this symbol and see where it leads to.

The elementary ideas of valence, which we are trying to develop, do not explain why atoms have different valences. Why should hydrogen have a valence of 1; oxygen, 2; nitrogen, 3; and carbon, 4? Also, why should the union between some elements be different from those between others?

For example, chlorine atoms combine with sodium atoms to form a stable solid compound, sodium chloride (common salt), in which there are no discrete molecules. In fact, the sodium is present as positively charged particles called sodium ions and the chloride as negatively charged particles called chloride ions. When sodium chloride dissolves in water these ions are able to move about independently. In contrast, chlorine atoms combine with each other to form a gas containing diatomic molecules Cl—Cl which, though it dissolves in water to a small extent, does not ionize like sodium chloride. We are going to need more than the simple straight-line valence-bond picture to explain differences like these.

Modern advances, especially in atomic physics, have resolved this enigma, and the chemist, with the help of the physicist, has been able to construct a satisfactory theory of valence. The modern ideas of valence, in addition to being intellectually satisfying, also possess the unique merit of not greatly upsetting the notion of organic chemists who still write their formulae with straight lines signifying valence bonds.

DIFFERENT TYPES OF VALENCE

We have seen that NaCl and the Cl–Cl molecule have different properties and that the older theory does not offer an explanation for this. It seems necessary to recognize two **different types** of valence, typified by the two examples cited. Again, why should we recognize only two types of valence and not several types, as may be presumably required to explain the combination in the case of several other molecules? Fortunately, it has been found that the two types mentioned above are two extreme types, and though there are a few cases in which some transition types must be recognized, generally speaking, the two types are enough to construct a satisfactory picture. It is known that the valence bond of the sodium chloride type frequently occurs in inorganic compounds and that the type of bond in the chlorine molecule is more universal and most of the organic compounds have such bonds. The Na–Cl bond is called an **electrovalent bond** and the Cl–Cl bond is called a **covalent bond**.

ELECTRONIC INTERPRETATION OF VALENCE

What is the origin of the differences between these two types of bonds? The modern theory of valence is intimately bound up with the picture of atomic structure as revealed by physicists. We have already presented an elementary idea of the modern theory of the structure of the atom and have drawn some pictures representing the atomic structure of some familiar elements. We have seen that the valence of an element is its combining capacity. We know that hydrogen is univalent; oxygen is bivalent; nitrogen is trivalent and carbon is tetravalent. Are there elements with zero valence, that is to say elements which do not combine with any other element? The atmosphere, in addition to the major constituents, nitrogen and oxygen, contains traces of five gaseous elements called helium, neon, argon, krypton, and xenon. These gases are called inert gases to indicate that they do not combine with any other element. In other words, their valence is zero. Let us study the arrangement of the electrons in the atoms of these gases:

ELEMENT	ARRANGEMENT OF ELECTRONS IN SEVERAL SHELLS				
	K	L	M	N	O
Helium	2				
Neon	2	8			
Argon	2	8	8		
Krypton	2	8	18	8	
Xenon	2	8	18	18	8

All of the above elements, with the exception of helium, have one thing in common—their outermost shells contain 8 electrons. Helium's outermost shell contains 2 electrons. We know that these elements have zero valence and show no tendency to combine or form valence bonds. Furthermore, they are the only elements that show this inertness to chemical combination. It is reasonable, therefore, to associate an arrangement of 8 electrons, called an **octet**, in the outermost shell of an atom with chemical stability or lack of chemical reactivity. According to the electronic theory of valence, when atoms combine they do so in order to acquire more stable electronic structures (for example, the structures of inert gas atoms) by some kind of electronic transaction and the valence bond results. The nature and magnitude of these transactions will, naturally, vary from element to element, depending upon the number of electrons the elements originally had in the outermost shell. This tendency of an atom to acquire an octet of valence electrons by some arrangement with another element which is also enabled to have a similar octet, in the bargain, is manifested as valence.

ELECTROVALENCE

Let us once again look at the structures of sodium chloride and chlorine. Let us draw a picture of the atomic arrangements in sodium and chlorine side by side and find how they each attain an octet of valence electrons by forming the compound sodium chloride. The sodium atom, as we have already seen, contains 2, 8, and 1 electrons respectively in the K, L, and M shells, and the chlorine atom has 2, 8, and 7 electrons in these shells. If the sodium atom can eliminate the electron in the M shell it will attain the structure typified by the inert gas neon and thus have a complete octet of electrons in its outermost shell. The chlorine atom has 7 electrons in its valence shell and is one short of the much coveted octet of electrons and naturally strives to acquire one electron and complete its octet. Here are two atoms, one trying to get rid of an extra electron, and the other trying to acquire an electron, both being motivated by a common objective, that of acquiring an octet. The problem seems to contain an ideal solution. When sodium and chlorine unite to form sodium chloride the extra unwanted single electron from the sodium atom is transferred to the chlorine atom, where it is eagerly wanted. The arrangement is mutually satisfactory, since both the atoms have thus been able to complete their octets. But during this transfer the sodium atom by losing an electron (which has a negative charge) has gained a positive charge. It is no longer the neutral sodium atom, but is a positively charged ion. Similarly, the chlorine atom after the transaction is richer by one electron, and hence gains a single negative charge to become a negative ion. We can picture the combination in the following manner, representing the nucleus as before, by a small circle and the electrons by crosses:

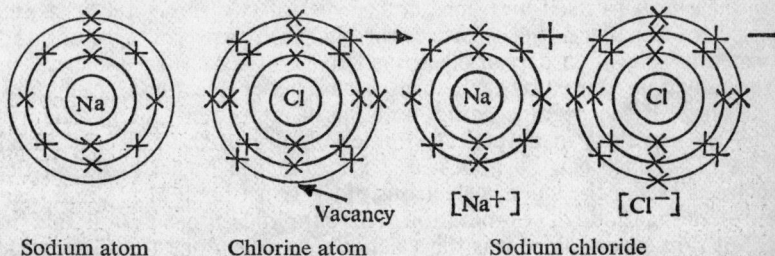

Sodium atom   Chlorine atom   Sodium chloride

Vacancy $[Na^+]$ $[Cl^-]$

Fig. 2

The sodium and the chloride ions thus formed hold on to each other by virtue of electrical attraction between their opposite charges. This type of bonding between atoms is called **electrovalence**.

COVALENCE

Let us now look at the state of affairs in the chlorine molecule Cl-Cl. It will be readily seen that the arrangement is different from that in NaCl (sodium chloride). There is no doubt that the valence bond in this case also is a manifestation of the tendency of the two chlorine atoms to complete their octets by forming the chlorine molecule. But the mechanism by which both the chlorine atoms attain this coveted state is different from that in the case of NaCl. The chlorine atom has seven electrons in its valence shell and exhibits a tendency to capture an electron. When two such chlorine atoms come together they can help each other out and both attain this happy state of eight by forming a close partnership, as pictured below:

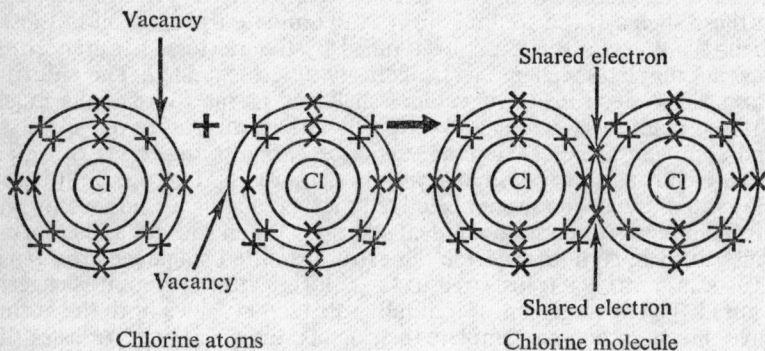

Vacancy

Shared electron

Cl + Cl → Cl Cl

Vacancy

Shared electron

Chlorine atoms

Chlorine molecule

Fig. 3

In the chlorine molecule each atom now contains one more electron than before, and thus has completed its octet. The important point to be emphasized is that in the formation of this type of bond there has been no transfer of electrons from one atom to the other. This is a case of mutual sharing of electrons. The chlorine molecule thus formed is a complete unit in itself and does not consist of two oppositely charged parts held together as in the case of the NaCl ion pair. This type of bond formed by sharing a pair of electrons between two atoms, one electron originally contributed by each atom, is called a **covalent bond**. We know that most of the organic compounds, as was pointed out in Chapter One, are covalent in nature. A simpler picture of these two essential types of valence can be obtained if we omit the nucleus and all the electrons except the valence electrons of the constituent atoms and then represent the mode of combination. This simplification is permissible (in all but rare cases), since it is the valence electrons alone that determine an atom's chemical behaviour. Thus in the case of the sodium chloride ion pair and the chlorine molecule we write the following, showing only the valence electrons:

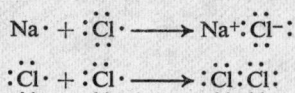

$$Na\cdot + :\overset{..}{\underset{..}{Cl}}\cdot \longrightarrow Na^+:\overset{..}{\underset{..}{Cl}}^-:$$

$$:\overset{..}{\underset{..}{Cl}}\cdot + :\overset{..}{\underset{..}{Cl}}\cdot \longrightarrow :\overset{..}{\underset{..}{Cl}}:\overset{..}{\underset{..}{Cl}}:$$

DATIVE (or CO-ORDINATE) COVALENCE

In addition to the above two types, there is a third type of valence called **dative covalence**. Let us consider the structure of two molecules like **ammonia** (NH_3) and **boron trichloride** (BCl_3). We have already seen that two electrons are needed to represent a bond, since in combination the valences of both atoms have to be satisfied. Thus the structure Cl–Cl implies that two electrons are shared between the chlorine atoms. The ammonia molecule is therefore $H:\overset{\overset{\displaystyle H}{..}}{\underset{..}{N}}:H$, and the boron tri-chloride molecule is $Cl:\overset{\overset{\displaystyle Cl}{..}}{B}:Cl$. As may be seen, the nitrogen atom has a pair of unshared electrons. It is willing to share this unused lone pair with the boron atom which has only six electrons around it in the BCl_3 molecule, and thus help the boron atom to complete its octet. The resulting bond is also a covalent bond, but with a difference. In the case of the chlorine molecule, each atom contributes equally, that is, one electron each for the formation of the bond. But in the case of the combination between the ammonia molecule and the boron trichloride molecule the bargain is all one-sided: the nitrogen atom donating two electrons for the formation of the bond, as indicated below, where the arrow points to the acceptor atom.

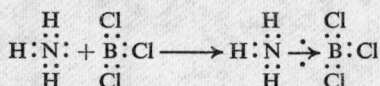

$$H:\overset{\overset{\displaystyle H}{..}}{\underset{..}{N}}: + \overset{\displaystyle Cl}{\underset{\displaystyle Cl}{B}}:Cl \longrightarrow H:\overset{\overset{\displaystyle H}{..}}{\underset{..}{N}} \rightarrow \overset{\displaystyle Cl}{\underset{\displaystyle Cl}{B}}:Cl$$

As a result of this donation of electrons, nitrogen gains a formal positive charge. It may be mentioned that the theory of valence outlined above was proposed independently by the American chemist G. N. LEWIS and the German chemist KOSSEL around 1916.

THE PERIODIC TABLE

Having understood the principal types of valence, we shall now consider the carbon atom more fully. We saw that the chlorine molecule is covalent in nature and that most of the organic compounds are also covalent. It seems pertinent to inquire into the reasons for the key position occupied by carbon in the list of elements. Owing principally to the labours of the Russian chemist MENDELEEF, all the elements have been arranged in a table known as the PERIODIC TABLE. The elements are arranged in the increasing order of their atomic weights and certain

PERIODIC TABLE OF ELEMENTS

PERIODS	GROUP I	GROUP II	TRANSITION ELEMENTS										GROUP III	GROUP IV	GROUP V	GROUP VI	GROUP VII	INERT Elem.
1	1 H 1.008																	2 He 4.003
2	3 Li 6.940	4 Be 9.02											5 B 10.82	6 C 12.010	7 N 14.008	8 O 16.000	9 F 19.00	10 Ne 20.183
3	11 Na 22.997	12 Mg 24.32											13 Al 26.97	14 Si 28.06	15 P 30.98	16 S 32.06	17 Ci 35.457	18 A 39.944
4	19 K 39.096	20 Ca 40.08	21 Sc 45.10	22 Ti 47.90	23 V 50.95	24 Cr 52.01	25 Mn 54.93	26 Fe 55.85	27 Co 58.94	28 Ni 58.69	29 Cu 63.57	30 Zn 65.38	31 Ga 69.72	32 Ge 30.98	33 As 74.91	34 Se 78.96	35 Br 79.916	36 Kr 83.7
5	37 Rb 85.48	38 Sr 87.63	39 Y 88.92	40 Zr 91.22	41 Nb 92.91	42 Mo 95.95	43 Tc (99)	44 Ru 101.7	45 Rh 102.91	46 Pd 106.7	47 Ag 107.88	48 Cd 112.41	49 In 114.76	50 Sn 118.70	51 Sb 121.76	52 Te 127.61	53 I 126.92	54 Xe 131.3
6	55 Cs 132.91	56 Ba 137.36	57 La 138.9 / 58 · 71	72 Hf 178.6	73 Ta 180.88	74 W 183.92	75 Re 186.31	76 Os 190.2	77 Ir 193.1	78 Pt 195.23	79 Au 197.2	80 Hg 200.6	81 Tl 204.39	82 Pb 207.2	83 Bi 209.00	84 Po (210)	85 At (210)	86 Rn (222)
7	87 Fr (223)	88 Ra 226.05	89 Ac (227) / 90 · 103															

RARE EARTH ELEM.

LANTHANIDE SERIES	57 La 138.92	58 Ce 140.13	59 Pr 140.92	60 Nd 144.27	61 Pm (145)	62 Sm 150.43	63 Eu 152.0	64 Gd 156.9	65 Tb 159.2	66 Dy 162.46	67 Ho 163.5	68 Er 167.2	69 Tm 169.4	70 Yb 173.04	71 - Lu 174.99
ACTINIDE SERIES	89 Ac (227)	90 Th 232.12	91 Pa 231	92 U 238.07	93 Np 237	94 Pu (242)	95 Am (243)	96 Cm (243)	97 Bk (245)	98 Cf (246)	99 Es (254)	100 Fm (255)	101 Md (256)	102 No (253)	103

NOTE: The numerals above the elements refer to atomic numbers and those below the elements refer to atomic weights.

Fig. 4

regularities appear. The periodic table is a help to systematization and comprehension. It is reproduced on page 16.

It will be seen in the second and third periods of the table that every eighth element has similar properties. Here we encounter once again the magic number eight. The inert group consists of the inert gases—helium, neon, etc.—having a valence of zero. The first group contains such similar elements as lithium, sodium, potassium, etc., all having a valence of 1. In the second group we find another family of similar elements—beryllium, magnesium, calcium, strontium, barium, etc.—all having a valence of 2. The same is true of the other groups. Let us consider the position of carbon in the periodic table. It is in the fourth group. The elements to the left of carbon in the periodic table are electropositive elements. That is to say, when they combine with other elements they exhibit a tendency to lose electrons to form positive ions or the positive ends of the molecules, as explained in the case of NaCl. The more easily an element allows an electron to be taken from it by the other element forming the compound, the more electropositive it is said to be. The elements to the left of carbon in the periodic table, that is, those in groups three, two, and one, become progressively more electropositive in nature. These elements are the elements which form bases typified by **sodium hydroxide**. Elements to the right of carbon, except the rare gases (that is, those in groups five, six, and seven), show a progressive increase in an electronegative nature. The elements from groups five to seven are elements which form acids typified by **hydrochloric acid**.

THE CARBON ATOM

The carbon atom in the middle of the periodic table in group four shows a unique character in being neither electropositive nor electronegative, and perhaps it is due to this electrically neutral character that carbon owes all its distinctiveness. The valence shell of carbon contains four electrons. To acquire an inert gas structure through electrovalence it must either lose all four electrons or capture four additional electrons, both processes being very difficult because of the intense electric field which would be produced round the resulting ion. The predominant characteristic of the carbon atom is to share its four valence electrons with other atoms to form four covalent bonds, e.g.:

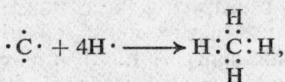

$$\cdot \overset{\cdot}{\underset{\cdot}{C}} \cdot + 4H \cdot \longrightarrow H : \overset{\overset{H}{\cdot \cdot}}{\underset{\underset{H}{\cdot \cdot}}{C}} : H,$$

thereby acquiring the stable structure of neon.

Vast numbers of organic compounds occur because of the stability of long chains of carbon atoms. There are two main reasons for their stability:

(1) a carbon atom forms covalent bonds of roughly the same strength with hydrogen, the halogens, nitrogen, oxygen, and sulphur as it does with other carbon atoms, so there is little tendency for a

C–C bond to break in order to form C–H, C–Halogen, C–N, C–O, or C–S bonds; and

(2) the outer electron shell of a carbon atom exerting four covalent bonds is completely filled with four pairs of electrons, all shared with other atoms.

Thus the carbon atom is resistant to chemical attack because it has neither an unshared pair of electrons on to which a reactant could attach itself nor room to accept a pair of electrons in dative covalence formation with a reactant.

POLAR AND NON-POLAR MOLECULES

We shall now treat the molecules as a whole and try to trace what influence the nature of the bond in a molecule has on its properties. Molecules can be roughly divided into two types, non-polar and polar. The following examples are illustrative:

$$:\overset{..}{\underset{..}{Cl}}:\overset{..}{\underset{..}{Cl}}: \qquad\qquad :\overset{..}{\underset{..}{I}}:\overset{..}{\underset{..}{Cl}}:$$

Chlorine (nonpolar)   Iodine chloride (polar)
Covalent

What is meant by the term polar molecule? Take the case of iodine chloride. We know that chlorine is more electronegative than iodine, and hence it exerts a stronger pull on the shared electron pair that constitutes the bond. In a sense, the sharing is not equal, and the shared electrons seem to belong a little more to the chlorine than the iodine. We can signify this state of affairs by inserting the shared electrons not midway between the iodine and chlorine but a little displaced towards the chlorine thus: $:\overset{..}{I}\ :\overset{..}{\underset{..}{Cl}}:$ or by marking the line representing the covalent bond with an arrowhead showing the direction in which the electrons are displaced: I $\longrightarrow$ Cl. This would amount to a partial transfer of electric charges from the iodine to the chlorine, the chlorine atom acquiring a formal negative charge and the iodine atom a formal positive charge. Hence one end of the ICl molecule is negative and the other end positive, both poles giving rise to what is called a **dipole**. The dipole moment of a molecule is a valuable research tool in problems of molecular structure. In a molecule that contains two poles the dipole moment is obtained by multiplying the distance between the poles by the electric charge on one of the poles (it is the same size for both poles, though of opposite sign). A highly polar molecule will have a high dipole moment, while non-polar molecules will have zero or negligible moments. The hydrogen molecule has no dipole moment, and thus reveals its symmetrical nature without any displacement of the electron pair towards either atom: H:H. The water molecule has a dipole moment of 1·85 units, and hence is assigned an angular or bent

structure thus: H $\overset{\textstyle O}{\diagup\diagdown}$ H. Coming to organic molecules like **carbon tetrachloride**, CCl_4, we find that it has no dipole moment, while chloroform, $CHCl_3$, has a dipole moment of 1·15 units. From this it can be concluded that the C–Cl bond is polar, but in CCl_4 the four polar valences balance one another, which is possible only if the molecule is symmetrical.

QUALITATIVE ANALYSIS OF ORGANIC COMPOUNDS

How does one set about determining the structure of a compound in practice? First of all, the substance in question must be obtained in a high degree of purity. The next step is a qualitative analysis whose objective is to determine the nature of the elements present in the compound. Tests for carbon and hydrogen in an organic compound are usually unnecessary, since it invariably contains them. If necessary, in other words, to see if it is organic, these two elements can be detected by heating a small amount of the substance with copper oxide (CuO), when the carbon of the organic compound is oxidized to carbon dioxide and the hydrogen to water. These products of oxidation are recognized in the usual manner, the former by its ability to turn lime-water milky and the latter by its ability to restore the blue colour to white anhydrous copper sulphate. Nitrogen, halogens, sulphur, etc., in organic compounds are detected by fusing the compound with a little sodium metal, extracting the product with water, and analysing the extract obtained. Under the conditions of the experiment the nitrogen of the organic compound is converted to **sodium cyanide**, the chlorine to **sodium chloride**, and the sulphur to **sodium sulphide**. The aqueous sodium fusion extract will contain these substances, and they can be identified by the usual methods of inorganic chemistry—the nitrogen by the production of the Prussian blue, the chloride as silver chloride, and so on.

QUANTITATIVE ANALYSIS

After detecting the elements in a particular compound one must proceed to a quantitative analysis to find the percentage of the several elements in the compound. A knowledge of the percentage composition will lead to the simplest formula, called the empirical formula of the compound. This quantitative process is called estimation. Carbon and hydrogen in an organic compound are estimated by allowing a small amount of the material to be oxidized, that is burnt up completely in a suitable apparatus in a current of oxygen. The products of oxidation— namely, carbon dioxide and water vapour—are trapped in separate absorption vessels which have been previously weighed, the carbon dioxide by potash bulbs and the water by tubes containing anhydrous calcium chloride or other water-absorbing reagents. The increase in weight of the potash bulbs and the calcium chloride tubes gives the weight of the carbon dioxide and water formed from the known weight of the compound. In the formation of carbon dioxide one atom of carbon on complete oxidation gives one molecule of carbon dioxide.

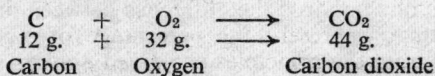

$$C \quad + \quad O_2 \quad \longrightarrow \quad CO_2$$

C		O_2		CO_2
12 g.	+	32 g.	$\longrightarrow$	44 g.
Carbon		Oxygen		Carbon dioxide

Writing out the equation and supplying the atomic weights, we find that 44 g. of carbon dioxide are produced from 12 g. of carbon. Hence the weight of carbon dioxide (increase in weight of potash bulb) multiplied by $\frac{12}{44}$ or $\frac{3}{11}$ will give the weight of carbon. Knowing the weight of the compound taken, we can calculate the percentage of carbon in the compound. Similarly in the case of water,

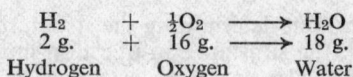

$$
\begin{array}{ccc}
H_2 & + & \tfrac{1}{2}O_2 \longrightarrow H_2O \\
2 \text{ g.} & + & 16 \text{ g.} \longrightarrow 18 \text{ g.} \\
\text{Hydrogen} & & \text{Oxygen} \qquad \text{Water}
\end{array}
$$

18 g. of water are obtained by the oxidation of 2 g. of hydrogen. Hence the weight of water obtained (increase in weight of the calcium chloride tubes) multiplied by $\frac{2}{18}$ or $\frac{1}{9}$ gives the weight of the hydrogen in the compound, from which the percentage of hydrogen is easily calculated.

When an organic substance containing nitrogen is heated in an atmosphere of carbon dioxide—in a combustion tube in the presence of copper oxide, which helps oxidation—the carbon and hydrogen of the compound are oxidized to CO_2 and H_2O respectively, and nitrogen in the elementary state is released. The volume of nitrogen can be measured, its weight computed, and the percentage of nitrogen in the compound determined. Since air contains nitrogen, all the air in the combustion tube must first be flushed out with carbon dioxide. The combustion is completed in an atmosphere of carbon dioxide, and the nitrogen released is collected in a measuring burette, called a nitrometer, after allowing it to bubble through potassium hydroxide solution. After applying the usual corrections the weight of nitrogen is calculated from the volume of nitrogen collected. Halogen and sulphur in organic compounds are determined by a method discovered by Carius. The principle of the method consists of oxidizing a known weight of the substance in a sealed tube with concentrated nitric acid to which some silver nitrate crystals have been added. The chlorine in the organic compound is converted into silver chloride, which is collected, washed, dried, and weighed. From the weight of silver chloride obtained, the percentage of chlorine in the compound can be determined. Sulphur is determined by heating the substance in a sealed tube with concentrated nitric acid alone. The sulphur is oxidized to sulphuric acid and the latter, precipitated as barium sulphate, is washed, dried, and weighed. From the weight of barium sulphate obtained, the weight of sulphur and the percentage of sulphur in the compound can be calculated:

1 g. atom of S (sulphur) gives 1 g. molecular weight of $BaSO_4$ (barium sulphate), i.e. 32 g. of sulphur gives 233 g. of barium sulphate.

The weight of barium sulphate multiplied by $\frac{32}{233}$ equals the weight of sulphur. There is no direct method for the estimation of oxygen. It is always found by difference. If the percentages of all the other elements do not total up to one hundred the difference between one hundred and the total is taken to represent the percentage of oxygen. A simple problem will illustrate the principles explained above.

PROBLEM I: 0·92 g. of an organic substance on combustion gave 1·76 g. of CO_2 and 1·08 g. of water. Calculate the percentage composition of the compound.

SOLUTION:

$$\text{Wt. of } CO_2 \text{ formed} = 1 \cdot 76 \text{ g.}$$

$$\text{Wt. of carbon} = 1 \cdot 76 \times \frac{3}{11} = 0 \cdot 48 \text{ g.}$$

$$\% \text{ of carbon} = \frac{0 \cdot 48}{0 \cdot 92} \times 100 = 52 \cdot 17$$

$$\text{Wt. of water formed} = 1 \cdot 08 \text{ g.}$$

$$\text{Wt. of hydrogen} = 1 \cdot 08 \times \frac{1}{9} = 0 \cdot 12 \text{ g.}$$

$$\% \text{ of hydrogen} = \frac{0 \cdot 12}{0 \cdot 92} \times 100 = 13 \cdot 04$$

The % of Carbon + % of Hydrogen = 52·17 + 13·04 = 65·21. This figure falls short of 100 by (100 − 65·21) or 34·79. Since no mention is made in the problem of any other element except carbon and hydrogen, this difference is taken to represent the % of oxygen.

Hence the % composition of the compound is:

C   52·17%

H   13·04%

O   34·79% (by difference)

DETERMINATION OF EMPIRICAL AND MOLECULAR FORMULAE

The percentage by weight of an element A in a compound $A_aB_bC_c$ is directly proportional to a (the number of atoms of A in one molecule of the compound) and to the atomic weight of A,

i.e.
$$\%A \propto a \times \text{At. wt. A}$$
$$\therefore a \propto \frac{\%A}{\text{At. wt. A}}$$

Hence $a : b : c$, the ratio of the numbers of atoms of elements A, B, and C which are present in a molecule of the compound, is the same as:

$$\frac{\%A}{\text{At. wt. A}} : \frac{\%B}{\text{At. wt. B}} : \frac{\%C}{\text{At. wt. C}}$$

This ratio can always be adjusted to a whole number ratio, generally by dividing through all the terms by the smallest term present. The result gives us the **empirical formula** of the compound: the simplest formula which expresses the atomic composition of the compound as a whole number ratio. Taking the data on percentage composition in PROBLEM I, let us work out the empirical formula of the compound. In doing this, we may tabulate the results in the following manner:

ELEMENT	% COMPOSITION	% COMPOSITION RESPECTIVE ATOMIC WEIGHT	FIG. IN COLUMN 3 LEAST NUMBER
C (At. wt. 12)	52·17	$\dfrac{52\cdot17}{12} = 4\cdot35$	$\dfrac{4\cdot35}{2\cdot17} = 2$ (nearly)
H (At. wt. 1)	13·04	$\dfrac{13\cdot04}{1} = 13\cdot04$	$\dfrac{13\cdot04}{2\cdot17} = 6$ (nearly)
O (At. wt. 16)	34·79	$\dfrac{34\cdot79}{16} = 2\cdot17$	$\dfrac{2\cdot17}{2\cdot17} = 1$

In column four the division by the least number, 2·17, is carried out to simplify the ratio 4·35 : 13·04 : 2·17 and arrive at a ratio of whole numbers. The simplest or the empirical formula of the compound is $C_2H_6O_1$ or simply C_2H_6O. The empirical formula gives the ratio of the several atoms present in a molecule, but not the actual number of atoms present in it. The actual number of atoms present in the molecule will give us the **molecular formula**. The molecular formula is a multiple of the empirical formula. Now we must determine what multiple of the empirical formula is the molecular formula. To determine this, a knowledge of the molecular weight of the compound is necessary. If the substance can be converted into the vapour without decomposition the absolute density of the vapour can be determined by suitable methods. The molecular weight of a gas (or vapour) is the number of grams which occupy 22·4 litres, measured under standard conditions, 760 mm. pressure and 0° C. From the experimentally determined value of the density, the molecular weight is easily calculated. If the substance is not volatile the molecular weight can be determined in solution by either the freezing-point or the boiling-point method. The principle of these methods may be outlined. A pure solvent boils at a fixed temperature under a given pressure. When a small quantity of a solute is dissolved in it and the boiling point is again determined the boiling point increases to a higher temperature. This elevation of the boiling point is proportional to the molecular concentration of the solute. By measuring the elevation in boiling point caused by dissolving a known weight of the substance in a solvent like acetone, chloroform, or water, the molecular weight can be computed. Similarly, when a solute is dissolved in a solvent the freezing point is depressed, and measurement of the depression of the freezing point caused by dissolving a known weight of the substance in a suitable solvent enables us to calculate the molecular weight of the substance.

If the molecular weight is known it is easy to calculate the molecular formula from the empirical formula. In the problem just completed the empirical formula is C_2H_6O. The molecular formula is a multiple of this and can be represented as $(C_2H_6O)_x$, with x representing a whole number to be determined. If we know that the absolute density of the compound in the vapour state is 0·0020491 g./ml. we can calculate the molecular formula thus:

$$\text{Mol. wt. of compound} = \text{Wt. of 22·4 litres at 760 mm. and 0° C.}$$
$$= 22,400 \times 0·0020491 = 45·9$$
$$(C_2H_6O)_x = 45·9$$

Supplying the atomic weights for carbon, hydrogen, and oxygen

$$(2 \times 12 + 6 \times 1 + 16 \times 1)_x = 45·9$$
$$46_x = 45·9$$
$$x = 1 \text{ (since } x \text{ must be a whole number)}$$

Hence the molecular formula is $(C_2H_6O)_1$. In this instance the empirical and molecular formula are the same. If x were any other number, for example, 2 or 3, the subscripts of the empirical formula would have to be multiplied by that number to give the molecular formula.

ISOMERISM

Having determined the molecular formula, the next job is to assign a structure to the compound. Here an unexpected problem crops up. As you know, the molecular formula of an inorganic compound completely defines it. That is to say, the formula $KClO_3$, once known, uniquely determines the properties of the substance in question, namely potassium chlorate, and there is no ambiguity about the matter. The situation is different in the case of organic compounds. Two or more organic substances can have the same molecular formula, but with different properties. For example, when the roll call of organic substances is taken and the formula C_2H_6O is called, two entities pop up claiming the same molecular formula. How are we to reconcile this? Compounds having the same molecular formula but with different properties are called **isomers**, and the phenomenon is called **isomerism**. Isomers differ in their structure, that is in the mode of linkage of the atoms within the molecule. We shall trace historically the development of the idea of isomerism, so frequently met in organic chemistry.

The first known case of isomerism, that of **silver cyanate** and **silver fulminate**, came to be noticed by two chemists, WOHLER, working under the great BERZELIUS in Sweden, and LIEBIG, working under GAY-LUSSAC in Paris. The whole affair began in a brisk quarrel and ended in a Damon and Pythias friendship. Liebig and Wohler were an inseparable pair in the further development of organic chemistry. Wohler was studying a white powder, silver cyanate, trying to find out why it decomposed when heated. He found that it contained silver, carbon, nitrogen, and oxygen and assigned it the formula AgCNO.

Liebig, who was interested in fireworks, prepared, at the risk of his eyesight, the slender colourless crystals of silver fulminate which when heated exploded violently. He found its formula to be AgCNO. Liebig rushed to the conclusion that Wohler was wrong in assigning the formula AgCNO to the cyanate. His own compound undoubtedly had the formula AgCNO, and its properties were very different from those described by Wohler. He wrote to Wohler in a vigorous strain, not concealing his feeling that Wohler must be an incompetent analyst. Wohler thought that Liebig was quite presumptuous. He could not understand how two different substances could have the same formula. Nevertheless, instead of ignoring Liebig, like a true man of science he re-examined the problem dispassionately and checked and rechecked his analysis of the cyanate. He also analysed the fulminate. He was surprised to find that both analyses were correct and that the two compounds had the same formula, AgCNO. This was indeed a puzzle, since it contradicted accepted notions.

Naturally, Wohler took his problem to his master, Berzelius, who named the phenomenon **isomerism** from the Greek *isos* meaning the same and *meros* meaning parts. Silver fulminate and silver cyanate are isomers containing equal parts of the same elements. How can this be? The same elements in the same proportions, yet the two substances are entirely different. Obviously, the properties of a substance seem to be influenced by some other factor in addition to the number and nature of the atoms constituting the molecule. This additional factor is the arrangement of the atoms within the molecule. The atoms in the fulminate are arranged differently from the atoms in the cyanate, just as a child with the same four matchsticks can arrange them to form the letters E or M. Even Berzelius could not have foreseen the wide prevalence of the phenomenon of isomerism among organic compounds.

It now becomes necessary to examine the disposition of the bonds in the case of organic molecules and to get a clear picture of their architecture. Carbon is almost invariably tetravalent, that is it has four valences. In organic chemistry carbon is always associated with the picture $-\overset{|}{\underset{|}{C}}-$ with four grappling hooks. This is the first and the most important point to be remembered in organic chemistry. A famous bridge player when asked for tips regarding the game replied, 'If you know how to count up to thirteen you can become a good bridge player.' He was referring to the length of the different suits of cards and emphasized the need for calculation. In a similar manner it may be said that if you are always aware of the fact that carbon has four valences, then you are likely to become thoroughly acquainted with the subject. Each of the bonds consists of a pair of electrons moving in some way between and around the two atomic nuclei joined by the bond. Once again we cannot know the precise details of the position and motion of the electrons, but we can assess relative probabilities of finding them at various places near the nuclei. We will have more to say about this later, but for the time being two points should be noted: (1) The electrons are more often in the region between the two nuclei than elsewhere: it is the attraction of the two negative electrons for the two positive nuclei which produces the bonding effect. (2) The pair of electrons in one covalent bond from a carbon atom repels the pairs in all the other bonds from that atom so that these bonds distribute themselves symmetrically round the carbon atom and as far away from each other as they can get. This causes the four valence bonds of carbon to point to the four corners of a regular tetrahedron.

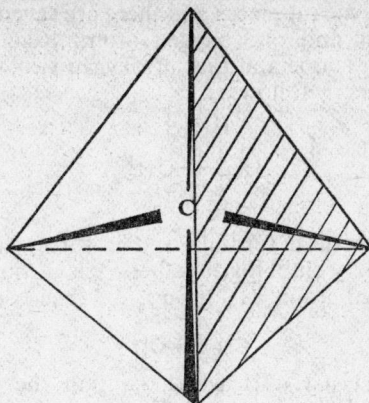

Fig. 5

Covalent bonds are rigid and directional, unlike the electrovalent bond. This rigidity and the directive nature of the covalent bond necessitate a careful consideration of the molecular architecture of organic compounds. For this reason, we shall have to study the mode of linkage of the atoms within the molecule, the proximity or otherwise of certain atoms or groups, and so on. As we develop the structural theory, we shall feel the necessity for extending our ideas from planar diagrams to three-dimensional models showing the spatial distribution of the atoms in the molecule.

Let us take the molecular formula C_2H_6O. It was previously noted that two substances had this formula. The organic chemist is essentially a builder of molecules. Just as an engineer erecting a building has to follow certain rules and procedures, so must the chemist follow certain prescribed rules in constructing molecules. The most important rule to be observed in the construction of molecules is that the valence of the elements constituting the molecule should be correctly portrayed. Carbon has a valence of four, oxygen has a valence of two, and hydrogen has a valence of one. The problem now resolves to this: with the same raw materials, that is two atoms of carbon, six of hydrogen, and one of oxygen, we must build two different buildings without transgressing the valence rule. Let us see how this can be done. First, consider the foundation of a building. Similarly, let us put down the foundation of the molecule and write out the two carbon atoms, without forgetting to give each carbon atom its due of four valences.

$$ -\overset{|}{\underset{|}{C}}- \qquad -\overset{|}{\underset{|}{C}}- $$

Now cement the foundation by joining one valence of a carbon atom to a valence of the other carbon atom, and you get the following picture:

$$ -\overset{|}{\underset{|}{C}}-\overset{|}{\underset{|}{C}}- $$

Now there are six vacant places and there are seven applicants, that is six hydrogen atoms and one oxygen atom. Remembering that the valence of hydrogen is one and that of oxygen two, erect the complete edifice which appears as follows:

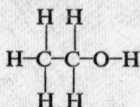

$$\begin{array}{ccc} & H & H \\ & | & | \\ H- & C- & C-O-H \\ & | & | \\ & H & H \end{array}$$

This is called a structural formula and is somewhat cumbersome to write. It is abridged by clubbing together each carbon atom along with its attached hydrogen atoms as a group. We then get the constitutional formula:

$$CH_3CH_2OH$$

Compounds like CH_3CH_2OH which contain the hydroxyl or $-OH$ group are called alcohols, and the particular one written above is called **ethyl alcohol.** It is a pleasant-smelling liquid which reacts with phosphorus pentachloride, sodium, and acids.

Another structure using the same two atoms of carbon, six of hydrogen, and one of oxygen can be built without violating the valence rules. It is shown as follows:

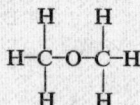

$$\begin{array}{ccc} H & & H \\ | & & | \\ H-C- & O- & C-H \\ | & & | \\ H & & H \end{array}$$

This can be abridged to CH_3OCH_3. This compound is a *gas* and is quite unreactive. Compounds like CH_3OCH_3, which contain an oxygen bridge between two carbon atoms, are called ethers, and this particular compound is **dimethyl ether.** If we compare the structures CH_3CH_2OH and CH_3OCH_3 we at once see the profound difference in their architecture. This difference in architecture is reflected in differences in their properties. Ethyl alcohol contains one methyl group CH_3 (a hall if you please) and dimethyl ether contains two such halls. Ethyl alcohol contains a smaller room $-CH_2-$, and there is no such room in the case of dimethyl ether. Again, in ethyl alcohol we have an $-OH$ window which we do not find in dimethyl ether. The two carbon atoms in the ether are joined through an oxygen bridge, while in the case of the alcohol the two carbon atoms are joined directly.

As the molecular complexity increases, the number of possible isomers increases rapidly. To give you an idea of this, it has been calculated that, theoretically, a compound with the molecular formula $C_{15}H_{32}$ can have 4,347 different isomers, and a compound with a formula $C_{40}H_{82}$ can possess no less than 62,491,178,805,831 isomers! By including other elements, the number of theoretically possible isomers having a given molecular formula will be several times the figures given above. Not all the isomers are known. Such astronomical figures give you an idea of the potential number of organic compounds, and the immensity of the domain of organic chemistry can stagger the imagination. Such considerations might easily lead you to despair. But much simplification has been brought about in the study of the subject by the concept of functional groups and the resultant classification of the vast number of compounds into families known as the **Homologous Series**.

HOMOLOGOUS SERIES

A **homologous series** is a group of related compounds in which the formula of each member differs from those of its preceding and succeeding members by one $-CH_2-$ group. Members of a homologous series have similar methods of preparation and similar chemical properties. The physical properties of the members change uniformly on ascending the series.

The functional groups behave as units in entering or leaving a molecule, and they confer characteristic properties on compounds containing them. Common functional groups include $-OH$ (hydroxy),

$\overset{\displaystyle H}{\underset{\displaystyle |}{-C}}=O$ (aldehydo), $\overset{\displaystyle |}{-C}=O$ (keto), $-COOH$ (carboxyl), $-NH_2$ (amino),

$-NO_2$ (nitro), etc. The various functional groups have specific characteristics, and if we know the particular functional group or groups in a given molecule, then the properties of the molecule can be readily inferred. This means that all the compounds with $-OH$ groups, the alcohols, will have properties in common. This is also true in the case of acids, ethers, etc. In addition, we must, in the case of more complex molecules, take into consideration the proximity of one functional group to another and the result of their interaction. Nevertheless, the idea of functional groups has certainly brought about simplicity and has transformed the study of a bewildering array of compounds into a logical, well-defined, and coherent whole.

SUMMARY

Consideration of the structure of organic compounds in terms of atoms, molecules, and valence. An atom is the smallest indivisible particle of an element. A molecule is a combination of atoms in a fixed ratio. The number and nature of such atoms may vary. An atom consists of a nucleus with a positive charge with electrons moving around it. Nucleus contains neutrons and protons. An electron is a particle with unit negative charge and no mass. A proton has unit positive charge and unit mass. A neutron has no charge but does have unit mass. Extranuclear electrons equal the number of positive charges on the nucleus and move in different 'shells'. Chemical properties and reactions involve electrons in the outermost shell. Valence has been defined as the combining capacity of elements. Electronic interpretation of valency as a manifestation of the tendency for each atom to attain a more stable state, e.g. eight electrons in the outermost shell. Characteristics of inert gas atoms. Chemical combination results from a redistribution of outermost electrons in two or more atoms by processes involving electron transfer (electrovalence) or electron sharing (covalence) or donation of electrons (dative covalence).

Qualitative and quantitative analysis of organic compounds. Determination of empirical and molecular formulae from percentage composition. The phenomenon of isomerism in which two or more compounds have the same molecular formula but exhibit different chemical or physical properties owing to difference in structure, for example, CH_3OCH_3 and CH_3CH_2OH. Classification of organic compounds into families on the basis of functional groups present. Definition of a homologous series as a class of compounds containing the same functional group or groups in which succeeding members differ in molecular formula by a $-CH_2-$ group. Gradation of physical properties within a homologous series. Important functional groups include $-OH$, $>C=O$, $-COOH$, $-NO_2$, $-NH_2$.

Problem Set No. 1

In the calculations below, use the following values of atomic weight: H 1, C 12, N 14, O 16, S 32, Cl 35·5.

1. Calculate the percentage composition of compounds having the following molecular formulae: (a) $C_4H_8O_2$; (b) C_4H_9Cl.
2. A certain substance contains 53·34% carbon, 15·56% hydrogen, and 31·10% nitrogen. Find the empirical formula for the substance.
3. A sample of a gas weighs 0·1977 g. and occupies a volume of 100 ml. at 0° and 760 mm. pressure. What is the molecular weight of the gas?
4. A certain substance contains 32·9% carbon, 4·1% hydrogen, 19·25% nitrogen, and the remainder consists of oxygen. Given the fact that the molecular weight is 146, find the molecular formula of the substance.

Check your answers with those in the answer section in back of the book.

PART II

ALIPHATIC COMPOUNDS

CHAPTER THREE

HYDROCARBONS

ALKANES

The simplest type of organic compounds are the **hydrocarbons**—those containing only carbon and hydrogen. Taking a carbon atom which has a valence of four and attaching four hydrogen atoms to it, we get the molecular formula CH_4 (or the structural formula $H-\overset{\displaystyle H}{\underset{\displaystyle H}{\overset{|}{\underset{|}{C}}}}-H$) for the simplest hydrocarbon called **methane**. Similarly, from two carbon atoms we derive the formula C_2H_6 corresponding to the structural formula $H-\overset{\displaystyle H}{\underset{\displaystyle H}{\overset{|}{\underset{|}{C}}}}-\overset{\displaystyle H}{\underset{\displaystyle H}{\overset{|}{\underset{|}{C}}}}-H$, also written as CH_3-CH_3, for the next hydrocarbon called **ethane**. A hydrocarbon with three carbon atoms has the molecular formula C_3H_8 corresponding to the structure $CH_3-CH_2-CH_3$ and is called **propane**. We can go on building a number of molecules in this series by adding one $^-CH_2$ group to the previous molecule. Thus **butane** would be $CH_3CH_2CH_2CH_3$ and **pentane** would be $CH_3CH_2CH_2CH_2CH_3$, etc. These hydrocarbons constitute a family or a homologous series and are known variously as saturated hydrocarbons, **paraffins**, or **alkanes**. They have the general formula C_nH_{2n+2}. The description **saturated hydrocarbons** signifies that each carbon atom is joined to four different carbon or hydrogen atoms by single covalent bonds, while the word *paraffin* is derived from the Latin—*parum affinis*, meaning slight affinity—to indicate the chemical unreactivity of this class of compounds. They can be prepared by similar methods, and their chemical properties are almost identical. But their physical properties show a gradation. The first four members of this series are gases; then we pass on to liquids with low boiling points, through liquids with high boiling points to solids with low melting points and then to solids with high melting points. A similar gradation in other physical properties like solubility, density, refractive index, viscosity, etc., is also noticed.

From the general formula of the paraffin C_nH_{2n+2} we can derive the univalent radical C_nH_{2n+1} by a removal of a hydrogen atom. This radical is called the **alkyl radical** and is usually represented by the symbol R. Thus R may stand for the methyl $^-CH_3$, ethyl $CH_3-CH_2{}^-$, or any other higher radical.

PREPARATION: The alkanes can be prepared by one of the following general methods: by reduction of alkyl halides (see Chapter Four) through nascent hydrogen produced by treatment with a zinc copper couple or aluminium amalgam and alcohol. This general method can be indicated by the following equation:

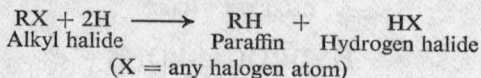

$$RX + 2H \longrightarrow RH + HX$$

Alkyl halide   Paraffin   Hydrogen halide

(X = any halogen atom)

Thus methane can be obtained from methyl iodide, ethane from ethyl iodide, and so on.

Another method suitable for the higher members of the series is the reaction known as the WURTZ SYNTHESIS after its discoverer. The word synthesis means that a compound containing a larger number of carbon atoms is prepared from one containing a smaller number. The Wurtz reaction brings about the union of two alkyl radicals by the action of metallic sodium on two molecules of the same alkyl halide or different alkyl halides as follows:

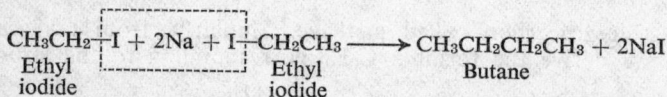

$$CH_3CH_2\text{---}I + 2Na + I\text{---}CH_2CH_3 \longrightarrow CH_3CH_2CH_2CH_3 + 2NaI$$

Ethyl iodide   Ethyl iodide   Butane

The abstraction of the two iodine atoms as sodium iodide by sodium gives rise to two ethyl radicals which combine to form butane. Similarly, the interaction of sodium with a mixture of methyl and ethyl iodide gives propane mixed with ethane and butane.

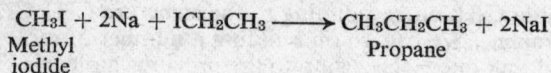

$$CH_3I + 2Na + ICH_2CH_3 \longrightarrow CH_3CH_2CH_3 + 2NaI$$

Methyl iodide   Propane

Methane can be made in the laboratory by treating aluminium carbide (Al_4C_3) with water.

$$Al_4C_3 + 12H_2O \longrightarrow 3CH_4 + 4Al(OH)_3$$

The above methods of preparation are used only when pure samples of specific alkanes are required. They are generally obtained by processing of naturally occurring petroleum, which is a mixture largely composed of alkanes.

PROPERTIES AND REACTIONS: Methane is also called marsh gas because it is found in swamps and marshes and is formed by the decomposition of organic matter brought about by certain bacteria. Mixed with air or oxygen, it is the fire-damp of coal mines that leads to explosions. The principal sources of methane are natural gas, which contains about 90% methane, and coal gas, containing about 40% of it. Ethane is also present in natural gas to a certain extent, and propane and butane are recovered from oilfields.

Methane and its homologues, though resistant to chemical change in comparison with other compounds, do undergo some reactions. The most widely known and frequently used reaction of the alkanes is the combustion of the gases or the vapours of the higher members in air or oxygen. In excess air they burn completely to carbon dioxide and water vapour; this is the reaction which occurs in the cylinders of petrol and diesel engines.

$$C_8H_{18} + 12\tfrac{1}{2}O_2 \longrightarrow 8CO_2 + 9H_2O$$
Octane

In a limited supply of air the carbon is not completely burned, and its white-hot particles make the flame luminous yellow:

$$C_{19}H_{40} + 10O_2 \longrightarrow 19C + 20H_2O$$

This is the reaction which shows up as a candle flame.

The alkanes react readily with chlorine and bromine (but not iodine) in the presence of sunlight, ultra-violet light, heat, or a catalyst, e.g.:

$$CH_3 \cdot CH_2 \cdot CH_3 + Cl_2 \xrightarrow{\text{U.V. light}} \begin{cases} CH_3 \cdot CH_2 \cdot CH_2Cl + HCl \quad \text{\textit{n}-Propyl chloride} \\[2ex] \underset{\substack{| \\ Cl}}{CH_3 \cdot CH \cdot CH_3} + HCl \quad \text{Isopropyl chloride} \end{cases}$$

Propane

$$\underset{\substack{| \\ CH_3}}{CH_3 \cdot CH \cdot CH_3} + Cl_2 \xrightarrow{300°C} \begin{cases} \underset{\substack{| \\ CH_3}}{CH_3 \cdot CH \cdot CH_2 \cdot Cl} + HCl \quad \text{Isobutyl chloride} \\[2ex] \underset{\substack{CH_3 \\ | \\ Cl}}{CH_3 \cdot C \cdot CH_3} + HCl \quad \textit{tert}\text{-Butyl chloride} \end{cases}$$

Isobutane

Since the alkane molecule is saturated and there is no valence bond left unused, chlorine can enter the molecule only by first rupturing a carbon–hydrogen bond. Thus the hydrogen atom severed from the alkane pairs off with one chlorine atom of the molecule of chlorine as hydrogen chloride and the alkyl radical unites with the other chlorine atom to form an alkyl chloride. In effect, one hydrogen atom of the alkane has been replaced by a chlorine atom, and the reaction is called a **substitution reaction.** Reactions of this type are typical of **saturated compounds.** Fluorine reacts explosively, and hence fluorine compounds are obtained indirectly. Thus the order of reactivity of the halogens is F > Cl > Br > I.

The first member of a homologous series often differs from the rest of the series. Methane shows this difference in its reactions with chlorine. In bright sunlight it reacts explosively with chlorine:

$$CH_4 + 2Cl_2 \longrightarrow C + 4HCl$$

In diffused daylight it undergoes progressive substitution and, one by one, all the hydrogen atoms are replaced by chlorine atoms—the stages becoming progressively easier:

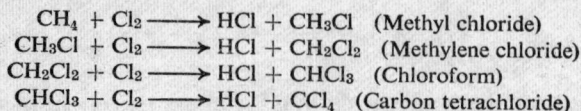

$$CH_4 + Cl_2 \longrightarrow HCl + CH_3Cl \quad \text{(Methyl chloride)}$$
$$CH_3Cl + Cl_2 \longrightarrow HCl + CH_2Cl_2 \quad \text{(Methylene chloride)}$$
$$CH_2Cl_2 + Cl_2 \longrightarrow HCl + CHCl_3 \quad \text{(Chloroform)}$$
$$CHCl_3 + Cl_2 \longrightarrow HCl + CCl_4 \quad \text{(Carbon tetrachloride)}$$

Methyl chloride is a gas used in refrigerators, methylene chloride is a solvent, chloroform an anaesthetic, and carbon tetrachloride a solvent. It is to be noted that the extent of substitution in methane is difficult to control and mixtures of the different chlorides result.

The alkanes, though unreactive towards even the strongest of acids at room temperature, react with nitric acid in the vapour phase at high temperatures. Nitroalkanes of the type RNO_2 are produced by replacement of a hydrogen atom with a nitro (NO_2) group—a process referred to as nitration. *n*-Butane, for example, when nitrated as above, gives 1-nitrobutane and 2-nitrobutane:

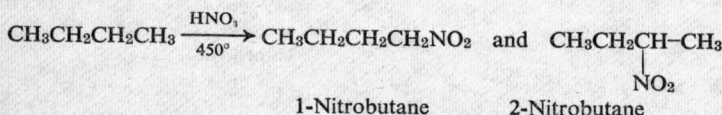

$$CH_3CH_2CH_2CH_3 \xrightarrow[450°]{HNO_3} CH_3CH_2CH_2CH_2NO_2 \quad \text{and} \quad CH_3CH_2\underset{\underset{NO_2}{|}}{C}H-CH_3$$

$$\text{1-Nitrobutane} \qquad\qquad \text{2-Nitrobutane}$$

The energy required for nitration is also sufficient to rupture the molecule of butane, and the above products are accompanied by smaller amounts of 1-nitropropane ($CH_3CH_2CH_2NO_2$), nitroethane ($CH_3CH_2NO_2$), and nitromethane (CH_3NO_2). Ethane and propane resemble methane in their properties, and no special mention need be made of them.

ISOMERISM OF HYDROCARBONS

In drawing a structural formula we do not usually attempt to show the three-dimensional shape of the molecule: for instance, instead of trying to represent the four valencies of each carbon atom pointing towards the corners of a regular tetrahedron, we just draw them like the arms of a cross.

Butane with the formula C_4H_{10} can exist in two isomeric forms. All four carbon atoms may be in a 'straight chain' or three of them may form a straight chain and the fourth one may form a branch or a fork. The two possible carbon skeletons are:

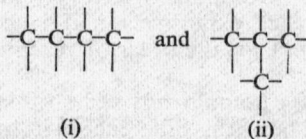

$$-\overset{|}{\underset{|}{C}}-\overset{|}{\underset{|}{C}}-\overset{|}{\underset{|}{C}}-\overset{|}{\underset{|}{C}}- \quad \text{and} \quad -\overset{|}{\underset{|}{C}}-\overset{|}{\underset{}{C}}-\overset{|}{\underset{|}{C}}-$$

$$\text{(i)} \qquad\qquad \text{(ii)}$$

Also, since a carbon atom can rotate freely about the axis of each of its bonds:

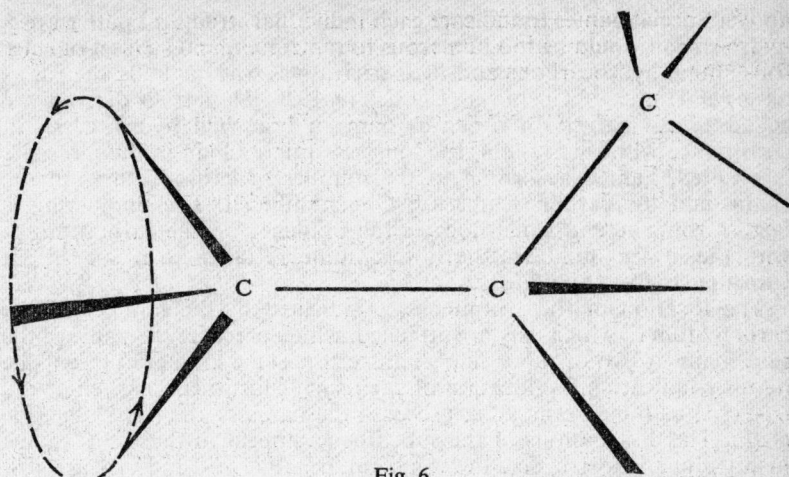

Fig. 6

then $\quad$ (structures) $\quad$ and $\quad$ (structures)

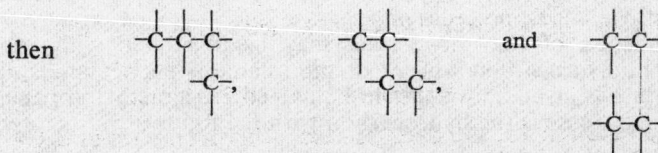

are not different structures but only different ways of drawing (i). Supplying the hydrogen atoms, we get the corresponding structural formulae:

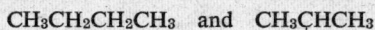

$$CH_3CH_2CH_2CH_3 \quad \text{and} \quad CH_3CHCH_3$$
$$\qquad\qquad\qquad\qquad\qquad\quad CH_3$$

Normal or *n*-Butane $\qquad$ Isobutane

The straight chain isomer is given the prefix *normal* and the branched chain the prefix *iso*. In the case of pentane three compounds can be built up as:

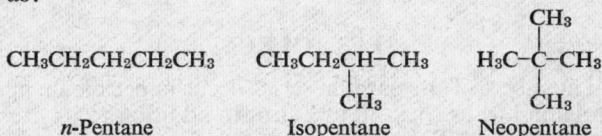

$$CH_3CH_2CH_2CH_2CH_3 \quad CH_3CH_2CH{-}CH_3 \quad H_3C{-}\overset{\displaystyle CH_3}{\underset{\displaystyle CH_3}{C}}{-}CH_3$$
$$\qquad\qquad\qquad\qquad\qquad\qquad CH_3$$

n-Pentane $\qquad$ Isopentane $\qquad$ Neopentane

NOMENCLATURE: The names methane, ethane, propane, and butane are accepted by the chemists because they have long been used to represent the first four members of the alkane series. Names of hydrocarbons beyond butane are derived from Greek numerals indicating the number of carbon atoms in the molecule. Thus the hydrocarbon C_5H_{12} is **pentane**, C_7H_{16} is **heptane**, and so on. The ending *ane* indicates an alkane. These generic names are related only to the molecular formula of the hydrocarbons and do not indicate their structures except when the prefixes *n*- and *iso*- are used in conjunction with them, as illustrated above. Special names to indicate each individual structural pattern of a hydrocarbon would be too numerous to remember, and so a set of rules for naming hydrocarbons and their derivatives was made by the International Union of Chemists (I.U.C.) in 1930. Names so derived are called I.U.C. names. In order to name a branched hydrocarbon in accordance with this system, the longest straight chain in the molecule is selected, named according to the number of carbon atoms in the chain, and the carbon atoms are then numbered. The numbering is begun from the end which locates the branches on the carbon atoms with the lowest numbers, the lowest numbers taken to mean the set with the smallest individual number.

The location of these branches is indicated by the number of the carbon atom to which they are attached, which precedes the name of the side chain. If two or more side chains of the same kind are present the number indicating the location of each such substituent, as well as *di-*, *tri-*, etc., as the case may be, precedes the name of the particular side chain. The designations of the substituents appear in the I.U.C. name in an alphabetical order, in which prefixes like *di-* and *tri-* are disregarded in fixing the order of appearance. In other words, ethyl appears before dimethyl, since *e* precedes *m*.

The application of the above rules may be illustrated with two examples. The hydrocarbon whose common name is neopentane and whose structure is given above should be named 2,2-dimethylpropane. A more complex case is the hydrocarbon with the following structure:

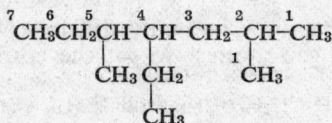

$$\overset{7}{C}H_3\overset{6}{C}H_2\overset{5}{C}H-\overset{4}{C}H-\overset{3}{C}H_2-\overset{2}{C}H-\overset{1}{C}H_3$$

$$\underset{CH_3}{\underset{|}{\ }} \ \underset{CH_2}{\underset{|}{\ }} \qquad \overset{1}{\underset{CH_3}{\underset{|}{\ }}}$$

$$\underset{CH_3}{\underset{|}{\ }}$$

The longest chain here is either of two identical seven-carbon chains. Numbering of the carbon atoms starts at the right end, so that the substituents have the smallest numerals to indicate their location. The I.U.C. name is 4-ethyl-2,5-dimethylheptane.

PETROLEUM

Several members of the paraffin series occur in petroleum and natural gas. Petroleum means rock-oil in Latin. In addition to its use as a fuel, petroleum is the chief starting material for the organic chemistry industry.

It is believed that animal and plant remains which were entrapped in the lower strata of the earth, in past geological ages, were subjected to pressure, high temperature, and some chemical influences which resulted in the production of petroleum. Depending upon the conditions, the elements—nitrogen, sulphur, oxygen, and hydrogen—in the organic matter were converted into inorganic gases, and most of the carbon was transformed to carbonaceous gases (natural gas), liquids (petroleum), and solids (coal, shale, etc.).

The leading petroleum-producing areas in the United States are in Texas, California, Kansas, Louisiana, Oklahoma, Wyoming, and New Mexico. Other petroleum-rich countries are Mexico, Venezuela, Arabia, and Russia.

In recent years oil prospecting beneath the coastal waters of the Gulf of Mexico and off the Californian coast has given an indication of extensive underwater oil deposits. A most encouraging flow of natural gas and oil is now being obtained from under the North Sea.

Petroleum is a dark-coloured emulsion of a large number of hydrocarbons along with various compounds containing sulphur, oxygen, and nitrogen, with sandy soil and water. The colour depends upon the region from which the petroleum is obtained, and varies from yellow through red to black. The crude petroleum is pumped to the refinery from the well and the emulsion is broken by steam coils and the water and sand removed. The oil is then subjected to fractional distillation and the crude oil is separated into several fractions or 'cuts', having different ranges of boiling points. The 'cuts' are collected with an eye to satisfying commercial demand. The following fractions are usually collected:

Boiling point	Name of the 'cut'
$-161°$ to $-88°$ C.	Natural gas (mostly CH_4)
$-42°$ to $0°$ C.	Bottle gas (C_3H_8 to C_4H_{10})
$20°$ to $70°$ C.	Petroleum ether ($C_5H_{12}-C_6H_{14}$)
$50°$ to $200°$ C.	Petrol or gasoline ($C_6H_{14}-C_{12}H_{26}$)
$200°$ to $300°$ C.	Paraffin or kerosene ($C_{12}H_{26}-C_{15}H_{32}$)
Above $300°$ C.	Gas oil, lubricating oil, and waxes ($C_{16}H_{34}-C_{24}H_{50}$)
Residue	Coke (Asphalt)

The net result of the fractionation process is that the original crude oil is separated into a series of products of gradually increasing boiling points. When the heavy lubricating oils are cooled paraffin wax and Vaseline (Petrolatum) separate out. Often, in practice, the distillation is stopped after the kerosene fraction, and the residue, known as 'reduced crude', is used for fuel and lubrication.

The next step is a process of desulphurization. The sulphur compounds present in the distillate, in addition to their objectionable odour, cause corrosion of containers. Further, if sulphur-containing oils are used in internal-combustion engines, sulphuric acid is produced and corrodes the engine. Hydrogen sulphide and some sulphur-containing compounds known as **mercaptans** are removed by washing the distillates with sodium hydroxide solution. Mercaptans not removed by this process are converted into odourless and non-corrosive substances by using a solution of sodium plumbite in the presence of added sulphur. This process is referred to in petroleum technology as 'sweetening'. We shall now consider the more important fractions, their treatment and uses.

Gaseous Hydrocarbons. Gaseous hydrocarbons contain one to four carbon atoms, which constitute about 2% of the oil used in furnaces and internal-combustion engines. The propane and butane are separated and compressed in the liquid form convenient for transport in cylinders. The compressed propane LPG (Liquid Propane Gas) or butane 'Gaz' becomes a gas once again when the pressure on it is released. An advantage of LPG is that it produces less exhaust fumes. The methane obtained from natural gas, when it is treated with steam at 1,000° C. in the presence of catalysts, produces hydrogen. This is a rich source of hydrogen for synthetic ammonia.

Petroleum Ether. Petroleum ether is used as a solvent and in dry cleaning, etc.

Petrol or Gasoline. The petrol fraction obtained by the direct distillation of petroleum is a poor motor fuel and must be refined. The principal objective of the refining process is to remove unsaturated hydrocarbons, which form gummy substances on standing, and to remove sulphur compounds. After the removal, additives, like anti-knocks, deoxidants, anti-rusting substances, lubricants, and so on, are introduced to serve specific purposes.

Because of the great importance of petrol in world economy, considerable research work has been done to improve fuels, standardize them, and ensure good performance in engines. The 'straight-run petrol' is not good enough to use as a fuel. Rigid standards are set for the evaluation of petrol: the Octane scale is used. 'Knocking' is due to pre-ignition of the fuel in the cylinder prior to sparking. The sudden explosion causes the cylinder to vibrate, and hence much of the fuel energy is wasted away as heat and mechanical energy. Hence knocking must be avoided to get the maximum power from the fuel and minimize wear and tear on the engine. It has been determined that branched-chain hydrocarbons have less tendency to knock than the straight-chain hydrocarbons. Catalysts like tetraethyl lead greatly reduce knocking. *n*-Heptane is found to be the worst offender as far as knocking is concerned, and hence it is rated 0. 2,2,4-Trimethylpentane (wrongly called *iso*-octane) has exceptionally high anti-knock characteristics, and is rated 100. The octane number of a fuel is the percentage of *iso*-octane in a mixture of *n*-heptane and *iso*-octane that gives the same knock properties as the fuel in question. For example, if a petrol has the same anti-knock properties in a test engine as a mixture of 75% *iso*-octane and 25% *n*-heptane the petrol has an octane rating of 75.

Due to the very heavy demand for petrol, it became necessary to find methods of supplementing the supply from oil wells. In the German Fischer–Tropsch process low-grade fuels were produced (synthetic gasoline) by hydrogenation of carbon monoxide. The two gases in the ratio of 2 : 1 were passed over a catalyst, usually cobalt with thoria as a promoter, under a pressure of 5–15 atmospheres at a temperature of 200° C. An American modification called Hydrocol or Synthol process produces 80-octane petrol. The other process, known as the Bergius process, depends upon the hydrogenation of coal dust. The original process required a high temperature of about 2,000° C. The I.C.I. have modified the process by carrying out the hydrogenation of a mixture of coal dust and oil at 450° C.

CRACKING

In addition to these (synthetic) methods for augmenting the supply of oil, an attempt was made to convert the less commercially useful high-boiling fractions into petrol. When the paraffin hydrocarbons are passed through heated tubes maintained at about 500–800° C. a thermal decomposition (pyrolysis or cracking) sets in. The large molecules are in a sense cracked and give rise to smaller molecules. The hydrocarbon vapours are led through a chamber, at 500° C., under pressure in the presence of such catalysts as silica or alumina. As a result of the cracking processes, the high-boiling hydrocarbons after kerosene are converted into a mixture of low-boiling hydrocarbons, olefins, or unsaturated compounds and deposits of coke. The cracking may be carried out by placing the catalyst on a fixed bed or a moving bed. The modern tendency is to use what is known as the fluid-catalyst technique. The finely powdered catalyst is mixed with air to make it flow like a liquid. The quality of cracked petrol is further improved by a reforming process. The reforming is done by passing the petrol, under pressure, rapidly over a catalyst in a steel pipe, kept heated. The time of contact is very short, approximately 15 seconds. This process converts the straight-chain hydrocarbons into branched-chain hydrocarbons, and thus contributes to an increase in the octane rating of the fuel.

PETROCHEMICALS

Petroleum, in addition to its use as a fuel, has recently become the raw material for the production of industrial chemicals. During the last twenty-five years hydrocarbons obtained from petroleum have been increasingly used as chemical intermediates. The chemicals that stem from petroleum as a raw material are conveniently called **petrochemicals**, and this infant industry, which has assumed such large dimensions, is called the **petrochemical industry**. Increasing amounts of chemicals processed from petroleum are supplementing available quantities from coal, coal tar, coke, and agricultural sources. Most of these chemicals are needed to manufacture products like synthetic rubber, resins, fibres, etc., and the petroleum hydrocarbons have indeed become the vital raw material, especially when other supplies of the chemicals are either inadequate or their sources are depleted.

The rapid growth of the industry can be gauged from the following figures. In 1920 no appreciable amount of chemicals was made from petroleum. Petroleum is now the chief starting material for the organic chemistry industry.

The paraffin hydrocarbons from petroleum are subjected to three types of reactions—**oxidation**, **halogenation**, and **nitration**. The oxidation of paraffins gives rise to alcohols, aldehydes, ketones, and acids. The chlorination of paraffins, either in the liquid or in the vapour phase, with or without catalysts, gives chlorine compounds, which in turn could be converted into alcohols or alkenes. The nitration of paraffins gives nitroparaffins which can be converted into amines, aldehydes, or ketones by methods to be discussed later.

ALKENES

The general formula for the alkenes (*olefines*) family of hydrocarbons is C_nH_{2n}. The alkanes, as has been previously pointed out, are saturated hydrocarbons in which the carbon atoms are united by single bonds. The alkenes are characterized by the presence of a double bond between two carbon atoms. If we write down the graphical formula of ethyl alcohol and imagine the removal of the elements of water from it, say by using sulphuric acid, which has a great affinity for water, we may picture the dehydration to proceed in the manner outlined below:

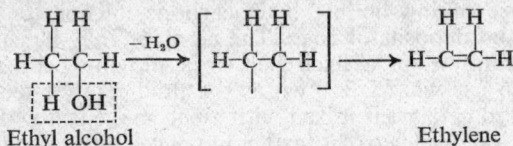

$$
\underset{\text{Ethyl alcohol}}{\begin{array}{c} H\ \ H \\ | \ \ | \\ H-C-C-H \\ |\ \ \ | \\ \overline{H\ \ OH} \end{array}} \xrightarrow{-H_2O} \left[\begin{array}{c} H\ \ H \\ | \ \ | \\ H-C-C-H \\ |\ \ \ | \end{array} \right] \longrightarrow \underset{\text{Ethylene}}{\begin{array}{c} H\ \ H \\ | \ \ | \\ H-C=C-H \end{array}}
$$

In the resulting compound two valences, one belonging to each carbon, have been released owing to the removal of H and ^{-}OH previously attached to them. In other words, in the resulting molecule two valences are free and are not used for purposes of combination with other elements. The free valences join together, and the structure of the resulting compound is written, as above, with a double bond between the two carbon atoms. This compound is called ethylene, and the series of hydrocarbons it heads is referred to as alkenes. Compounds in which there are one or more double bonds (or triple bonds) are referred to as **unsaturated compounds**. The unsaturation here is with respect to valence and denotes a state of affairs where all the valences of the carbon atom have not been used for combination with other elements. The names of unsaturated hydrocarbons end in the termination *ene*. Thus, alkene is the family name and ethylene (ethene is the I.U.C. name) (C_3H_6), propene, and butene (C_4H_8) are the names of the first three members of the series.

Because ethylene was found to react with chlorine or bromine to give oily addition products it was called 'Olefiant gas' or oil-forming gas. From this name the word olefine has come to denote substances related to ethylene. The alkenes can be regarded as being derived from the paraffins by the abstraction of two atoms of hydrogen, to form a double bond.

PREPARATION: Ethylene can be prepared from ethyl alcohol by treating it with sulphuric acid at 170° or phosphoric acid at 200° or catalytically by alumina at 350°, as shown below:

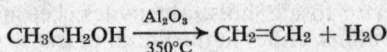

$$
CH_3CH_2OH \xrightarrow[350°C]{Al_2O_3} CH_2{=}CH_2 + H_2O
$$

Other alcohols can also be treated likewise to give different alkenes in varying yields. Alkenes can also be prepared by the removal of hydrogen halide from alkyl halides thus:

$$\underset{\substack{||| \\ HHBr}}{H-C-C-C-H} \xrightarrow[\text{(Alcoholic)}]{+KOH} \underset{\substack{|\\H}}{\overset{\substack{HHH\\|||}}{H-C-C=C-H}} + KBr + H_2O$$

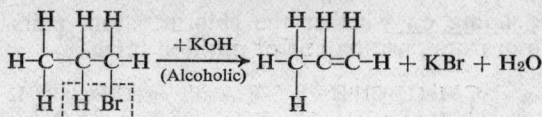

An acid, hydrogen bromide, has to be removed; hence the use of an alkali like potassium hydroxide. The reaction is best accomplished in alcoholic solution. However, ethylene cannot be prepared from an ethyl halide in this way. This process, referred to as **dehydrohalogenation**, is often used to introduce a double bond into an organic compound. The prefix *de-* means removal of, and hence dehydrohalogenation signifies the removal of hydrogen halide. The removal of the elements of water from ethyl alcohol to produce ethylene is similarly referred to as dehydration. We shall later discuss processes called dehydrogenation (removal of hydrogen), decarboxylation (removal of a carboxyl group), and deamination (removal of an amino group).

Industrially, ethylene and propene are prepared by cracking ethane and propane respectively in the presence of catalysts, usually chromium or molybdenum oxides supported on aluminium oxide, as shown below.

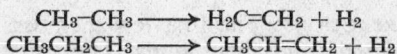

$$CH_3-CH_3 \longrightarrow H_2C=CH_2 + H_2$$
$$CH_3CH_2CH_3 \longrightarrow CH_3CH=CH_2 + H_2$$

Applied to higher alkanes, the method is less useful since complex mixtures of alkenes result.

NOMENCLATURE: Alkenes are named under the I.U.C. rules by slight modifications of the rules laid down for naming alkanes. The longest chain containing the double bond is selected, and the carbon atoms numbered from the end which gives the smallest numbers to the double-bonded carbon atoms. The location of the double bond is specified by referring to the number of the doubly bound carbon atom nearest to carbon atom 1 and the ending *ene* is used to indicate an olefin. For example, in the structure below, the longest chain containing the double bond is numbered as indicated and the compound named 3-ethyl-4-methyl-hexene-1:

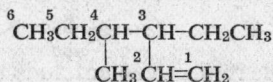

$$\overset{6}{C}H_3\overset{5}{C}H_2\overset{4}{C}H-\overset{3}{C}H-CH_2CH_3$$
$$\underset{\substack{|\\CH_3}}{}\underset{\substack{2\\CH_3}}{|}\overset{1}{C}H=CH_2$$

SAYTZEFF'S RULE: From a study of the dehydration and the dehydrohalogenation of several alcohols and alkyl halides, an empirical rule known as SAYTZEFF'S RULE has been enunciated. According to this rule, in these reactions the hydrogen is preferentially eliminated from the carbon atom with the smaller number of hydrogen atoms. Let us take a simple illustration. Supposing we have a compound with the structure shown below:

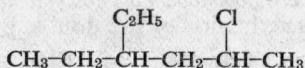

$$\underset{\substack{}}{CH_3-CH_2-\overset{\substack{C_2H_5\\|}}{C}H-CH_2-\overset{\substack{Cl\\|}}{C}H-CH_3}$$

Dehydrohalogenation of this compound can produce two isomeric alkenes, depending on whether the chlorine atom pairs off with a hydrogen atom from position one or position three.

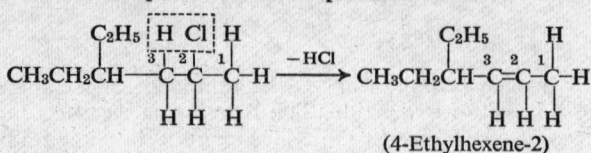

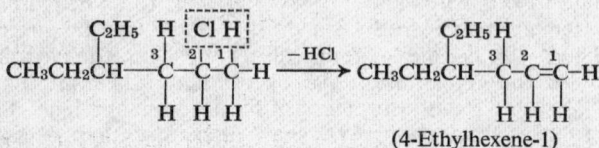

$$CH_3CH_2CH \overset{C_2H_5}{\underset{H}{|}} \overset{H \ Cl}{\underset{H}{|}} \overset{H}{\underset{H}{|}} \xrightarrow{-HCl} CH_3CH_2CH \overset{C_2H_5}{\underset{H}{|}} = C \overset{H}{\underset{H}{|}} - C - H$$

(4-Ethylhexene-2)

$$CH_3CH_2CH \overset{C_2H_5}{\underset{H}{|}} \overset{H \ Cl \ H}{\underset{H}{|}} \xrightarrow{-HCl} CH_3CH_2CH \overset{C_2H_5 H}{\underset{H}{|}} - C = C - H$$

(4-Ethylhexene-1)

Saytzeff's rule states that the first alternative is favoured and 4-ethyl-hexene-2 is the principal product, though the other product may be formed in smaller amounts. Saytzeff's rule seems to be a case of 'poor becoming poorer'; the carbon atom, having fewer hydrogen atoms, loses one more hydrogen atom.

PROPERTIES AND REACTIONS: The first few members are gases, higher members being liquids and then solids. The usual gradation in the physical properties—boiling point, viscosity, density, etc.—observable in the homologous series of alkanes is noted here also.

Chemically, the alkenes are much more reactive than the alkanes. Although the double bond between two carbon atoms is a stronger link than a single bond, it is not twice as strong: the second bond formed is weaker than the first, making it more vulnerable to attack by suitable reagents. The unsaturated carbon atoms, each with a valence which has not been fully utilised for combination, tend to become saturated under fairly mild conditions. For example, if chlorine is mixed with ethylene two chlorine atoms become linked to the doubly bonded carbon atoms, as follows:

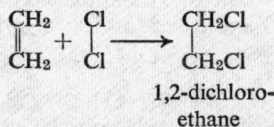

$$\overset{CH_2}{\underset{CH_2}{\|}} + \overset{Cl}{\underset{Cl}{|}} \longrightarrow \overset{CH_2Cl}{\underset{CH_2Cl}{|}}$$

1,2-dichloro-ethane

This type of reaction, where one reactant adds on to another reactant to give a single product, is known as an **addition reaction** and is characteristic of unsaturated compounds. Indeed, it forms the basis of one of the standard practical tests for unsaturation. The compound being tested is shaken with a red-brown solution of bromine (another halogen like chlorine) in chloroform or carbon tetrachloride. The bromine will add on to the compound if it is unsaturated and the red-brown colour will disappear rapidly. Hydrogen can be added to alkenes to furnish alkanes in the presence of nickel as a catalyst. This process is referred to as hydrogenation or reduction of the double bond, as shown below:

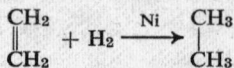

$$\overset{CH_2}{\underset{CH_2}{\|}} + H_2 \xrightarrow{Ni} \overset{CH_3}{\underset{CH_3}{|}}$$

Hydrogen halide gases add on to ethylene in a similar way:

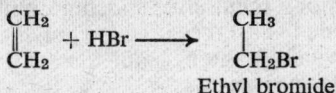

$$\begin{array}{c} CH_2 \\ \| \\ CH_2 \end{array} + HBr \longrightarrow \begin{array}{c} CH_3 \\ | \\ CH_2Br \end{array}$$

Ethyl bromide

Ethylene is a symmetrical alkene, that is the portions on either side of the double bond in the structure of ethylene are the same. Only a single product can result from the addition of the elements of a halogen acid HX to the double bond. In the case of an unsymmetrical alkene like propene, such an addition with hydrogen bromide may occur in two ways:

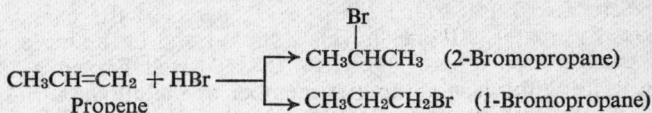

$$CH_3CH{=}CH_2 + HBr \underset{\text{Propene}}{\overset{}{\longrightarrow}} \begin{cases} \overset{Br}{\underset{|}{CH_3CHCH_3}} \quad \text{(2-Bromopropane)} \\ CH_3CH_2CH_2Br \quad \text{(1-Bromopropane)} \end{cases}$$

MARKOWNIKOFF enunciated an empirical rule named after him which enables one to predict the course of addition of an unsymmetrical reagent like a hydrogen halide to an unsymmetrical alkene. According to Markownikoff the negative part of such an addendum (the halogen in the above case) attaches itself to that carbon atom which is poorer in hydrogen. In propene, carbon atom 1 has two hydrogen atoms and carbon atom 2 has only one hydrogen atom. The rule, therefore, predicts the formation of 2-bromopropane, which is actually the product.

The alkenes are readily oxidized when treated with potassium permanganate. The term oxidation in a narrow sense is used to refer to a reaction whereby a substance is enriched in oxygen. Ethylene, for example, gives ethylene glycol, a dihydroxy compound when oxidized with potassium permanganate in neutral or mildly alkaline solution, as shown below:

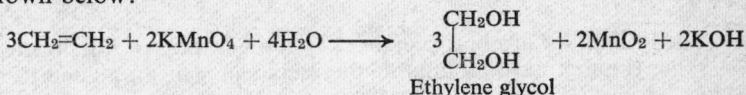

$$3CH_2{=}CH_2 + 2KMnO_4 + 4H_2O \longrightarrow 3\begin{array}{c} CH_2OH \\ | \\ CH_2OH \end{array} + 2MnO_2 + 2KOH$$

Ethylene glycol

During the oxidation, the purple colour of the permanganate solution disappears and the reaction constitutes a test, known as BAEYER'S TEST, to detect unsaturation in any compound. More drastic conditions of oxidation cause rupture of the molecule at the site of the double bond.

Alkenes react with ozone (O_3) to give ozonides. The latter when decomposed with water furnish ketones or aldehydes:

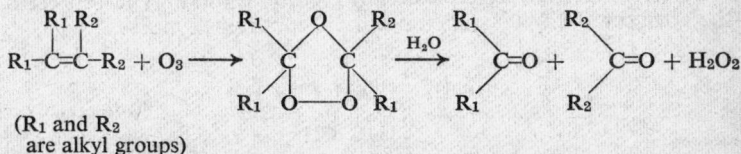

$$R_1{-}\overset{\overset{R_1 \quad R_2}{|\quad|}}{C}{=}\overset{}{C}{-}R_2 + O_3 \longrightarrow \begin{array}{c} R_1 \quad O \quad R_2 \\ \diagup \ \diagup \ \diagdown \\ C \qquad C \\ \diagup \quad \diagdown \diagup \quad \diagdown \\ R_1 \quad O{-}O \quad R_1 \end{array} \xrightarrow{H_2O} \begin{array}{c} R_1 \\ \diagdown \\ C{=}O \\ \diagup \\ R_1 \end{array} + \begin{array}{c} R_2 \\ \diagdown \\ C{=}O \\ \diagup \\ R_2 \end{array} + H_2O_2$$

(R₁ and R₂ are alkyl groups)

The position of a double bond in an unknown, unsaturated compound is often determined by forming its ozonide, treating it with water, and identifying the products.

Just as alkenes form addition compounds with many other substances, they may unite among themselves. Such a combination of a substance with itself or a related substance is called **polymerization**. The discussion of this reaction, which is the basis of the synthetic rubber industry, is deferred to Chapter Nineteen.

ALKYNES OR ACETYLENES

Alkynes or acetylenes have the general formula C_nH_{2n-2}, and compared with a paraffin have four fewer hydrogen atoms. Hence the alkynes are also unsaturated like the alkenes, but the extent of unsaturation is greater. Alkenes have a double bond in the molecule and the alkynes a triple bond. Acetylene (C_2H_2), well known as the gas used in welding torches, is the first member of the alkyne series and is represented by the structural formula $H-C\equiv C-H$. This means that six electrons are shared between the two carbon atoms, and the equivalent electronic structure is H:C:::C:H. There are four unsatisfied valences, two on each of the triply bonded carbon atoms, and, as is to be expected from our knowledge of the alkenes, the alkynes are very reactive.

PREPARATION: Alkynes are generally prepared by the removal of two moles of hydrogen halide from organic dihalides containing a halogen atom on each of two adjacent carbon atoms. The reagent used to effect this dehydrohalogenation is alcoholic potassium hydroxide, as in the case of alkenes:

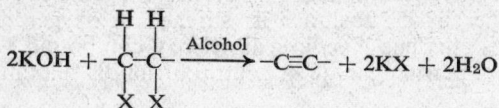

$$2KOH + \begin{matrix} H & H \\ | & | \\ -C-C- \\ | & | \\ X & X \end{matrix} \xrightarrow{\text{Alcohol}} -C\equiv C- + 2KX + 2H_2O$$

Since such dihalides are best prepared by the addition of a halogen to an alkene, the method amounts to the conversion of an alkene to an alkyne.

A cheap and convenient method of making acetylene consists of treating calcium carbide with water:

$$CaC_2 + 2H_2O \longrightarrow C_2H_2 + Ca(OH)_2$$

NOMENCLATURE: Alkynes are named in the same way as alkenes except that the termination *yne* is used. For example, the alkyne represented by the structure below is called 3-ethyl-4-methylhexyne-1:

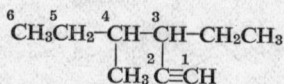

$$\overset{6}{C}H_3\overset{5}{C}H_2-\overset{4}{C}H-\overset{3}{C}H-CH_2CH_3$$
$$| \quad 2 | \quad 1$$
$$CH_3 \quad C\equiv CH$$

REACTIONS: Alkynes give addition products with most of the reagents that add on to alkenes, the only difference being that the addition takes place twice over. Bromine, for example, adds to acetylene as follows:

$$\text{HC}{\equiv}\text{CH} + \text{Br}_2 \longrightarrow \overset{\overset{\displaystyle \text{Br} \quad \text{Br}}{|\quad\;\;|}}{\text{HC}{=}\text{CH}} \overset{\text{Br}_2}{\longrightarrow} \overset{\overset{\displaystyle \text{Br} \quad \text{Br}}{|\quad\;\;|}}{\underset{\underset{\displaystyle \text{Br} \quad \text{Br}}{|\quad\;\;|}}{\text{H}-\text{C}-\text{C}-\text{H}}}$$

1,1,2,2-Tetra-
bromoethane

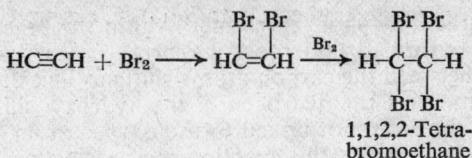

The product after the addition of the first mole of bromine is still unsaturated, and hence adds on a second mole. Because it is more reactive than bromine, chlorine reacts explosively with acetylene on mixing, producing black smoke and hydrogen chloride gas:

$$\text{C}_2\text{H}_2 + \text{Cl}_2 \longrightarrow 2\text{C} + 2\text{HCl}$$

Dichloroethylene can be made by absorbing acetylene in antimony pentachloride and then distilling the addition compound:

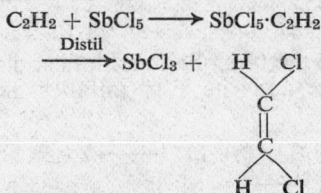

$$\text{C}_2\text{H}_2 + \text{SbCl}_5 \longrightarrow \text{SbCl}_5{\cdot}\text{C}_2\text{H}_2$$

$$\overset{\text{Distil}}{\longrightarrow} \text{SbCl}_3 + \quad \underset{\underset{\displaystyle \text{H} \quad\quad \text{Cl}}{\underset{\displaystyle \text{C}}{\|}}}{\overset{\overset{\displaystyle \text{H} \quad\quad \text{Cl}}{\diagdown\;\;\diagup}}{\text{C}}}$$

Tetrachloroethane, or 'Westron', $\text{CHCl}_2{\cdot}\text{CHCl}_2$ is formed if excess antimony pentachloride is used.

The addition of hydrogen chloride to acetylene produces vinyl chloride —an important chemical in the manufacture of plastics.

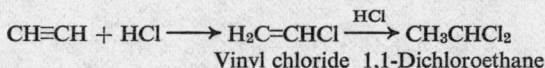

$$\text{CH}{\equiv}\text{CH} + \text{HCl} \longrightarrow \text{H}_2\text{C}{=}\text{CHCl} \overset{\text{HCl}}{\longrightarrow} \text{CH}_3\text{CHCl}_2$$

Vinyl chloride 1,1-Dichloroethane

The addition of acetic acid to acetylene similarly furnishes vinyl acetate, also used in the manufacture of plastics.

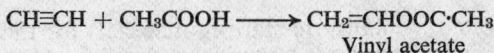

$$\text{CH}{\equiv}\text{CH} + \text{CH}_3\text{COOH} \longrightarrow \text{CH}_2{=}\text{CHOOC}{\cdot}\text{CH}_3$$

Vinyl acetate

Acetylene adds on hydrogen cyanide in the presence of a barium cyanide–carbon catalyst to give acrylonitrile, a product used in industry to make certain types of synthetic rubber substitutes and also to make the synthetic fibre orlon (see Chapter Nineteen).

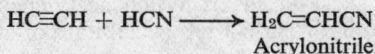

$$\text{HC}{\equiv}\text{CH} + \text{HCN} \longrightarrow \text{H}_2\text{C}{=}\text{CHCN}$$

Acrylonitrile

When acetylene is bubbled into an aqueous solution containing sulphuric acid and mercuric sulphate, water adds to the triple bond:

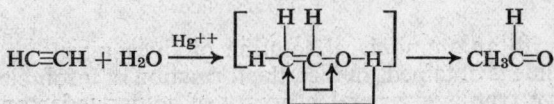

$$\text{HC}{\equiv}\text{CH} + \text{H}_2\text{O} \overset{\text{Hg}^{++}}{\longrightarrow} \left[\overset{\overset{\displaystyle \text{H} \quad \text{H}}{|\quad|}}{\text{H}-\text{C}{=}\text{C}-\text{O}-\text{H}} \right] \longrightarrow \text{CH}_3\overset{\overset{\displaystyle \text{H}}{|}}{\text{C}}{=}\text{O}$$

The unsaturated alcohol represented within the bracket is unstable, and as soon as formed undergoes a rearrangement in structure. The double bond between the carbon atoms shifts to link the oxygen atom, and at the same time the hydrogen of the hydroxyl shifts to the other carbon atom, the atom not linked to the oxygen atom. The product is **acetaldehyde.** In industry this reaction is used in the manufacture of acetic acid by oxidation of acetaldehyde. Acetylene adds on **arsenic trichloride** ($AsCl_3$) in the presence of anhydrous aluminium chloride as catalyst to give Lewisite, an effective war gas. It is less persistent than mustard gas (see Chapter Eight), but acts more rapidly. Inhalation of Lewisite vapour for ten minutes at a concentration of 0·12 mg. per litre of air is fatal. The reaction is represented as follows:

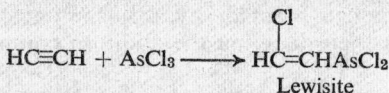

$$HC \equiv CH + AsCl_3 \longrightarrow \overset{\displaystyle Cl}{HC = CHAsCl_2}$$
Lewisite

In the presence of a mixture of ammonium and cuprous chloride, one molecule of acetylene adds itself to another molecule to give vinyl acetylene.

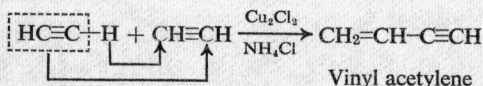

$$HC \equiv C-H + CH \equiv CH \xrightarrow[NH_4Cl]{Cu_2Cl_2} CH_2 = CH-C \equiv CH$$
Vinyl acetylene

Vinyl acetylene is an important product in the manufacture of Neoprene, a synthetic rubber (see Chapter Nineteen).

With formaldehyde, acetylene gives propargyl alcohol which is a mono-addition product, and 2-butyne-1,4-diol which is a diaddition product:

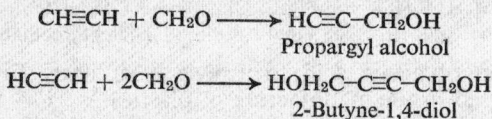

$$CH \equiv CH + CH_2O \longrightarrow HC \equiv C-CH_2OH$$
Propargyl alcohol

$$HC \equiv CH + 2CH_2O \longrightarrow HOH_2C-C \equiv C-CH_2OH$$
2-Butyne-1,4-diol

These unsaturated alcohols are starting materials for the manufacture of the industrially important chemicals, **acrylic acid** and **butadiene.** Several new reactions of acetylene were discovered by the German chemist REPPE during the Second World War. They have added to the importance of acetylene in industry.

The hydrogen atoms in acetylene are acidic and can be replaced by such metals as copper and silver. When acetylene is passed through cuprous ammonium hydroxide a red precipitate of cuprous acetylide is obtained:

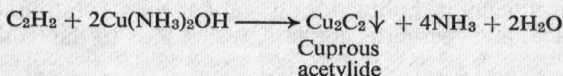

$$C_2H_2 + 2Cu(NH_3)_2OH \longrightarrow Cu_2C_2 \downarrow + 4NH_3 + 2H_2O$$
Cuprous
acetylide

Similarly, with an ammoniacal solution of silver nitrate the insoluble silver acetylide is obtained. In fact, the formation of insoluble acetylides of the above type is a general property of compounds containing an acetylenic hydrogen atom, that is a hydrogen atom attached to a triply bonded carbon atom. The acetylides are somewhat explosive in the dry state.

Of purely theoretical interest is the polymerization of acetylene on passing through a red-hot tube of benzene.

$$3C_2H_2 \longrightarrow C_6H_6$$

The polymerization can be envisaged as being due to the successive addition of three molecules of acetylene to one another.

SUMMARY

Alkanes or Paraffins: General formula C_nH_{2n+2}; preparation: (i) Reduction of RX to RH. (ii) Wurtz synthesis: $RX + R'X \xrightarrow{Na} R-R' + 2NaX$ (R and R' may be same or different). Methane by action of water on aluminium carbide. Reactions: Chlorination of methane gives CH_3Cl, CH_2Cl_2, $CHCl_3$, and CCl_4. Nitration in the vapour phase. Isomerism among alkanes. Two butanes and three pentanes. I.U.C. nomenclature for naming alkanes.

Petroleum. Occurrence of alkanes in petroleum and natural gas. Economic importance as a fuel and as a raw material for the manufacture of a number of chemicals. Theories of the formation of petroleum —probably formed by the chemical action on plant remains entrapped in the lower strata of the earth. Fractionation of petroleum. The importance of the 'cut' known as petrol or gasoline—standardization of petrol—idea of octane number. Methods of supplementing petrol by cracking of heavy oils or by synthetic methods which include Fischer–Tropsch process of hydrogenation of carbon monoxide and hydrogenation of coal.

Alkenes. General formula C_nH_{2n}. Preparation of alkenes from ethyl alcohol by dehydration with sulphuric acid at 170° or by dehydrohalogenation of alkyl halides. Industrial method involves cracking of ethane. Nomenclature of alkenes. Saytzeff's rule to predict alkene formed in dehydration and dehydrohalogenation. Hydrogen eliminated from the carbon atom with a smaller number of hydrogen atoms (poor becoming poorer). Reactions: addition of X_2, H_2, HX. Markownikoff's rule whereby the negative part of an addendum becomes linked to the carbon atom with the fewest hydrogen atoms. Oxidation of double bond with potassium permanganate. Baeyer's test. Ozonides and their use in locating position of double bond.

Alkynes. General formula C_2H_{2n-2}. Contain $-C{\equiv}C-$. Prepared by treating 1,2-dihalides with alcoholic potassium hydroxide. Acetylene prepared in the laboratory from calcium carbide. Addition of Br_2, HCN, HCl, CH_3COOH, $AsCl_3$ to acetylene. Hydration of acetylene to acetaldehyde. Self-addition to give vinyl acetylene. Condensation with formaldehyde. Formation of copper and silver acetylides. Trimerization of acetylene to benzene.

Problem Set No. 2

1. Write the structural formula for each of the hydrocarbons with the molecular formula C_6H_{14} and name them according to the I.U.C. system.
2. Four structural formulae are given below along with a proposed name for each of them. Correct the names given, wherever necessary.

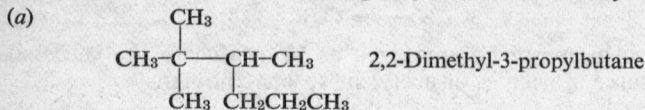

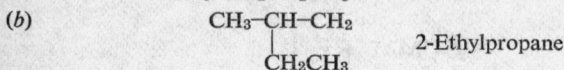

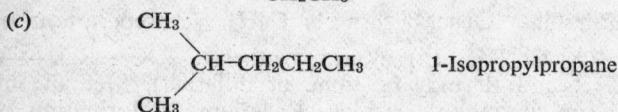

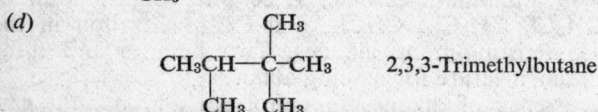

(a)

$$CH_3$$
$$CH_3-C\ \ \ \ CH-CH_3 \qquad \text{2,2-Dimethyl-3-propylbutane}$$
$$CH_3 \ \ CH_2CH_2CH_3$$

(b)

$$CH_3-CH-CH_2$$
$$\qquad\qquad\qquad \text{2-Ethylpropane}$$
$$CH_2CH_3$$

(c)

$$CH_3$$
$$\quad CH-CH_2CH_2CH_3 \qquad \text{1-Isopropylpropane}$$
$$CH_3$$

(d)

$$CH_3$$
$$CH_3CH-C-CH_3 \qquad \text{2,3,3-Trimethylbutane}$$
$$CH_3 \ \ CH_3$$

3. A mixture of methyl iodide and ethyl iodide is heated with sodium. What are the products that may be formed?
4. Give the names of the products of the following reactions:

   (a) Treatment of one mole of acetylene with one mole of bromine.
   (b) Treatment of one mole of methylacetylene (propyne) with two moles of hydrogen chloride.

5. How can the following transformations be carried out?

   (a) Ethanol to acetylene.
   (b) Ethanol to 1,1-dibromoethane.
   (c) Ethylene to 1,1,2,2-tetrachloroethane.

CHAPTER FOUR

HALOGEN DERIVATIVES OF THE PARAFFINS

ALKYL HALIDES

The mono-halogen derivatives of the alkanes, that is those which contain one halogen atom instead of a hydrogen atom of the alkanes, are known as alkyl halides. They are given the general formula RX, in which R is an alkyl radical and X is a halogen. The C–Cl bond is a polar bond in which the chlorine atom is held more loosely to the carbon than are the hydrogens. The importance of the alkyl halides arises from the fact that the halogen atom in them is a centre of reactivity in an otherwise indifferent alkane chain. Hence, starting from alkyl halides, by suitably bundling away the halogen (which is rather loosely held) by treatment with metallic sodium or compounds of potassium or silver, a variety of compounds can be prepared.

NOMENCLATURE: The binary system of naming, so common in inorganic chemistry, is employed for designating simple organic compounds like the alkyl halides. These are named as if they were compounds of two elements—one the inorganic halogen and the other the organic radical. For example, CH_3Cl is methyl chloride, CH_3CH_2I is ethyl iodide (cf. with NaCl, which is sodium chloride). Compounds with more complex structures are named according to the I.U.C. rules, choosing the chain containing the halogen as the longest chain and numbering it to give the smallest number to the carbon atom attached to the halogen atom. Thus $BrCH_2CH_2CH_2Br$ is 1,3-dibromopropane, $ClCH_2CH=CHCH_3$ is

$$CH_2Br$$
$$|$$

1-chloro-2-butene, and $CH_3CH_2\overset{|}{\underset{|}{C}}-CH_3$ is 1-bromo-2,2-dimethylbutane,

$$CH_3$$

etc.

PREPARATION: Alkyl halides can be prepared by the halogenation of the alkanes, but usually several different products are obtained, depending upon the extent of halogenation. It is cumbersome to separate these products. The alkyl halides are, in general, prepared from the corresponding alcohols which are readily available. The hydroxyl group in an alcohol can be replaced by a halogen using phosphorus halides or thionyl chloride. Thus methyl iodide can be prepared by carefully adding iodine in small amounts to methyl alcohol and red phosphorus kept in a flask. The phosphorus iodide produced *in situ* reacts with the alcohol giving methyl iodide.

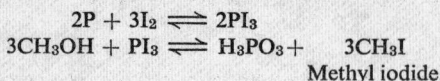

$$2P + 3I_2 \rightleftharpoons 2PI_3$$
$$3CH_3OH + PI_3 \rightleftharpoons H_3PO_3 + \quad 3CH_3I$$
Methyl iodide

Similarly, ethyl chloride can be prepared by adding phosphorus trichloride to ethyl alcohol:

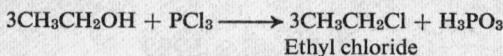

$$3CH_3CH_2OH + PCl_3 \longrightarrow 3CH_3CH_2Cl + H_3PO_3$$
Ethyl chloride

With thionyl chloride the reaction is represented as follows:

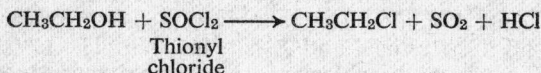

$$CH_3CH_2OH + SOCl_2 \longrightarrow CH_3CH_2Cl + SO_2 + HCl$$
Thionyl
chloride

PROPERTIES: The alkyl halides, excepting the first few members, are pleasant-smelling, heavy liquids immiscible with water. Their chemical properties centre round the mobility of the halogen atom which can be replaced by different groups giving rise to a variety of compounds. The reactivities increase in the order Cl, Br, I because the strength of the C–X bond is greatest for chlorine (and so least easily broken) and smallest for iodine (and so most easily broken). We shall illustrate this statement with a few examples, using ethyl iodide as a representative.

(I). When ethyl iodide is treated with sodium in an ethereal solution two molecules unite to give butane. This is the Wurtz synthesis, previously mentioned.

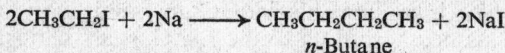

$$2CH_3CH_2I + 2Na \longrightarrow CH_3CH_2CH_2CH_3 + 2NaI$$
n-Butane

(II). Ethyl iodide when heated with salts and compounds resembling salts gives products in which the ethyl group is exchanged for the metal of the salt. For example, with alcoholic potassium cyanide it gives ethyl cyanide (also called propionitrile).

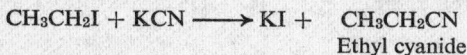

$$CH_3CH_2I + KCN \longrightarrow KI + \quad CH_3CH_2CN$$
Ethyl cyanide

With silver nitrite, nitroethane is formed.

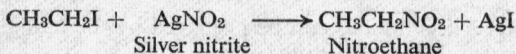

$$CH_3CH_2I + \quad AgNO_2 \longrightarrow CH_3CH_2NO_2 + AgI$$
Silver nitrite   Nitroethane

With silver salts of organic acids, ethyl esters result.

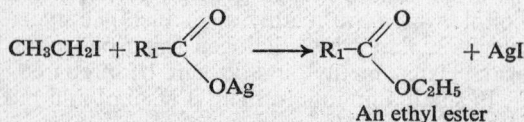

$$CH_3CH_2I + R_1-C{\scriptsize\begin{matrix}O\\OAg\end{matrix}} \longrightarrow R_1-C{\scriptsize\begin{matrix}O\\OC_2H_5\end{matrix}} + AgI$$
An ethyl ester

The reaction with sodium alkoxides serves as the basis for preparing ethers (referred to as Williamson's reaction).

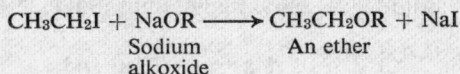

$$CH_3CH_2I + NaOR \longrightarrow CH_3CH_2OR + NaI$$

Sodium   An ether
alkoxide

When heated with potassium hydrogen sulphide an unpleasant-smelling product called ethyl mercaptan is obtained:

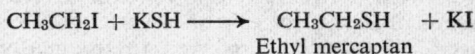

$$CH_3CH_2I + KSH \longrightarrow CH_3CH_2SH + KI$$

Ethyl mercaptan

Ethyl mercaptan serves as the starting material for the preparation of a hypnotic called sulphonal. Ethyl mercaptan is also added to natural gas to serve as a warning agent in case a leak develops. In most of the exchange reactions just discussed the iodides are more reactive than bromides, which in turn react faster than chlorides.

(III). The alkyl halides undergo a reaction referred to as hydrolysis, which is often encountered in organic chemistry. Essentially, hydrolysis consists of the splitting of a compound by water into two components formed by the addition of hydrogen and hydroxyl respectively. Ethyl iodide, for instance, undergoes hydrolysis to give ethyl alcohol and hydriodic acid.

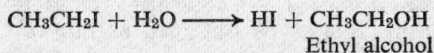

$$CH_3CH_2I + H_2O \longrightarrow HI + CH_3CH_2OH$$

Ethyl alcohol

This reaction proceeds faster in the presence of a base which neutralizes the acid formed.

$$HI + NaOH \longrightarrow NaI + H_2O$$

Experimentally, it has been established that a primary halide of the type RCH_2X hydrolyses faster than a secondary halide of the type R_1R_2CHX. Tertiary halides of the type $R_1R_2R_3CX$ are more reactive than secondary halides and sometimes even more reactive than primary halides. These differences in reactivity have been attributed to differences in the mechanism of hydrolysis of primary and secondary halides, on the one hand, and tertiary halides, on the other hand.

(IV). Alkyl halides above ethyl halides are converted to alkenes when boiled with a solution of potassium hydroxide in ethyl alcohol. Aqueous base is less efficient in this reaction and promotes hydrolysis. For example, propyl iodide furnishes propylene with alcoholic potassium hydroxide.

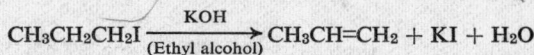

$$CH_3CH_2CH_2I \xrightarrow[\text{(Ethyl alcohol)}]{KOH} CH_3CH=CH_2 + KI + H_2O$$

This type of reaction is referred to as an elimination reaction, since a molecule of hydrogen iodide is eliminated from ethyl iodide. Tertiary halides are more likely to undergo the elimination reaction than secondary halides, which in turn are more reactive than primary halides.

(V). Alkyl halides react with magnesium in dry ether giving alkyl magnesium halides, also known as Grignard reagents because the reaction was discovered by the French chemist VICTOR GRIGNARD. For example, ethyl iodide gives ethyl magnesium iodide.

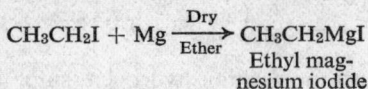

$$CH_3CH_2I + Mg \xrightarrow[\text{Ether}]{\text{Dry}} CH_3CH_2MgI$$
$$\text{Ethyl mag-}$$
$$\text{nesium iodide}$$

Ether is most often the preferred solvent for preparing a Grignard reagent. By means of Grignard reagents a variety of organic substances can be prepared. They will be referred to frequently in succeeding chapters.

USES OF ALKYL HALIDES: The several reactions of alkyl halides previously discussed indicate the different types of transformation that they could be subjected to, and hence are useful intermediates in synthetic work. Methyl bromide, methyl chloride, and ethyl chloride are used as refrigerants. Ethyl chloride is also used as an anaesthetic.

POLYHALOGEN DERIVATIVES

Of the dihalogen derivatives of methane, only methylene dichloride has been put to use. It is used as a solvent for extraction purposes and also as a refrigerant. It may be obtained as one of the products of chlorination of methane or by reduction of carbon tetrachloride.

Of the trihalogen compounds, mention should be made of chloroform ($CHCl_3$), bromoform ($CHBr_3$), and iodoform (CHI_3). These compounds are made by the action of the halogen concerned on heating with ethyl alcohol in the presence of sodium hydroxide. The halogen reacts with sodium hydroxide to give sodium hypohalite, which oxidizes ethyl alcohol to acetaldehyde. The latter is substituted by a halogen and then cleaved as follows:

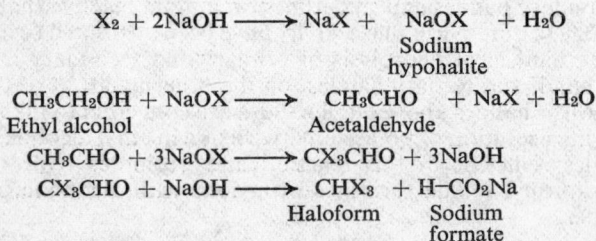

$$X_2 + 2NaOH \longrightarrow NaX + \underset{\substack{\text{Sodium} \\ \text{hypohalite}}}{NaOX} + H_2O$$

$$\underset{\text{Ethyl alcohol}}{CH_3CH_2OH} + NaOX \longrightarrow \underset{\text{Acetaldehyde}}{CH_3CHO} + NaX + H_2O$$

$$CH_3CHO + 3NaOX \longrightarrow CX_3CHO + 3NaOH$$

$$CX_3CHO + NaOH \longrightarrow \underset{\text{Haloform}}{CHX_3} + \underset{\substack{\text{Sodium} \\ \text{formate}}}{H-CO_2Na}$$

Acetone and, in fact, methyl ketones of the type $RCOCH_3$ similarly furnish haloforms when treated with sodium hypohalite. With acetone, for example, the reaction is as follows:

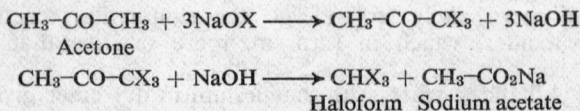

$$\underset{\text{Acetone}}{CH_3-CO-CH_3} + 3NaOX \longrightarrow CH_3-CO-CX_3 + 3NaOH$$

$$CH_3-CO-CX_3 + NaOH \longrightarrow \underset{\text{Haloform}}{CHX_3} + \underset{\text{Sodium acetate}}{CH_3-CO_2Na}$$

The reaction is known as the haloform reaction. Since iodoform is a yellow crystalline substance with a characteristic smell and melting point, the iodoform reaction serves as a test for ethyl alcohol, acetone, and in fact methyl ketones of the general formula $RCOCH_3$.

Chloroform is a colourless liquid with a sweet taste. It is made industrially by the haloform reaction and also by reduction of carbon tetrachloride with iron and dilute acid.

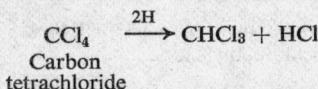

$$CCl_4 \xrightarrow{2H} CHCl_3 + HCl$$
$$\text{Carbon}$$
$$\text{tetrachloride}$$

Chloroform for anaesthesia must be protected from exposure to air or oxygen which in the presence of sunlight produces a poisonous gas called phosgene.

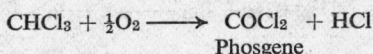

$$CHCl_3 + \tfrac{1}{2}O_2 \longrightarrow COCl_2 + HCl$$
$$\text{Phosgene}$$

Consequently, chloroform is usually stored in brown bottles and also contains small amounts of ethyl alcohol, which destroys traces of phosgene that may be formed. Iodoform, though formerly used as an antiseptic for wounds, has been supplanted by other substitutes.

Carbon Tetrachloride (CCl_4). Carbon tetrachloride is prepared industrially by exhaustive chlorination of methane and also by the reaction of chlorine with carbon disulphide in the presence of a catalyst like iodine.

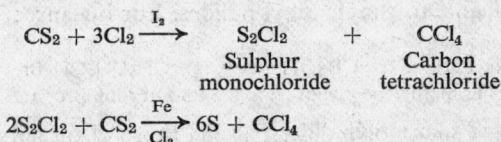

$$CS_2 + 3Cl_2 \xrightarrow{I_2} S_2Cl_2 + CCl_4$$
$$\qquad\qquad \text{Sulphur} \qquad \text{Carbon}$$
$$\qquad\qquad \text{monochloride} \quad \text{tetrachloride}$$

$$2S_2Cl_2 + CS_2 \xrightarrow[Cl_2]{Fe} 6S + CCl_4$$

The sulphur chloride formed in the first reaction above is treated further with carbon disulphide to produce more carbon tetrachloride. The sulphur is also used up and converted to carbon disulphide:

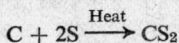

$$C + 2S \xrightarrow{Heat} CS_2$$

Although carbon tetrachloride is an excellent fire extinguisher, its use for this purpose around the home has been urgently discouraged for two reasons: its vapours are seriously toxic, and the generation of extremely poisonous phosgene also accompanies this use. Moreover, the use of carbon tetrachloride as a solvent around the home is equally dangerous and similarly discouraged.

Freon. Another completely halogenated derivative of methane, of commercial importance, is dichlorodifluoromethane (CCl_2F_2), commonly known as Freon. It is non-inflammable, non-toxic, and widely used as a refrigerant and as the propellant in aerosol canisters. Freon is prepared in industry from carbon tetrachloride and antimony fluoride:

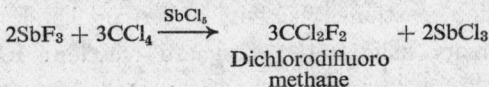

$$2SbF_3 + 3CCl_4 \xrightarrow{SbCl_5} 3CCl_2F_2 + 2SbCl_3$$
$$\text{Dichlorodifluoro}$$
$$\text{methane}$$

Fluorocarbons. Hydrocarbons in which all the hydrogen atoms are replaced by fluorine are used increasingly as lubricants and electrical insulators because of their remarkable stability to most chemicals. Teflon and Kel–F are two such fluorocarbon plastics made respectively by polymerization of tetrafluoroethylene and trifluorochloroethylene:

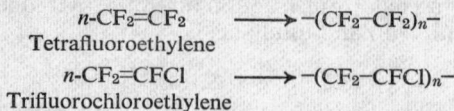

$$n\text{-}CF_2=CF_2 \longrightarrow -(CF_2-CF_2)_n-$$
Tetrafluoroethylene

$$n\text{-}CF_2=CFCl \longrightarrow -(CF_2-CFCl)_n-$$
Trifluorochloroethylene

Polyhalogen Compounds with More Than One Carbon Atom. The addition of halogens to alkenes gives halides with halogens attached to adjacent carbon atoms. For example:

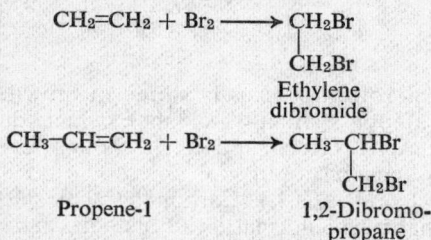

$$CH_2=CH_2 + Br_2 \longrightarrow \begin{array}{c} CH_2Br \\ | \\ CH_2Br \end{array}$$
Ethylene
dibromide

$$CH_3-CH=CH_2 + Br_2 \longrightarrow \begin{array}{c} CH_3-CHBr \\ | \\ CH_2Br \end{array}$$
Propene-1   1,2-Dibromo-
propane

Halides with halogens on different carbon atoms can be prepared by methods applicable to simple alkyl halides. For instance:

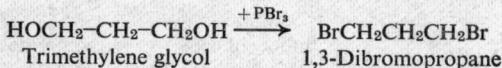

$$HOCH_2-CH_2-CH_2OH \xrightarrow{+PBr_3} BrCH_2CH_2CH_2Br$$
Trimethylene glycol   1,3-Dibromopropane

The reactions of such polyhalides parallel those of monohalides.

Ethylene chloride is used in large amounts to make the rubber substitute 'Thiokol'. It also serves as a source of vinyl chloride from which vinyl resins are made. Ethylene dibromide is a component of petrol. It reacts with the lead of tetraethyl lead to give volatile lead bromide, which is removed in the exhaust gases.

SUMMARY

Alkyl halides are monohalogen derivatives of alkanes. Prepared from ROH by reacting with red phosphorus and halogen, $SOCl_2$, PX_5, or PX_3 (X = halogen). Reactions: (I) Wurtz reaction. (II) Exchange reactions with potassium cyanide, silver nitrite, silver salt of an organic acid, sodium alkoxides, and potassium hydrogen sulphide. (III) Hydrolysis in the presence of alkali to ROH. Reactivity of a primary halide (RCH_2X) is greater than that of a secondary halide (R_1R_2CHX) which is less reactive than a tertiary halide ($R_1R_2R_3CX$). (IV) Alkali in ethanol promotes elimination reaction, that is $RCH_2CH_2X \rightarrow RCH=CH_2$. In such elimination reactions, reactivity of tertiary halides > secondary halides > primary halides. (V) Grignard reaction: $RX + Mg \xrightarrow{\text{Ether}} RMgX$. Uses of alkyl halides.

Polyhalogen Derivatives. Haloforms made by the action of halogen on ethanol or acetone in the presence of NaOH. Reaction consists in initial oxidation to acetaldehyde, halogenation, and then rupture of the product. Conversion of acetone to haloform is similar. Iodoform test for ethyl alcohol. Chloroform made industrially by the haloform reaction or reduction of carbon tetrachloride. Uses of chloroform and iodoform. Industrial preparation of carbon tetrachloride from methane or from carbon disulphide. Uses of carbon tetrachloride. Freon (CCl_2F_2) used in refrigerators made from carbon tetrachloride and antimony trifluoride. Teflon and Kel-F made from tetrafluoroethylene and trifluorochloroethylene, respectively.

1,2-Dihalides obtained by the addition of halogen to double bond, for example, ethylene → ethylene dichloride. 1,3-Dihalides from 1,3-dihydroxy compounds. Uses of ethylene dichloride and dibromide.

Problem Set No. 3

1. Write the structural formula for: (*a*) 1,5-dichloropentane; (*b*) 2-chloro-2-methylbutane; (*c*) 1-bromo-2-heptene; (*d*) 1-bromo-2-ethyl-2,3-dimethylbutane.
2. How are the following prepared?

    (*a*) 2-Bromopropane from 1-propanol.
    (*b*) 2-Nitropropane from 1-propanol.
    (*c*) *n*-Butane from ethanol.
    (*d*) Methyl ethyl ether from ethanol and methanol.
    (*e*) Butyne-1 from 1-bromobutane.
3. Would you prepare 1-chlorobutane by addition of hydrogen chloride to butane-1? Give reasons.
4. Complete the following equations:

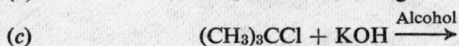

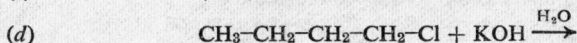

    (*a*) $\quad CH_3CH_2CH_2Br + NaCN \longrightarrow$

    (*b*) $\quad CH_3CHBrCH_3 + Mg \longrightarrow$

    (*c*) $\quad (CH_3)_3CCl + KOH \xrightarrow{\text{Alcohol}}$

    (*d*) $\quad CH_3-CH_2-CH_2-CH_2-Cl + KOH \xrightarrow{H_2O}$

CHAPTER FIVE

ALCOHOLS AND ETHERS

CLASSIFICATION OF ALCOHOLS

Alcohols may be considered either as alkyl derivatives of water or hydroxyl derivatives of paraffins, as indicated:

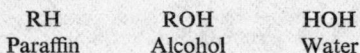

RH	ROH	HOH
Paraffin	Alcohol	Water

Their characteristic functional group is the hydroxyl or ^-OH group. The alcohols are further classified into monohydric, dihydric, trihydric, and polyhydric alcohols, depending on whether they contain one, two, three, or several ^-OH groups in the molecule. In the case of polyhydric alcohols, it has been found that more than one hydroxyl group cannot be attached to the same carbon atom. A system containing two hydroxyl groups attached to the same carbon atom is unstable and eliminates water in the same manner as carbonic acid.

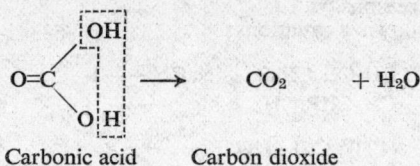

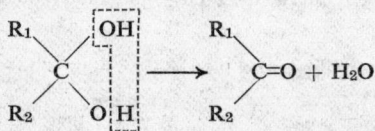

$$O=C \begin{matrix} OH \\ O\,H \end{matrix} \longrightarrow CO_2 + H_2O$$

Carbonic acid   Carbon dioxide

$$\begin{matrix} R_1 \\ R_2 \end{matrix} C \begin{matrix} OH \\ O\,H \end{matrix} \longrightarrow \begin{matrix} R_1 \\ R_2 \end{matrix} C=O + H_2O$$

The alcohols are named after the paraffins from which they are derived. Thus from methane we get the name methyl alcohol for CH_3OH, from ethane we get ethyl alcohol for CH_3CH_2OH and so on. The above are common names generally used for alcohols with less than five carbon atoms. The I.U.C. system of designation is preferred for alcohols with five or more carbon atoms. In this system the name is compounded from the stem name of the longest chain of carbon atoms in the molecule and the termination *ol*. The position of ^-OH is indicated by a numeral. For example, $\overset{4}{C}H_3\overset{3}{C}H_2\overset{2}{C}H_2\overset{1}{C}H_2OH$ is 1-butanol,

56

$$\overset{5}{C}H_3\overset{4}{C}H_2\overset{3}{C}H_2\overset{2}{C}HOH\overset{1}{C}H_3 \text{ is 2-pentanol, and } \overset{4}{C}H_3\overset{3}{C}H_2\overset{2}{\underset{\underset{CH_3}{|}}{\overset{\overset{1 CH_3}{|}}{C}}}-OH \text{ is 2-}$$

methyl-2-butanol and so on. The alcohols are further classified into primary, secondary, and tertiary alcohols. The functional groups contained by each are:

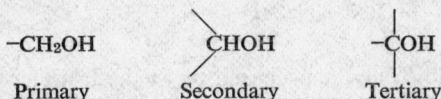

$$-CH_2OH \qquad \diagdown\!\!\!\!\underset{\diagup}{C}HOH \qquad -\underset{|}{\overset{|}{C}}OH$$

  Primary   Secondary   Tertiary

Before proceeding to give a general account of the preparation and properties of alcohols a special feature in the structure of alcohols should be mentioned. The electronic formula of an alcohol can be written as $R \!:\! \overset{..}{O} \!:\! H$, to indicate the distribution of electrons around the oxygen atom. When hydrogen is linked to one of the highly electronegative atoms, such as oxygen or nitrogen, the pair of electrons constituting the bond are not shared equally by both atoms, but spend a larger proportion of the time nearer the more electronegative atom. In other words, the bond has appreciable ionic nature, with the hydrogen atom carrying a fractional positive charge indicated by $\delta+$ and the oxygen or nitrogen atom carrying a fractional negative charge indicated by $\delta-$. The situation may be represented somewhat as follows:

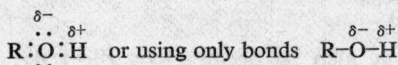

$$R \!:\! \overset{\delta-}{\underset{..}{\overset{..}{O}}} \!:\! \overset{\delta+}{H} \quad \text{or using only bonds} \quad R-\overset{\delta-}{O}-\overset{\delta+}{H}$$

The hydrogen atom of an alcohol molecule, with a fractional positive charge, will naturally exert an electrostatic attraction on the negatively charged oxygen atom of a second molecule of the alcohol thus:

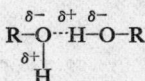

$$R-\overset{\delta-}{O}\cdots\overset{\delta+}{H}-\overset{\delta-}{O}-R$$
$$\underset{\overset{\delta+}{|}}{}$$
$$H$$

This attraction binds the two molecules together weakly and results in a type of cohesion referred to as association, which is brought about here by hydrogen bonding. The hydrogen bond or the hydrogen bridge is conventionally shown by dots and the stronger covalent bonds by dashes as usual. In fact, several molecules of an alcohol may form a large aggregate by association, as follows:

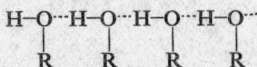

$$\underset{R}{H-O}\cdots\underset{R}{H-O}\cdots\underset{R}{H-O}\cdots\underset{R}{H-O}\cdots$$

When the boiling points of alkanes and alcohols of comparable molecular weight are compared (for example, *n*-hexane, C_6H_{14}, b.p. 68·7° and *n*-amyl alcohol, $C_5H_{11}OH$, b.p. 138·1°) the alcohols invariably are higher-boiling liquids. Association of different alcohol molecules as explained above causes an increase in their boiling points. Water is also highly associated and has a higher boiling point than unassociated hydrogen compounds of comparable molecular weight (NH_3, H_2S, etc.). The lower alcohols are very soluble in water, and the solubility diminishes as the molecular weight increases.

PREPARATION: (1) Alcohols may be prepared by the hydrolysis of alkyl halides either with aqueous alkali or silver oxide suspended in water:

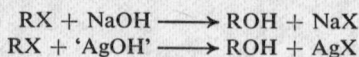

$$RX + NaOH \longrightarrow ROH + NaX$$
$$RX + \text{'AgOH'} \longrightarrow ROH + AgX$$

(2) The alcohols may be prepared by the hydration of alkenes. This is done by the addition of sulphuric acid to the double bond and hydrolysis of the resulting alkyl sulphuric acid.

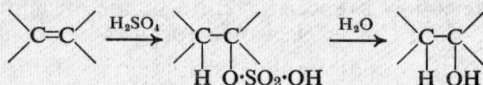

(3) The Grignard synthesis is capable of wide application for the preparation of different types of alcohols. The essential step is the addition of an alkyl magnesium halide (RMgX) to the carbonyl group (C=O) of a second component, followed by hydrolysis.

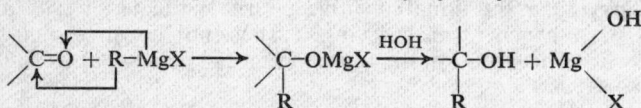

(4) Alcohols may also be obtained by the reduction of aldehydes, ketones, and acids:

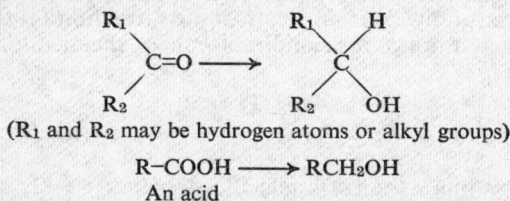

(R₁ and R₂ may be hydrogen atoms or alkyl groups)

$$R-COOH \longrightarrow RCH_2OH$$
An acid

Methods (3) and (4) will be dealt with in detail in Chapter Six.

Certain industrially important alcohols, like methanol and ethanol, have specific methods by which they are manufactured.

METHANOL

Methanol (Methyl Alcohol, Wood Spirit) was obtained until 1923 by the destructive distillation of wood, which gives a volatile product that is known as pyroligneous acid. Pyroligneous acid contains water, methyl alcohol (2–3%), acetone (0·5%), and acetic acid (about 10%). When the vapours of pyroligneous acid were passed into boiling milk of lime [$Ca(OH)_2$], the acetic acid was retained as the non-volatile calcium acetate. The vapours of water, acetone, and methyl alcohol which came over were condensed and the resulting liquid concentrated as far as possible by fractional distillation. Most of the water was removed by the fractionation, since it has a much higher boiling point (100° C.) than either acetone (56° C.) or methyl alcohol (65° C.). The small amount of water in the acetone–alcohol mixture was removed by letting it stand over quicklime. Fractionation, using long columns, gave an almost complete separation of acetone from methyl alcohol. The two compounds can also be separated by taking advantage of the fact that methyl alcohol alone forms a solid compound with anhydrous calcium chloride of the formula $CaCl_2,4CH_3OH$, and acetone is unaffected by the calcium chloride.

Methanol is now prepared synthetically by the hydrogenation of carbon monoxide, using a pressure of about 200 atmospheres. A mixture of water-gas and additional hydrogen is used:

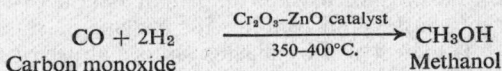

$$CO + 2H_2 \xrightarrow[350-400°C.]{Cr_2O_3-ZnO\ catalyst} CH_3OH$$

Carbon monoxide   Methanol

ETHANOL

Ethanol (Ethyl Alcohol, Alcohol) is commonly known simply as alcohol, and is the intoxicating principle in beer, wine, whisky, etc. It is prepared on a large scale for use in many fields.

The chief industrial methods of preparation are hydration of ethylene, previously mentioned, and fermentation of molasses or grain.

The fact that sugars on fermentation, under the influence of yeast, give alcohol has been known for quite some time. There is a reference in the *Rigveda*, an ancient Hindu religious text, to 'soma juice' which presumably was some kind of fermented juice containing ethyl alcohol. The name fermentation is applied to a process by means of which complex organic substances are split up into simpler ones by the agency of ferments or **enzymes**. Enzymes are catalysts of biological origin which bring about such reactions as oxidation, hydrolysis, reduction, etc., in a surprisingly efficient manner. The substances thus decomposed into simpler units are referred to as **substrates**. The enzymes are characterized by extreme specificity of action, and each enzyme has an optimum range of temperature for activity. We can illustrate the specific nature of an enzyme by a simple example. Both cane sugar and malt sugar (maltose) have the same molecular formula, $C_{12}H_{22}O_{11}$. The enzyme maltase, obtained from yeast, hydrolyses maltose to two molecules of glucose. But it is without action on cane sugar, whose chemical structure is closely related to, but different from, that of maltose. In order to hydrolyse cane sugar, another enzyme called invertase is required. This specific relationship between an enzyme and its substrate was compared to the relationship between a lock and its key by the great German chemist EMIL FISCHER.

Ethyl alcohol can be produced by the fermentation of a sugar solution and yeast. The sugar solution to be fermented is prepared and yeast is added to it. The temperature is kept at about 27° C. After a few hours the fermented liquid is filtered and the alcohol distilled off. The following enzymic reactions take place one after the other:

$$C_{12}H_{22}O_{11} + H_2O \xrightarrow[\text{invertase}]{\text{Hydrolysis by}} C_6H_{12}O_6 + C_6H_{12}O_6$$
$$\text{Glucose} \quad \text{Fructose}$$

$$C_6H_{12}O_6 \xrightarrow[\text{zymase}]{\text{Fermentation by}} 2C_2H_5OH + 2CO_2\uparrow$$

The alcohol thus obtained can be concentrated, and the small amounts of water still retained can be removed by suitable methods.

Sugar is much too costly a substance to be used as raw material for the production of alcohol. Industrially, alcohol is made from the 'black strap molasses' that remains as a residue at sugar refineries after crystallizable cane sugar has been removed from the cane juice. Alternatively, potatoes and rice, which are rich in starch, may be used as raw materials. The potatoes are mashed with water and malt is added to the mash. Malt is germinating barley and contains the enzyme diastase. At about 50° C. the starch is hydrolysed by diastase into maltose or malt sugar. It should be noted that starch is a complex substance with the empirical formula $C_6H_{10}O_5$, and hence its molecular formula may be written $(C_6H_{10}O_5)_n$, where n is a large number not known with certainty. The first step of the process can be represented by the equation:

$$2(C_6H_{10}O_5)_n + n\text{-}H_2O \xrightarrow[\text{from malt}]{\text{Diastase}} n\text{-}C_{12}H_{22}O_{11}$$
$$\text{Maltose}$$

The liquid thus obtained is cooled to ordinary temperature and yeast is then added. The enzymes, maltase, and zymase, present in yeast successively break down the maltose molecule, as depicted below, to give alcohol.

$$C_{12}H_{22}O_{11} + H_2O \xrightarrow{\text{Maltase}} 2C_6H_{12}O_6$$
$$\text{Glucose}$$

$$C_6H_{12}O_6 \xrightarrow{\text{Zymase}} 2C_2H_5OH + CO_2\uparrow$$
$$\text{Ethyl alcohol}$$

The alcohol obtained by any of the fermentation methods is exceedingly dilute, and concentration is effected by prolonged fractionation in tall fractionating columns. The commercial alcohol thus obtained contains 95·57% by weight of alcohol and 4·43% by weight of water, and is a constant-boiling mixture with a boiling point of 78·2° C. Since this boiling point is only slightly lower than that of absolute (i.e. 100%) ethyl alcohol (78·3°), separation from water cannot be effected by fractional distillation. Separation can be achieved, however, by a process known as **Azeotropic Distillation**, which consists of adding a suitable amount of benzene to the mixture and distilling off, successively, constant-boiling mixtures of the components. In order to separate a mixture of two substances, it seems odd to add a third substance to the mixture, but the method works. A constant-boiling mixture like the commercial alcohol referred to above is called an azeotrope. Azeotropes which have two components are called binary azeotropes.

Azeotropes which have three components are called ternary azeotropes. The boiling points of benzene, water, and ethyl alcohol are 80° C., 100° C., and 78·3° C., respectively. The three components form a ternary azeotrope with a boiling point of 64·8° C.; benzene and alcohol form a binary azeotrope with a boiling point of 68·2° C. So when a mixture of 95% alcohol and benzene is distilled the above ternary azeotrope distils first, followed by the binary azeotrope, and the final fraction (78·3° C.) is absolute ethyl alcohol.

The 95% alcohol obtained by fractionation from the fermentation process is also referred to as rectified spirit. In order to render commercial alcohol unfit for human consumption, it is denatured by addition of some poisonous and unpalatable substances, like methyl alcohol and pyridine.

BUTANOL

1-Butanol (*n*-Butyl Alcohol), $CH_3CH_2CH_2CH_2OH$, is obtained along with acetone by the fermentation of carbohydrates with the help of bacteria. The process uses the organism *Clostridium Acetobutylicum*, and was developed in 1911 by WEIZMANN in an effort to supply acetone needed for compounding the explosive Cordite. Since the end of the First World War, the process has been used mainly to supply 1-butanol used in quick-drying automobile lacquers.

REACTIONS OF ALCOHOLS: **(I) With Sodium.** Metallic sodium reacts vigorously with methanol, giving hydrogen and sodium methoxide. The reaction parallels the reaction of sodium with water.

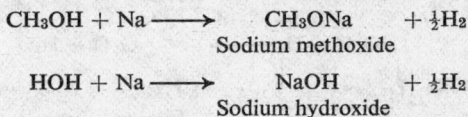

$$CH_3OH + Na \longrightarrow CH_3ONa + \tfrac{1}{2}H_2$$
Sodium methoxide

$$HOH + Na \longrightarrow NaOH + \tfrac{1}{2}H_2$$
Sodium hydroxide

Ethyl alcohol reacts with sodium in a similar manner, giving sodium ethoxide. Sodium ethoxide and sodium methoxide or, generally speaking, the alkoxides, are useful in organic synthesis. For instance, Williamson's synthesis of ethers makes use of alkoxides. Higher alcohols also react with sodium, but with decreasing readiness as the molecular weight increases. This observation can be explained by saying that as the size of the hydrocarbon residue increases, the functional ^-OH group becomes an increasingly minor part of the whole, and thus becomes a none too easy target for the sodium atom.

(II) **Alkene Formation.** The dehydration of alcohols by different reagents gives alkenes as has been described. It has been found that tertiary alcohols are more easily dehydrated than secondary alcohols, which in turn are dehydrated more readily than primary alcohols. This generalization is of practical utility in fixing the optimum conditions for the preparation of any desired alkene.

(III) **Oxidation.** Methyl, ethyl, and several higher alcohols are oxidized by such reagents as potassium permanganate and potassium dichromate. Bearing in mind that methane and ethane are not attacked

by the above reagents, it may be inferred that the introduction of the
$-OH$ group confers susceptibility to oxidation and that the hydroxyl-
bearing carbon atom in methanol and ethanol becomes oxidized.

A primary alcohol on oxidation gives rise to a substance called an
aldehyde (from alcohol dehydrogenated). The aldehydes are easily
oxidized further to become compounds called acids. Ethyl alcohol, for
example, gives on stepwise oxidation acetaldehyde and acetic acid.

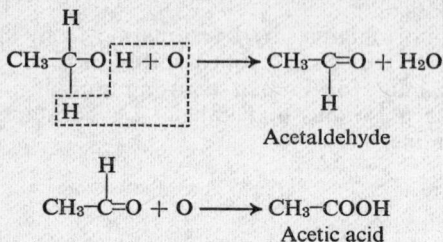

$$CH_3-\underset{\underset{H}{|}}{\overset{\overset{H}{|}}{C}}-O\,|H + O| \longrightarrow CH_3-\underset{H}{\overset{|}{C}}=O + H_2O$$

Acetaldehyde

$$CH_3-\underset{H}{\overset{|}{C}}=O + O \longrightarrow CH_3-COOH$$

Acetic acid

It will be noticed that both the aldehyde and the acid formed contain
the same number of carbon atoms as the parent alcohol. A secondary
alcohol on oxidation gives a ketone. The ketones are less susceptible to
further oxidation than aldehydes. But if powerful oxidizing agents are
used the molecule of the ketone is split up and acids are formed with a
smaller number of carbon atoms than the ketone possesses. Isopropyl
alcohol, for example, gives acetone as the initial product of oxidation.
Acetone is then split into acetic acid containing two carbon atoms and
carbonic acid ($CO_2 + H_2O$) containing one carbon atom.

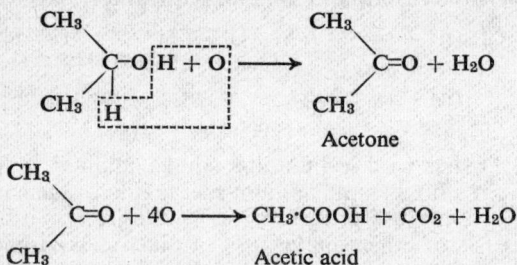

$$\begin{matrix} CH_3 \\ \end{matrix} \begin{matrix} \\ C-O\,|H + O| \\ \end{matrix} \longrightarrow \begin{matrix} CH_3 \\ \end{matrix} \begin{matrix} \\ C=O + H_2O \\ \end{matrix}$$

Acetone

$$\begin{matrix} CH_3 \\ \\ CH_3 \end{matrix} C=O + 4O \longrightarrow CH_3\cdot COOH + CO_2 + H_2O$$

Acetic acid

Tertiary alcohols are resistant to oxidation. Drastic oxidation first
gives an alkene, which is then oxidized successively to a ketone and acids
of a smaller number of carbon atoms. Tertiary butyl alcohol, for
example, is dehydrated to *iso*butylene, which is then oxidized to acetone,
acetic acid, etc.

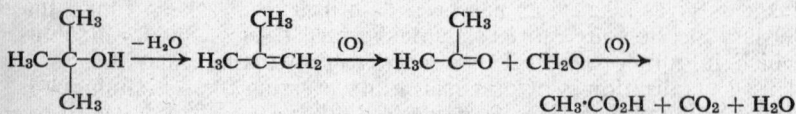

$$H_3C-\underset{CH_3}{\overset{CH_3}{\underset{|}{\overset{|}{C}}}}-OH \xrightarrow{-H_2O} H_3C-\underset{}{\overset{CH_3}{\overset{|}{C}}}=CH_2 \xrightarrow{(O)} H_3C-\underset{}{\overset{CH_3}{\overset{|}{C}}}=O + CH_2O \xrightarrow{(O)}$$
$$CH_3\cdot CO_2H + CO_2 + H_2O$$

A tertiary alcohol contains a hydroxylated carbon atom which carries
no hydrogen atom, and this factor is responsible for the resistance it
offers to oxidation. Normal conditions of oxidation in neutral and

alkaline media (KMnO$_4$) do not affect tertiary alcohols, which can therefore be distinguished from primary and secondary alcohols. Since primary and secondary alcohols give aldehydes and ketones which are easily distinguished from each other, the oxidation of alcohols serves to distinguish between the different types of alcohols.

(IV) **Esterification.** Acetic acid, CH$_3$COOH, reacts with ethyl alcohol under suitable conditions to give a sweet-smelling substance called ethyl acetate, the other product being water. The reaction, referred to as esterification, can be represented by the following equation:

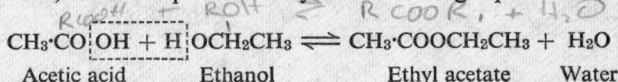

$$CH_3 \cdot CO \mid OH + H \mid OCH_2CH_3 \rightleftharpoons CH_3 \cdot COOCH_2CH_3 + H_2O$$

$$\text{Acetic acid} \qquad \text{Ethanol} \qquad\qquad \text{Ethyl acetate} \qquad \text{Water}$$

This reaction parallels the fundamental reaction of an acid with a base, the so-called neutralization reaction of inorganic chemistry.

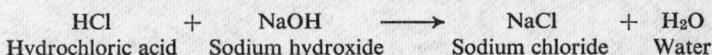

$$HCl + NaOH \longrightarrow NaCl + H_2O$$

$$\text{Hydrochloric acid} \quad \text{Sodium hydroxide} \qquad \text{Sodium chloride} \quad \text{Water}$$

In the case of neutralization, an acid and an alkali react to give salt and water. In esterification, which is the organic counterpart of neutralization, an organic acid reacts with an alcohol (which, though not an alkali, does contain the $^-$OH group) to give water and an ester. The ester obtained from ethyl alcohol and acetic acid is called ethyl acetate, just as the salt obtained from sodium hydroxide and sulphuric acid is called sodium sulphate.

But the analogy is only superficial and there are deep-seated differences between the two reactions. Esterification is a slow and reversible process between covalent molecules, as indicated by the two oppositely pointing arrows, and neutralization is an instantaneous and irreversible reaction between two ions. That is to say, in the case of the neutralization reaction, it is not possible to revert to the original acid and alkali by allowing the salt to react with water, since this reaction does not take place. The ester, however, can be hydrolysed by water to give the acid and the alcohol. Esterification is a reversible equilibrium reaction.

According to the Law of Mass Action, the use of an excess of either acid or alcohol as well as the removal of water formed should increase the yield of ester. Therefore one method of esterification is to mix the dry acid and the alcohol (one of them in excess, usually the alcohol), add a small quantity of sulphuric acid, and reflux the mixture for a few hours. On pouring the mixture, after the reaction is over, into water, the ester separates as an oily layer and is removed and distilled. Another method consists of saturating a mixture of the acid and alcohol with dry hydrogen chloride, refluxing, and working up the mixture as before. The inorganic acids used above function as catalysts and hasten the attainment of equilibrium conditions in esterification.

(V) **With Phosphorus Halides.** Alcohols react with phosphorus halides or with thionyl chloride to form alkyl halides (see pages 49 and 50).

DIHYDRIC ALCOHOLS

The dihydric alcohols, containing two hydroxyl groups in the molecules, are called **Diols** or **Glycols**. They can be prepared by oxidation of an alkene with permanganate. As has been pointed out, ethylene on oxidation with potassium permanganate gives ethylene glycol.

(1) Ethylene glycol can be readily prepared in the laboratory by refluxing ethylene dichloride with dilute sodium carbonate solution.

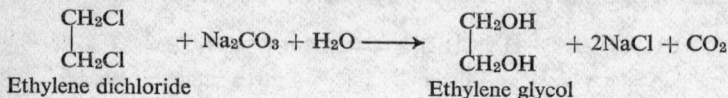

$$\begin{array}{c} CH_2Cl \\ | \\ CH_2Cl \end{array} + Na_2CO_3 + H_2O \longrightarrow \begin{array}{c} CH_2OH \\ | \\ CH_2OH \end{array} + 2NaCl + CO_2$$

Ethylene dichloride   Ethylene glycol

(2) A commercial method is the direct oxidation of ethylene to ethylene oxide and hydrolysis of the oxide.

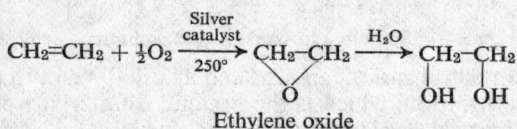

$$CH_2=CH_2 + \tfrac{1}{2}O_2 \xrightarrow[250°]{\text{Silver catalyst}} \underset{O}{CH_2-CH_2} \xrightarrow{H_2O} \underset{OH \quad OH}{CH_2-CH_2}$$

Ethylene oxide

The three-membered ring in ethylene oxide is very reactive. It adds on water as above and also ammonia to give ethanolamine, as follows:

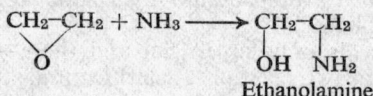

$$\underset{O}{CH_2-CH_2} + NH_3 \longrightarrow \underset{OH \quad NH_2}{CH_2-CH_2}$$

Ethanolamine

The ethanolamine formed may attach to one more molecule of ethylene oxide to give diethanolamine.

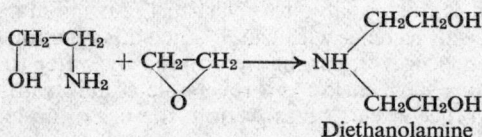

$$\underset{OH \quad NH_2}{CH_2-CH_2} + \underset{O}{CH_2-CH_2} \longrightarrow NH\begin{array}{l} CH_2CH_2OH \\ \\ CH_2CH_2OH \end{array}$$

Diethanolamine

With higher fatty acids the ethanolamines form soaps which are excellent emulsifying agents and are used extensively in the manufacture of shaving creams and lotions, insecticide sprays, furniture polishes, and dry-cleaning preparations.

With two hydroxyl groups present in the molecule, the formation of hydrogen bonds between the molecules proceeds more readily than in the case of monohydric alcohols. Hence it is not at all surprising that ethylene glycol is a thick syrupy liquid with a high boiling point (197° C.). When the glycol is converted into its mono- or di- ethers or esters the possibility of the hydrogen bonding is decreased and these compounds have lower boiling points than glycol.

Glycols have the same chemical properties as the alcohols described previously. However, owing to the presence of two $-OH$ groups, they have the additional property of the formation of cyclic or ring compounds. Thus, when ethylene glycol is heated with dilute sulphuric acid 1,4-dioxane is formed.

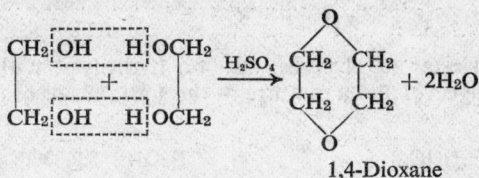

$$CH_2\overbrace{OH \quad H}OCH_2 \xrightarrow{H_2SO_4} \underset{O}{\overset{O}{\begin{array}{c} CH_2 \quad CH_2 \\ CH_2 \quad CH_2 \end{array}}} + 2H_2O$$

1,4-Dioxane

This is a case of intermolecular dehydration. **Dioxane** is a valuable solvent because it dissolves many organic substances and is also completely miscible with water. **Ethylene glycol,** when it is slowly introduced into a mixture of nitric and sulphuric acid, gives glycol dinitrate, a pale, yellow, highly explosive liquid, used with **glyceryl trinitrate** (nitroglycerine) in the manufacture of dynamite.

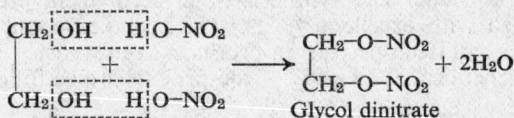

$$CH_2\overbrace{OH \quad H}O-NO_2 \longrightarrow \begin{array}{c} CH_2-O-NO_2 \\ CH_2-O-NO_2 \end{array} + 2H_2O$$

Glycol dinitrate

As can be seen from the above equation, glycol dinitrate is an ester of an inorganic acid.

By reaction with ethyl alcohol and sulphuric acid, ethylene glycol may be converted into its monoethyl ether (ethylcellosolve) or diethyl ether.

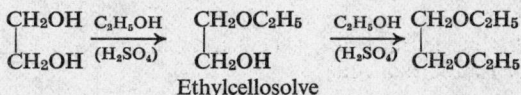

$$\begin{array}{c} CH_2OH \\ CH_2OH \end{array} \xrightarrow[(H_2SO_4)]{C_2H_5OH} \begin{array}{c} CH_2OC_2H_5 \\ CH_2OH \end{array} \xrightarrow[(H_2SO_4)]{C_2H_5OH} \begin{array}{c} CH_2OC_2H_5 \\ CH_2OC_2H_5 \end{array}$$

Ethylcellosolve

The monoalkyl ethers are given the trade name 'Cellosolves' and are used extensively in the manufacture of lacquers and varnishes. Cellosolve acetate, shown below, is obtained by acetylating cellosolve, and is a valuable solvent in the manufacture of lacquers, motion-picture films, and related materials. About 75% of the ethylene glycol produced is used as an anti-freeze agent.

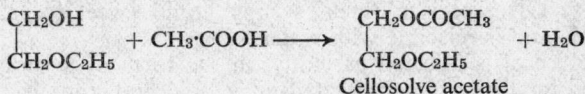

$$\begin{array}{c} CH_2OH \\ CH_2OC_2H_5 \end{array} + CH_3 \cdot COOH \longrightarrow \begin{array}{c} CH_2OCOCH_3 \\ CH_2OC_2H_5 \end{array} + H_2O$$

Cellosolve acetate

At one time most accidents involving dynamite resulted from its freezing. By addition of ethylene glycol dinitrate, the freezing point of dynamite is lowered, thus minimizing the risk of explosion.

TRIHYDRIC ALCOHOLS

Compounds containing three hydroxyl groups are called **triols,** the simplest being **1,2,3-propane-triol,** known to every chemist as **glycerol** (from the Greek *glykys* meaning sweet) or glycerin. Glycerol is commercially produced by the hydrolysis of fats.

$$\text{Fats} + \text{Alkali} \longrightarrow \text{Soap} + \text{Glycerol}$$

A fat is a triester of glycerol, and on hydrolysis with sodium hydroxide gives glycerol and a mixture of the sodium salt of the fatty acids called soap.

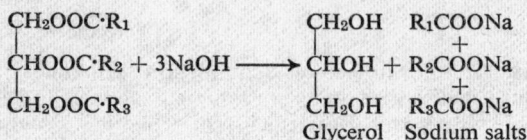

$$
\begin{array}{l}
CH_2OOC{\cdot}R_1 \\
| \\
CHOOC{\cdot}R_2 + 3NaOH \\
| \\
CH_2OOC{\cdot}R_3
\end{array}
\longrightarrow
\begin{array}{ll}
CH_2OH & R_1COONa \\
| & + \\
CHOH + & R_2COONa \\
| & + \\
CH_2OH & R_3COONa \\
\text{Glycerol} & \text{Sodium salts}
\end{array}
$$

Glycerol is a by-product in the soap industry and is recovered by suitable means. The modern method of industrial production of glycerol starts from propylene. The synthesis of glycerol from propylene has been developed with excellent yields, which makes possible the production of glycerol from petroleum products at low cost. Here, then, is another petrochemical. The steps of the process are as follows:

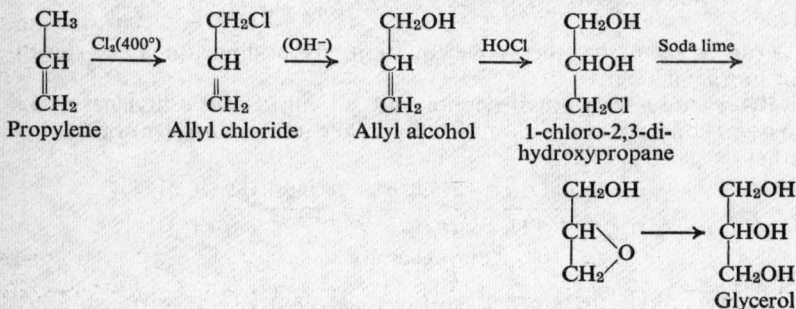

$$
\begin{array}{l}
CH_3 \\
| \\
CH \\
\| \\
CH_2
\end{array}
\xrightarrow{Cl_2(400°)}
\begin{array}{l}
CH_2Cl \\
| \\
CH \\
\| \\
CH_2
\end{array}
\xrightarrow{(OH^-)}
\begin{array}{l}
CH_2OH \\
| \\
CH \\
\| \\
CH_2
\end{array}
\xrightarrow{HOCl}
\begin{array}{l}
CH_2OH \\
| \\
CHOH \\
| \\
CH_2Cl
\end{array}
\xrightarrow{\text{Soda lime}}
$$

Propylene   Allyl chloride   Allyl alcohol   1-chloro-2,3-di-
                                                        hydroxypropane

$$
\begin{array}{l}
CH_2OH \\
| \\
CH{\diagdown} \\
\quad\quad O \\
CH_2{\diagup}
\end{array}
\longrightarrow
\begin{array}{l}
CH_2OH \\
| \\
CHOH \\
| \\
CH_2OH \\
\text{Glycerol}
\end{array}
$$

It is to be noted that in the chlorination of propylene at 400° C. substitution takes place on the terminal carbon atom in preference to addition to the double bond. At the high temperature used, the addition product is unstable and reverts to the unsaturated system.

PROPERTIES: Glycerol is a thick, syrupy liquid sweet to taste. It is hygroscopic and is miscible with water.

When glycerol is injected carefully, in the form of a spray, into a well-cooled mixture of concentrated nitric acid and sulphuric acid the nitric acid ester of glycerol, namely, glyceryl trinitrate (wrongly called nitroglycerine), is obtained.

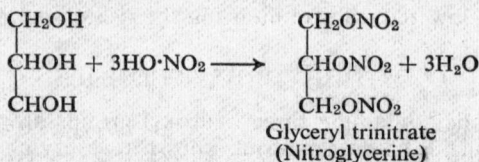

$$
\begin{array}{l}
CH_2OH \\
| \\
CHOH + 3HO{\cdot}NO_2 \\
| \\
CHOH
\end{array}
\longrightarrow
\begin{array}{l}
CH_2ONO_2 \\
| \\
CHONO_2 + 3H_2O \\
| \\
CH_2ONO_2
\end{array}
$$

Glyceryl trinitrate
(Nitroglycerine)

Nitroglycerine is an oil which crystallizes on cooling. The liquid is sensitive to shock and detonates violently when subjected to slight shock, essentially because the molecule has more than enough oxygen to convert the carbon and hydrogen to oxides and to liberate nitrogen.

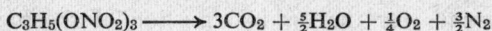

$$C_3H_5(ONO_2)_3 \longrightarrow 3CO_2 + \tfrac{5}{2}H_2O + \tfrac{1}{4}O_2 + \tfrac{3}{2}N_2$$

The frozen material to which some glycol dinitrate has been added is less sensitive to shock. The use of nitroglycerine as an explosive involved much risk until Nobel discovered the practical explosive dynamite. A mixture of nitroglycerine and **kieselguhr**, a siliceous earth, while retaining the explosive properties of the former, was found to be less sensitive to shock and could be transported with safety. To set off the explosion, a detonator containing mercuric fulminate [$Hg(ONC)_2$] or lead azide (PbN_6) is used. Nitroglycerine is also an essential ingredient, along with guncotton (cellulose trinitrate, an ester similar to nitroglycerine in structure), of other explosive formulations like Nobel's Blasting Gelatin and Cordite. Medicinally, small amounts of nitroglycerine are used in the treatment of heart disease (*angina pectoris*), because taken internally it causes dilation of the blood vessels and a corresponding decrease in blood pressure.

Glycerol is a valuable chemical that has many uses in both war and peace. In addition to being used in the production of explosives, glycerol is used as a moistening agent for tobacco, as a softening agent for cellophane films, in cosmetics, in food products, as a commercial solvent, and in the manufacture of plastics known as 'alkyd resins'.

POLYHYDRIC ALCOHOLS

Polyhydric alcohols contain several hydroxyl groups in the molecule and may be obtained by the reduction of sugars. **Sorbitol**, for example, is obtained by the reduction of glucose:

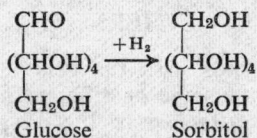

$$\begin{array}{ccc} CHO & & CH_2OH \\ | & +H_2 & | \\ (CHOH)_4 & \longrightarrow & (CHOH)_4 \\ | & & | \\ CH_2OH & & CH_2OH \\ Glucose & & Sorbitol \end{array}$$

It occurs in many fruits, particularly in the berries of the mountain ash. **Mannitol** is a **stereoisomer** of sorbitol (see page 110), that is an isomer in which the −OH groups have a different spatial arrangement from those of sorbitol and occurs widely distributed in nature. It has been isolated from moulds, algae, the sap of the manna tree, and many other plants. The spatial structure of these alcohols will be considered in the chapter on sugars.

The practical method of determining the number of hydroxyl groups in a polyhydric alcohol should be mentioned here. The molecular weight of the polyhydric alcohol is first determined. Then it is completely acetylated by boiling with acetic anhydride and the acetylated product is purified. Each original $-OH$ group has now been converted into the acetoxy group:

$$R(OH)_n \longrightarrow R(OOCCH_3)_n$$

in which R stands for the residual portion of the molecule. Thus ethylene glycol gives the diacetate on acetylation.

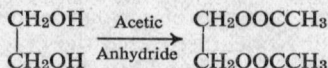

$$\begin{array}{ccc} CH_2OH & \text{Acetic} & CH_2OOCCH_3 \\ | & \xrightarrow{\hspace{1cm}} & | \\ CH_2OH & \text{Anhydride} & CH_2OOCCH_3 \end{array}$$

During acetylation the hydrogen atom (atomic weight 1) of every hydroxyl group has been replaced by an acetyl group ($CH_3CO = 43$) and the molecular weight of the polyacetyl derivative exceeds that of the original polyhydroxy compound by a multiple of 42 (i.e. $43 - 1$). The molecular weight of the purified polyacetyl derivative is determined. The gain in molecular weight divided by 42 will give the number of hydroxyl groups in the compound. This estimation of the number of hydroxyl groups is important, especially in the study of carbohydrates. The following problem is worked out to illustrate the method.

PROBLEM II: A substance with the molecular formula $C_5H_{12}O_4$ on treatment with acetic anhydride (exhaustive acetylation) gives a product whose molecular weight is 304. Calculate the number of hydroxyl groups in the substance.

SOLUTION:

Mol. wt. of the original substance $C_5H_{12}O_4$ is found by supplying atomic weights. ($12 \times 5 + 1 \times 12 + 16 \times 4 = 136$)
Mol. wt. of the acetylated product $= 304$
Gain in molecular weight $= 304 - 136 = 168$
No. of hydroxyl groups in the substance $= \frac{168}{42} = 4$ (Four)

ETHERS

The ethers are compounds with the general formula $R-O-R'$, in which R and R' may be the same or different alkyl radicals. They are named after the radicals attached to the oxygen atom. If both the radicals are the same, simple ethers are obtained, and if they are different, mixed ethers result. Thus we have CH_3OCH_3 dimethyl ether, a simple ether, and $CH_3OCH_2CH_3$, methyl ethyl ether, a mixed ether. Ethers may be considered as being derived from water (HOH) by substitution of the two hydrogen atoms by alkyl groups or more suitably as derived from alcohols $R-OH$ by replacement of the hydrogen atom with a member of the alkyl group. As a result of this replacement, association of molecules due to hydrogen bonding of the type encountered with alcohols is not possible and ethers have nearly the same boiling points as alkanes of comparable molecular weight. For instance, diethyl ether and *n*-pentane of nearly the same molecular weight boil respectively at $34 \cdot 5°$ and $36°$.

PREPARATION: While studying alkenes it was noted that ethylene could be obtained by the dehydration of ethyl alcohol with sulphuric acid, as follows:

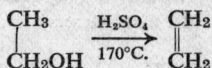

$$\underset{CH_2OH}{\overset{CH_3}{|}} \xrightarrow[170°C.]{H_2SO_4} \underset{CH_2}{\overset{CH_2}{||}}$$

If we could remove a molecule of water from two molecules of alcohol, as shown below, we could expect to get diethyl ether, known simply as ether.

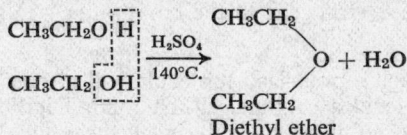

$$\begin{array}{c} CH_3CH_2O\,\boxed{H} \\ CH_3CH_2\,\boxed{OH} \end{array} \xrightarrow[140°C.]{H_2SO_4} \begin{array}{c} CH_3CH_2 \\ O + H_2O \\ CH_3CH_2 \end{array}$$
Diethyl ether

This is indeed the case, and both ethylene and ether can be obtained from ethyl alcohol and sulphuric acid by suitable variation of the conditions of the reaction. The initial product formed in the reaction between ethyl alcohol and sulphuric acid is ethyl hydrogen sulphate.

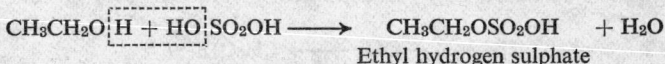

$$CH_3CH_2O\,\boxed{H} + \boxed{HO}\,SO_2OH \longrightarrow CH_3CH_2OSO_2OH + H_2O$$
Ethyl hydrogen sulphate

This is a simple case of esterification. The fate of this intermediate compound depends upon the conditions of the reactions it is subjected to. At 170° C., when an excess of sulphuric acid is used, the ethyl hydrogen sulphate decomposes to give ethylene, as shown below.

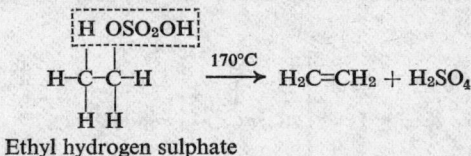

$$\underset{\underset{H\,\,H}{|\,\,\,\,|}}{H-C-C-H} \xrightarrow{170°C} H_2C=CH_2 + H_2SO_4$$
Ethyl hydrogen sulphate

At 140° C., in the presence of excess of alcohol, the ethyl hydrogen sulphate reacts with a molecule of the alcohol to give ether, as shown below:

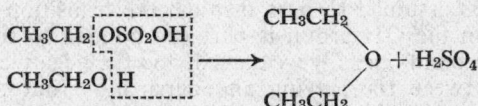

$$\begin{array}{c} CH_3CH_2\,\boxed{OSO_2OH} \\ CH_3CH_2O\,\boxed{H} \end{array} \longrightarrow \begin{array}{c} CH_3CH_2 \\ O + H_2SO_4 \\ CH_3CH_2 \end{array}$$

It will be noted that the molecule of sulphuric acid is regenerated. In a sense, the sulphuric acid may be looked upon as a catalyst. The regenerated sulphuric acid may react with another quantity of alcohol, reforming ethyl hydrogen sulphate, which, in turn, combines with more alcohol and produces an additional quantity of ether. The process, theoretically, should be a continuous one, and is referred to as a continuous etherification process. In practice, however, some of the sulphuric acid is lost in side reactions.

From a theoretical and structural point of view, Williamson's synthesis of ether, referred to under alkyl halides, is important. It is capable of being applied to the preparation of an ether of any desired structure, both simple and mixed. Thus diethyl ether can be prepared from ethyl iodide and sodium ethoxide.

$$C_2H_5 \;|\; I + Na \;|\; OC_2H_5 \longrightarrow C_2H_5OC_2H_5 + NaI$$

The mechanism of the reaction can be understood from a consideration of the reaction between an alkyl halide and an alkali, like sodium hydroxide. We have seen that in the compound CH_3I the C–I bond is polarized in the direction $\overset{\delta+}{H_3C} \rightarrow \overset{\delta-}{I}$ as revealed by dipole moment measurements. Such a polarized molecule will attract (or repel) electrically charged molecules or ions. When methyl iodide is heated with sodium hydroxide solution hydrolysis of the alkyl halide occurs, as shown below:

$$CH_3I + Na^+ + OH^- \longrightarrow CH_3OH + Na^+ + I^-$$

By cancelling the Na^+ in both sides of the equation the equation takes the form of a reaction between a polarized molecule and an ion.

$$\overset{\delta+}{H_3C} \rightarrow \overset{\delta-}{I} + OH^- \longrightarrow CH_3OH + I^-$$

Because of mutual repulsion of like charges, the $^-OH^-$ ion cannot be expected to attack the carbon from the side of the iodine atom carrying a partial negative charge. It would be more logical to assume that the ^-OH ion will be attracted to the positively charged carbon atom on the side away from the iodine atom, that is on the rear side, as illustrated below:

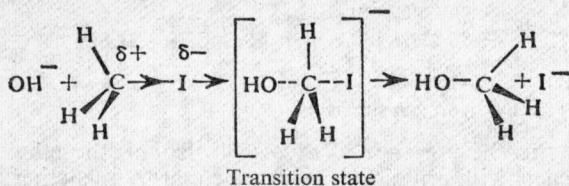

Transition state

The reaction is assumed to pass through the transition state pictured above, wherein the OH group is partially bonded to the carbon, the iodine–carbon bond partially weakened, and the negative charge evenly distributed between the leaving and departing groups. Williamson's synthesis of ethers similarly involves the attack of an alkoxide ion on an alkyl halide:

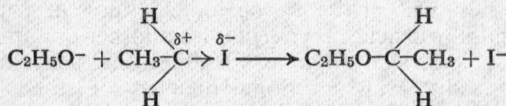

PROPERTIES: As a class, the ethers are stable compounds. Hence they are used as solvents, that is, as reaction media, just as water is a reaction medium in most inorganic reactions. In the case of an ether having the linkage C–C–O–C–C, we encounter a chain of carbon atoms interrupted by an oxygen atom. Since we know ethers are stable compounds, the oxygen bridge must be strong. Compared, however, with the C–C link, the C–O–C link is weaker because of the following important factor. The electronic formula of an ether, written below, shows that the oxygen atom has two unshared pairs of electrons, and hence offers points of attack to electron-seeking reagents. The ethers, therefore, form salts with acids:

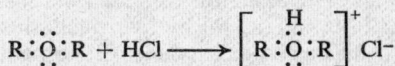

$$R : \overset{..}{\underset{..}{O}} : R + HCl \longrightarrow \left[R : \overset{H}{\underset{..}{O}} : R \right]^+ Cl^-$$

These salts, comparable to ammonium salts $[NH_4^+]X^-$ and hence called Oxonium salts, are stable only at low temperatures ($-100°$ C.).

The usefulness of ether as a solvent arises from its ability to form active intermediate complexes with dissolved reagents. Though ether cannot combine with itself to form hydrogen bonds, like water and alcohol, it can combine with electron acceptors by donation of either or both the lone pairs of electrons on the oxygen atom. For example, any alkyl magnesium halide (Grignard reagent) exists as a dietherate of the structure below:

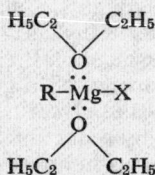

$$\begin{array}{ccc} H_5C_2 & & C_2H_5 \\ & O & \\ & \overset{..}{\underset{..}{}} & \\ R-Mg-X & \\ & O & \\ & \overset{..}{\underset{..}{}} & \\ H_5C_2 & & C_2H_5 \end{array}$$

Shared pair of electrons link Mg to each O atom

The vapours of the lower members are highly inflammable in the air. Otherwise, in general, ethers are inert. However, on exposure to air ethers slowly form explosive peroxides. They combine with an oxygen atom thus:

$$R-O-R + [O] \longrightarrow R-O-O-R$$

The distillation of diethyl ether that has been stored for a long time has frequently resulted in violent explosions due to concentration of the less volatile peroxide. The presence of peroxides in ether can be detected by the liberation of iodine when a sample is shaken with acidified potassium iodide. Ether containing peroxide must be washed with ferrous chloride or some other reducing agent and dried over calcium chloride before it is distilled. Ordinarily, as a precautionary measure, ether which is sold in bottles contains iron wire, which inhibits peroxide formation in storage.

When an ether is heated with hydriodic acid the carbon-to-oxygen linkage is broken and a mixture of alcohols and alkyl iodides, as shown below, is formed.

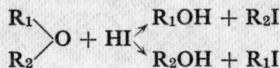

$$\begin{array}{l} R_1 \\ \diagdown \\ O + HI \diagup \begin{array}{l} R_1OH + R_2I \\ \\ R_2OH + R_1I \end{array} \\ R_2 \diagup \end{array}$$

Industrially important ethers, like ethylene oxide, dioxane, and cello-solves, are dealt with in the section on ethylene glycol.

SUMMARY

Alcohols contain $-OH$ group. Classification into primary (RCH_2OH), secondary (R_1R_2CHOH), and tertiary ($R_1R_2R_3COH$) alcohols. Association of alcohols due to hydrogen bonding and its effect on boiling points. Preparation: (1) Hydrolysis of alkyl halides. (2) Hydration of alkenes, e.g. ethylene to ethyl alcohol. (3) From Grignard reagents and carbonyl compounds. (4) Reduction of acids, ketones, and aldehydes. Manufacture of alcohols: methanol by hydrogenation of carbon monoxide. Ethanol by (1) hydration of ethylene; (2) fermentation of molasses and grain. Enzymes involved in alcoholic fermentation. Azeotropic distillation with benzene to make absolute alcohol. Butanol by fermentation of carbohydrate with *Clostridium acetobutylicum*. Reactions: formation of alkoxides, olefines, oxidation, and esterification. Oxidation useful in determining the classification of an alcohol. Esterification catalysed by mineral acids.

Ethylene Glycol (a dihydric alcohol). Preparation: (1) From ethylene dichloride. (2) From ethylene via ethylene oxide. Ethanolamine from ethylene oxide and its uses. Dioxane from ethylene glycol. Glycol dinitrate as a constituent of dynamite. Cellosolves (CH_2OH-CH_2OR), cellosolve acetates and their uses. Ethylene glycol as an anti-freeze agent. Glycerol (glycerine) a trihydric alcohol. Preparation: (1) Saponification of fats and oils. (2) From propene. Nitroglycerine an ester. Dynamite and its composition. Other uses of glycerine. Mannitol and sorbitol are hexahydric alcohols obtained by reduction of sugars. Method of determination of number of hydroxyls in an alcohol.

Ethers. Formula ROR'. Simple and mixed ethers. Preparation: (1) Diethyl ether from ethanol by treatment with sulphuric acid at 140°. (2) Williamson's synthesis involving an alkoxide and an alkyl halide. Mechanism of alkaline hydrolysis of alkyl halides. Ether as a solvent and its use in Grignard reactions. Peroxide formation and how to prevent it. Cleavage of ethers with hydriodic acid.

Problem Set No. 4

1. Write down the structures of all the possible isomers having the molecular formula $C_4H_{10}O$ and name them.
2. How would you distinguish between butanol-1 and butanol-2?
3. How are the following conversions accomplished?
   (a) Methanol to Propanol-2.
   (b) Ethanol to Butanol-2.
4. A compound with the molecular formula $C_4H_8O_4$ is changed into one with the formula $C_{10}H_{14}O_7$ by treatment with acetic anhydride. How many alcoholic groups are there in the original compound?
5. An ether on refluxing with hydriodic acid gave a mixture of methyl iodide and isopropyl iodide. What is the structure of the ether?

CHAPTER SIX

ALDEHYDES AND KETONES

The compounds we shall deal with in this chapter are closely related to the alcohols of the previous chapter, on the one hand, and on the other, they are related to the acids we shall be studying in the next chapter. A primary alcohol $R \cdot CH_2OH$ can be oxidized (dehydro-genated) to an aldehyde $R-\overset{\overset{\displaystyle H}{|}}{C}=O$ and the aldehyde oxidized further to an acid $R-\overset{\overset{\displaystyle O}{||}}{C}-OH$. These compounds are all related, but differ in their degree of oxidation. The reverse change, that is, starting from an acid through the aldehyde to the alcohol, can also be accomplished by reduction, though the acids are somewhat resistant to reduction. We can represent the set of changes as follows:

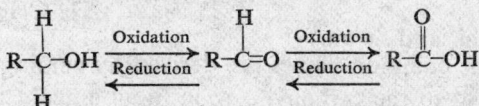

$$R-\underset{\underset{\displaystyle H}{|}}{\overset{\overset{\displaystyle H}{|}}{C}}-OH \underset{\text{Reduction}}{\overset{\text{Oxidation}}{\rightleftharpoons}} R-\overset{\overset{\displaystyle H}{|}}{C}=O \underset{\text{Reduction}}{\overset{\text{Oxidation}}{\rightleftharpoons}} R-\overset{\overset{\displaystyle O}{||}}{C}-OH$$

Similarly, the secondary alcohols on oxidation furnish ketones which can be reduced back to secondary alcohols:

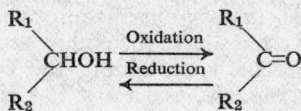

$$\overset{R_1}{\underset{R_2}{>}}CHOH \underset{\text{Reduction}}{\overset{\text{Oxidation}}{\rightleftharpoons}} \overset{R_1}{\underset{R_2}{>}}C=O$$

The aldehydes have the general formula $R-\overset{\overset{\displaystyle H}{|}}{C}=O$, and the ketones are represented by $R_1-\overset{\overset{\displaystyle O}{||}}{C}-R_2$. Both of them contain a common structural feature, that is, the $C=O$ bond, in which a carbon atom is joined to an oxygen atom through a double bond. This $C=O$ combination is called a carbonyl group and is one of the important functional groups. In an aldehyde the carbonyl group carries a hydrogen atom, whereas in the case of a ketone there is no hydrogen atom attached to the carbonyl group. Hence aldehydes and ketones have certain similarities and also certain differences in properties. The properties and reactions of aldehydes and ketones are determined by the polarized nature of the

74   *Organic Chemistry Made Simple*

carbonyl bond C=O, in which the oxygen atom, because it is more electronegative than the carbon atom, has a greater share of the common electrons, as indicated below in structure (3), which is intermediate between structures (1) and (2):

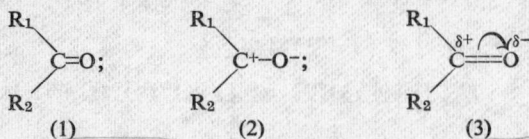

$$R_1R_2C{=}O; \qquad R_1R_2C^+{-}O^-; \qquad R_1R_2C\overset{\delta+}{=}\overset{\delta-}{O}$$

    (1)   (2)   (3)

The dipole moments of acetaldehyde, CH_3CHO (2·7 units), and acetone, CH_3COCH_3 (2·85 units), are considerably larger than would be expected if the carbonyl group possessed a purely double covalent bond.

NOMENCLATURE: **Aldehydes.** The common names for the aldehydes are formed from the stem names of the acids they give on oxidation. For example, HCHO is formaldehyde corresponding to H–COOH (formic acid), CH_3CHO is acetaldehyde corresponding to CH_3COOH (acetic acid), etc. The I.U.C. names for aldehydes end in -*al* and are derived by the application of the same rules as are applied for designating alcohols. Since an aldehyde group cannot be anywhere else except at carbon atom 1, its position is usually not indicated in the I.U.C. name. For example, CH_3CH_2CHO is propanal, $CH_3{-}CH{-}CH{-}CHO$ is

                                            CH_3 $CH_2{-}CH_3$

3-ethylpentanal, and so on.

Ketones. The common names are derived by naming the two alkyl groups attached to the carbonyl carbon in an alphabetical order and adding the ending ketone. Thus, $CH_3CH_2COCH_3$ is ethyl methyl ketone, $CH_3CHCOC_4H_9$ (n) is n-butyl isopropyl ketone, and so on. In the I.U.C. system the ketones have the ending -*one* after the stem name of the longest carbon chain containing the carbonyl group. Thus $CH_3CH_2COCH_3$ is butanone-2, and $CH_3CH_2CH{-}COCH_3$ is 3-methyl-pentanone-2 and so on.

PREPARATION: (1) Aldehydes and ketones may be prepared by the oxidation of primary and secondary alcohols respectively. The usual oxidizing agents used are:

(*a*) Dilute potassium dichromate and sulphuric or acetic acid. Concentrated potassium permanganate is much too powerful, and consequently oxidation proceeds to the further stage of acid. For example, ethyl alcohol can be oxidized, under the conditions stated, to acetaldehyde:

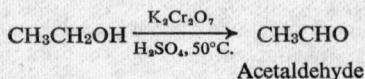

$$CH_3CH_2OH \xrightarrow[H_2SO_4,\,50°C.]{K_2Cr_2O_7} CH_3CHO$$

                                  Acetaldehyde

(b) Atmospheric oxygen and silver or platinum catalyst:

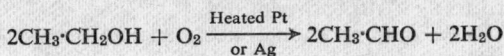

$$2CH_3 \cdot CH_2OH + O_2 \xrightarrow[\text{or Ag}]{\text{Heated Pt}} 2CH_3 \cdot CHO + 2H_2O$$

(2) Alcohols can be dehydrogenated by passing their vapours over a copper catalyst at 300° C.:

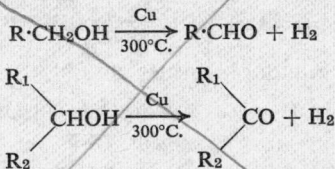

$$R \cdot CH_2OH \xrightarrow[\text{300°C.}]{\text{Cu}} R \cdot CHO + H_2$$

$$\begin{array}{c} R_1 \\ \diagdown \\ CHOH \\ \diagup \\ R_2 \end{array} \xrightarrow[\text{300°C.}]{\text{Cu}} \begin{array}{c} R_1 \\ \diagdown \\ CO + H_2 \\ \diagup \\ R_2 \end{array}$$

(3) Acid chlorides R—COCl on reduction with hydrogen, in the presence of palladium supported on barium sulphate, give the aldehydes.

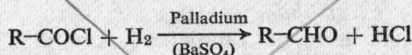

$$R-COCl + H_2 \xrightarrow[\text{(BaSO}_4\text{)}]{\text{Palladium}} R-CHO + HCl$$

A sulphur-containing poison, such as barium sulphate, is used along with the catalyst to prevent the aldehyde obtained from getting further reduced to an alcohol. An acid chloride contains a very reactive chlorine atom, and the poison does not interfere with the replacement of this chlorine by the hydrogen. The technique of this reduction was developed by ROSENMUND, and hence is known as the Rosenmund Reduction.

(4) Thermal decomposition of metal (usually Ca or Mg) salts of acids gives ketones. For example, calcium acetate when pyrolysed (heated strongly) decomposes to acetone and calcium carbonate.

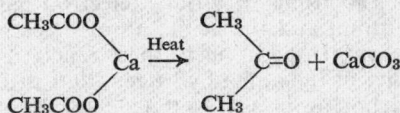

$$\begin{array}{c} CH_3COO \\ \diagdown \\ Ca \\ \diagup \\ CH_3COO \end{array} \xrightarrow{\text{Heat}} \begin{array}{c} CH_3 \\ \diagdown \\ C=O + CaCO_3 \\ \diagup \\ CH_3 \end{array}$$

Cyclic ketones are similarly prepared by thermal decomposition of salts of dibasic acids, as follows:

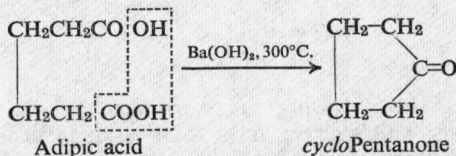

$$\begin{array}{c} CH_2CH_2CO\,OH \\ | \\ CH_2CH_2\,COOH \end{array} \xrightarrow{\text{Ba(OH)}_2,\,300°C.} \begin{array}{c} CH_2-CH_2 \\ | \qquad\quad C=O \\ CH_2-CH_2 \end{array}$$

Adipic acid   *cyclo*Pentanone

(5) A recent industrial method, originating in Germany, is known as the OXO-reaction. This reaction consists of the addition of carbon monoxide and hydrogen to the double bond of alkenes using a cobalt catalyst at a temperature of about 300° C. and a pressure of 200 atmospheres.

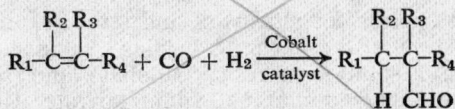

$$\begin{array}{c} R_2\ R_3 \\ |\ \ | \\ R_1-C=C-R_4 + CO + H_2 \end{array} \xrightarrow[\text{catalyst}]{\text{Cobalt}} \begin{array}{c} R_2\ R_3 \\ |\ \ | \\ R_1-C-C-R_4 \\ |\ \ | \\ H\ \ CHO \end{array}$$

The reaction results in the addition of H and ⁻CHO groups, and hence is also referred to as hydroformylation.

(6) *Special Methods*. Hydration of acetylene, as mentioned in Chapter Three (pages 45 and 46), in the presence of mercuric salts is an industrial method for the preparation of acetaldehyde:

$$CH{\equiv}CH + HOH \xrightarrow[HgSO_4]{H_2SO_4} CH_3CHO$$

The chlorination of ethanol furnishes chloral.

$$CH_3CH_2OH + 4Cl_2 \longrightarrow \underset{\text{Chloral}}{CCl_3 \cdot CHO} + 5HCl$$

Besides chlorination, oxidation of the ⁻CH₂OH group to ⁻CHO occurs simultaneously.

Acetone is an important by-product during the fermentation of carbohydrates by *Clostridium acetobutylicum* to produce butanol-1. Most of the acetone used in commerce is obtained from propene, a petrochemical.

$$\underset{\text{Propene}}{CH_3CH{=}CH_2} \xrightarrow[H_2O]{\text{Catalyst}} \underset{\text{Propanol-2}}{CH_3CH(OH)CH_3} \xrightarrow{[O]} CH_3{-}CO{-}CH_3$$

REACTIONS: The polarized carbonyl bond is the vulnerable spot in aldehydes and ketones. This unsaturated group, like the olefinic double bond, takes part in addition reactions, but the substances that add to the carbonyl group are quite different from those which add to an alkene. In most of these addition reactions of aldehydes and ketones it will be noticed that the basic atoms or groups (electron pair donors) join the carbonyl carbon atom, and the acidic atoms (electron-pair acceptors) join the oxygen atom. In addition to these reactions, mention may be made of the activity of the alpha hydrogen, that is, the hydrogen atom attached to the carbon atom next to the aldehyde group. The polarized carbonyl group has an inductive effect on the alpha carbon, and thus increases the reactivity of the alpha hydrogen atom:

(I) **Oxidation.** Aldehydes are easily oxidized to acids, especially in alkaline solutions. **Tollen's reagent** (a solution of silver nitrate in excess ammonium hydroxide) oxidizes aldehydes to acids, and in the process is reduced and gives a mirror of metallic silver. The reagent may be considered to contain silver oxide, and the reaction can be represented by the following equation:

$$R{\cdot}CHO + Ag_2O \longrightarrow R{\cdot}COOH + 2Ag$$

This reaction is specific for aldehydes and serves to distinguish them from ketones. Another test reagent is **Fehling's solution**, which consists of two parts: (*a*) a solution of copper sulphate, and (*b*) a solution of sodium hydroxide and sodium potassium tartrate (Rochelle's Salt).

When these two solutions are mixed an alkaline solution of a soluble copper tartrate complex is formed which may be considered a solution of cupric oxide. Aldehydes reduce Fehling's solution to a red precipitate of cuprous oxide.

$$R \cdot CHO + 2CuO \longrightarrow R \cdot COOH + Cu_2O$$

(II) **Reduction.** Reduction of aldehydes and ketones leads to primary and secondary alcohols respectively. Either catalytic or chemical methods of reduction are applicable. The former method involves reduction with hydrogen gas at atmospheric or higher pressures in the presence of suitable catalysts like Pt, Pd, Ni, etc.

A versatile reagent for the reduction of carbonyl compounds, introduced in 1947, is **lithium aluminium hydride**, obtained by treating lithium hydride with aluminium chloride, as shown below.

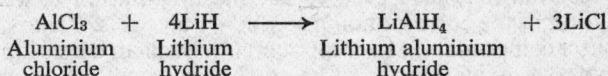

$$\underset{\substack{\text{Aluminium} \\ \text{chloride}}}{AlCl_3} + \underset{\substack{\text{Lithium} \\ \text{hydride}}}{4LiH} \longrightarrow \underset{\substack{\text{Lithium aluminium} \\ \text{hydride}}}{LiAlH_4} + 3LiCl$$

The reagent suspended in dry diethyl ether reacts with a carbonyl compound at room temperature to give an alcoholate, essentially according to the equation:

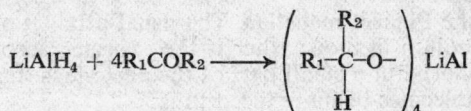

$$LiAlH_4 + 4R_1COR_2 \longrightarrow \left(\begin{array}{c} R_2 \\ | \\ R_1-C-O- \\ | \\ H \end{array} \right)_4 LiAl$$

Decomposition of this alcoholate with water furnishes free alcohol.

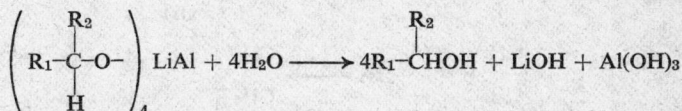

$$\left(\begin{array}{c} R_2 \\ | \\ R_1-C-O- \\ | \\ H \end{array} \right)_4 LiAl + 4H_2O \longrightarrow 4R_1-\overset{\overset{\textstyle R_2}{|}}{C}HOH + LiOH + Al(OH)_3$$

Carbon–carbon double bonds if present in the starting carbonyl compound are not affected.

Another elegant method used to bring about the same kind of reduction is called the MEERWEIN-PONNDORF–VERLEY reduction. The principle of the method can be understood from the following considerations. In the presence of aluminium isopropoxide: $Al[OCH(CH_3)_2]_3$, or aluminium t-butoxide: $Al[OC(CH_3)_3]_3$, alcohols and carbonyl compounds undergo a reversible reaction as follows:

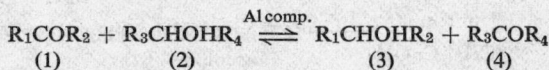

$$\underset{(1)}{R_1COR_2} + \underset{(2)}{R_3CHOHR_4} \overset{\text{Al comp.}}{\rightleftharpoons} \underset{(3)}{R_1CHOHR_2} + \underset{(4)}{R_3COR_4}$$

The R's are numbered for purposes of reference and may denote alkyl radicals or hydrogen atoms. Essentially, the carbonyl compound (1) is converted into its corresponding alcohol (3) and the alcohol (2) is transformed into its corresponding carbonyl compound (4). The reaction is reversible and the equilibrium may be shifted to any desired side by increasing the molar ratio of one of the reactants or by removing a product as it is formed.

For converting the carbonyl compound (1) to the alcohol (3), isopropyl alcohol is used as the alcohol (2) in the general equation given above, and a large ratio of (2) to (1) is obtained by using the isopropyl alcohol as the reaction medium itself. The special merit of this method is that only the carbonyl group is reduced, leaving double bonds, if any, in the carbonyl compound unaffected. Crotyl alcohol, for example, can be obtained from crotonaldehyde in excellent yields:

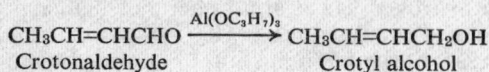

$$CH_3CH=CHCHO \xrightarrow{Al(OC_3H_7)_3} CH_3CH=CHCH_2OH$$
$$\text{Crotonaldehyde} \qquad\qquad \text{Crotyl alcohol}$$

In this connexion it may be of interest to note that the above reversible reaction may also be used for the preparation of ketones by the oxidation of the corresponding secondary alcohols. That is, the idea is to convert compound (2) to compound (4). Such an oxidation is called the OPPENAUER oxidation. In order to shift the equilibrium far to the right, in this case, acetone is used as solvent. The Oppenauer oxidation has been of particular use in the oxidation of steroid and other cyclic compounds containing a secondary alcoholic group.

The chemical reduction of a ketone often takes another course referred to as the **Pinacol reduction**. The usual product of reduction of acetone with sodium in moist ether is the expected secondary alcohol (*iso*propyl alcohol), but a small part of the acetone is transformed into pinacol, two molecules being reduced thus:

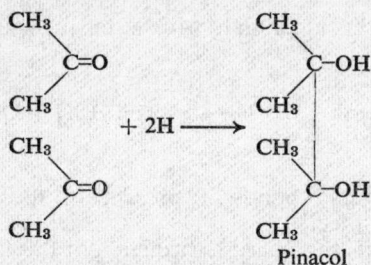

The use of magnesium amalgam instead of sodium improves the yield of pinacol. Pinacol is interesting because of the peculiar intramolecular rearrangement that it undergoes on being heated with dilute sulphuric acid to give a ketone called pinacolone, shown below.

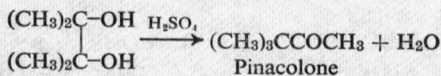

$$\begin{array}{c} (CH_3)_2C-OH \\ | \\ (CH_3)_2C-OH \end{array} \xrightarrow{H_2SO_4} (CH_3)_3CCOCH_3 + H_2O$$
$$\qquad\qquad\qquad \text{Pinacolone}$$

(III) Auto-oxidation and Reduction. Formaldehyde, trimethyl acetaldehyde, and other aldehydes like benzaldehyde, which do not have a hydrogen on the carbon atom next to the carbonyl group, the α carbon atom, undergo intermolecular oxidation and reduction when warmed with a concentrated sodium hydroxide solution. This reaction is known as CANNIZZARO reaction. Taking formaldehyde for an example, the reaction may be formulated as follows:

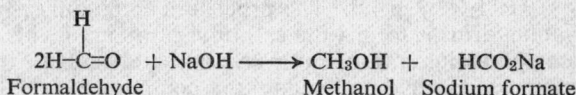

$$2H\underset{|}{\overset{H}{C}}=O + NaOH \longrightarrow CH_3OH + HCO_2Na$$

Formaldehyde   Methanol   Sodium formate

One molecule of formaldehyde is oxidized at the expense of the other.

(IV) **Substitution at the Alpha Position.** Halogens like chlorine and bromine substitute in the α position, where the α hydrogen is rendered mobile by the inductive effect of the polarized carbonyl group. Thus, with butyraldehyde and chlorine we get α-chlorobutyraldehyde.

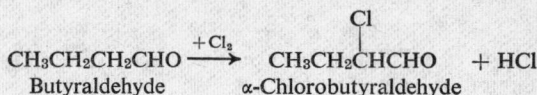

$$CH_3CH_2CH_2CHO \xrightarrow{+Cl_2} CH_3CH_2\overset{\overset{\displaystyle Cl}{|}}{C}HCHO + HCl$$

Butyraldehyde   α-Chlorobutyraldehyde

The preparation of chloral referred to earlier also involves α-chlorination.

(V) **Addition to the** $\diagdown$CO **Group.** A number of reagents add to the $\diagdown$C=O in such a way that the negative fragment of the addendum goes to the carbon atom of the carbonyl bond, with the positive fragment becoming linked to the oxygen atom. The mechanism will be quite clear when we bear in mind the polarized structure of the carbonyl bond. The positively charged carbon atom of the carbonyl group is open to attack by a **nucleophilic reagent**: a reagent with an unshared pair of electrons, looking for an electron-deficient nucleus with which to share them. This leaves the oxygen atom with a negative charge which it neutralizes by combining with a hydrogen nucleus.

Let us take the addition of hydrogen cyanide to form a cyanohydrin, as an example. First a cyanide ion combines with the carbonyl carbon atom:

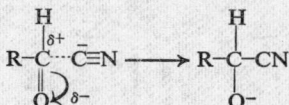

$$\underset{\overset{\|}{O}\overset{}{\delta-}}{\overset{\overset{\displaystyle H}{|}}{R-\overset{\delta+}{C}}\cdots C\equiv N} \longrightarrow \underset{O^-}{\overset{\overset{\displaystyle H}{|}}{R-\overset{|}{C}-CN}}$$

Then the negative oxygen atom removes a proton from another molecule of hydrogen cyanide or a water molecule:

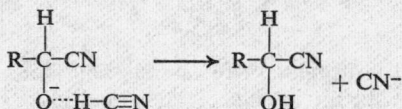

$$\underset{\overset{|}{O}\cdots H-C\equiv N}{\overset{\overset{\displaystyle H}{|}}{R-\overset{|}{C}-CN}} \longrightarrow \underset{OH}{\overset{\overset{\displaystyle H}{|}}{R-\overset{|}{C}-CN}} + CN^-$$

The overall reaction for acetone, for instance, is:

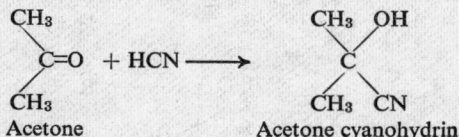

$$\underset{CH_3}{\overset{\displaystyle CH_3}{}}C=O + HCN \longrightarrow \underset{CH_3 \quad CN}{\overset{\displaystyle CH_3 \quad OH}{}}C$$

Acetone   Acetone cyanohydrin

This reaction is useful for lengthening a carbon chain by a single carbon atom since the $-CN$ group can be hydrolysed to a $-COOH$ group or reduced to a $-CH_2NH_2$ group. The procedure commonly employed to prepare cyanohydrins is to mix the carbonyl compound with a solution of sodium or potassium cyanide in water and add acid.

With ammonia, aldehydes give addition compounds called aldehyde ammonias (formaldehyde is an exception), while ketones give condensation products.

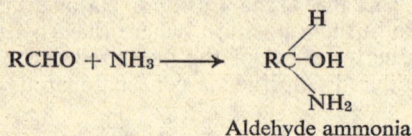

$$RCHO + NH_3 \longrightarrow RC\overset{H}{\underset{NH_2}{|}}OH$$

Aldehyde ammonia

Formaldehyde gives a condensation product, according to the equation, known as hexamethylenetetramine.

$$6CH_2O + 4NH_3 \longrightarrow (CH_2)_6N_4 + 6H_2O$$

Hexamine

The four nitrogen atoms are at the four corners and the six CH_2 groups are along the six edges of a regular tetrahedron:

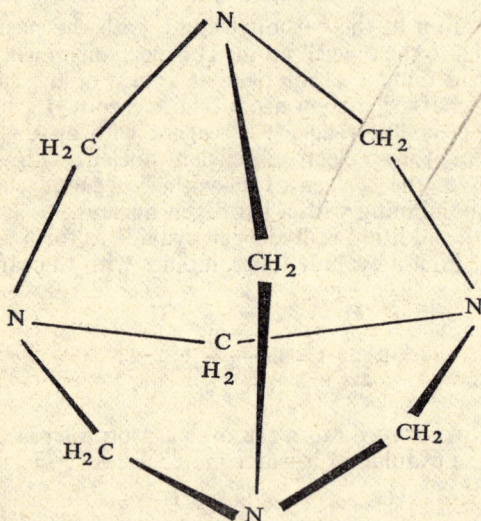

This is also called **urotropine** or **hexamine**, and is used as urinary antiseptic. Nitration with nitric acid of hexamine gives the high explosive known as cyclonite or RDX.

Sodium bisulphite gives crystalline addition products with aldehydes and ketones.

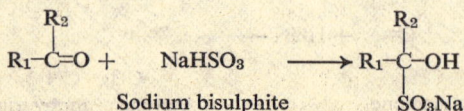

$$R_1\overset{R_2}{\underset{}{|}}C=O + NaHSO_3 \longrightarrow R_1\overset{R_2}{\underset{SO_3Na}{|}}C-OH$$

Sodium bisulphite

Since the reaction is reversible, an excess of a saturated solution of bisulphite is used to displace the equilibrium to the right. Since the aldehydes and ketones can be easily regenerated from either by decomposition with sodium carbonate or acid, the bisulphite addition compounds are useful for isolation and purification of aldehydes and ketones.

(VI) **Addition of Grignard Reagents to the Carbonyl Group.** Alkyl and aryl magnesium halides (Grignard reagents) add readily to C=O of ketones and aldehydes to give addition compounds containing magnesium halide which are decomposed with acids to give alcohols. Aldehydes give secondary alcohols and ketones give tertiary alcohols.

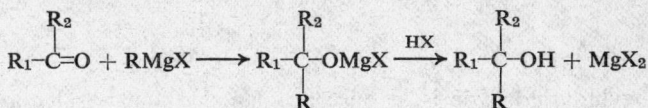

$$\underset{R_1}{\overset{R_2}{|}}C\!=\!O + RMgX \longrightarrow R_1\!-\!\overset{R_2}{\underset{R}{\overset{|}{C}}}\!-\!OMgX \xrightarrow{HX} R_1\!-\!\overset{R_2}{\underset{R}{\overset{|}{C}}}\!-\!OH + MgX_2$$

R and R_1 = alkyl or aryl groups
R_2 = alkyl or aryl groups or hydrogen

With formaldehyde as the carbonyl compound in the reaction, primary alcohols are obtained:

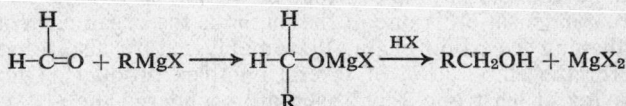

$$H\!-\!\overset{H}{\overset{|}{C}}\!=\!O + RMgX \longrightarrow H\!-\!\overset{H}{\underset{R}{\overset{|}{C}}}\!-\!OMgX \xrightarrow{HX} RCH_2OH + MgX_2$$

This reaction is of wide applicability. Only when R_1, R_2, and R are bulky or branched groups, the linking of R to the carbon atom of the carbonyl group becomes difficult.

(VII) **Condensation Reactions.** A condensation reaction is one in which two molecules combine to form a larger one, with the elimination of a simpler substance, usually water. Reagents like **hydroxylamine, phenylhydrazine, semicarbazide**, etc., add on to the carbonyl bond—the addition product subsequently losing one molecule of water. For example, the reaction of carbonyl compounds with hydroxylamine (NH_2OH) to give compounds called oximes can be written:

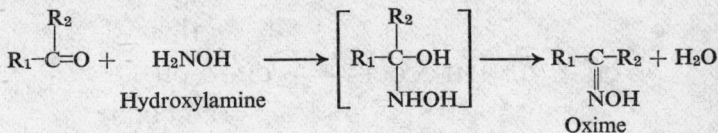

$$\underset{R_1}{\overset{R_2}{|}}C\!=\!O + \underset{\text{Hydroxylamine}}{H_2NOH} \longrightarrow \left[R_1\!-\!\overset{R_2}{\underset{NHOH}{\overset{|}{C}}}\!-\!OH\right] \longrightarrow \underset{\underset{\text{Oxime}}{NOH}}{R_1\!-\!C\!-\!R_2} + H_2O$$

Similarly, phenylhydrazine gives a phenylhydrazone with a carbonyl compound:

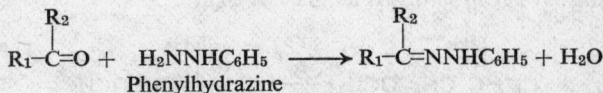

$$\underset{R_1}{\overset{R_2}{|}}C\!=\!O + \underset{\text{Phenylhydrazine}}{H_2NNHC_6H_5} \longrightarrow \overset{R_2}{\underset{}{R_1\!-\!C}}\!=\!NNHC_6H_5 + H_2O$$

Aldehydes and ketones react similarly with semicarbazide ($NH_2NHCONH_2$) to give derivatives known as semicarbazones. The above derivatives, oximes, phenylhydrazones, and semicarbazones are crystalline compounds and aid in the isolation, purification, and characterization of the aldehydes and ketones. Furthermore, many of them can be decomposed under suitable conditions to furnish the parent carbonyl compounds.

✗ (VIII) **Aldol 'Condensation'.** When acetaldehyde is warmed with a dilute solution of sodium hydroxide, β-**hydroxybutyraldehyde**, known as **'aldol'**, is obtained.

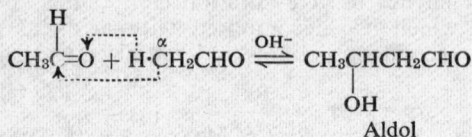

$$\overset{\text{H}}{\underset{}{\text{CH}_3\text{C}=\text{O}}} + \text{H·CH}_2\text{CHO} \overset{\text{OH}^-}{\rightleftharpoons} \text{CH}_3\text{CHCH}_2\text{CHO}$$
$$\underset{\text{OH}}{}$$
Aldol

β-hydroxybutyraldehyde is at once an aldehyde and an alcohol, and hence the name **aldol**. The aldol molecule has reactive hydrogen atoms on the α carbon atom next to the ‾CHO group, and hence the reaction can be repeated indefinitely, leading to compounds with high molecular weights. Since no simple substance is eliminated, this reaction is really addition, not condensation, but that is what it is commonly called. The aldol condensation is one of the methods the organic chemist has for lengthening a carbon chain. Presumably nature also adopts this method for the elaboration of several complex products. Only aldehydes having at least one α-hydrogen atom undergo the above aldol-type condensation. For instance, **trimethylacetaldehyde**, $(\text{CH}_3)_3\text{C}‾\text{CHO}$, and **benzaldehyde**, $\text{C}_6\text{H}_5\text{CHO}$, do not give aldols. Aldol on heating or with mineral acids loses a molecule of water, giving an unsaturated aldehyde called crotonaldehyde:

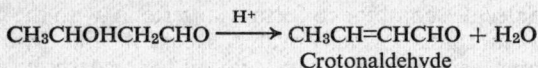

$$\text{CH}_3\text{CHOHCH}_2\text{CHO} \overset{\text{H}^+}{\longrightarrow} \text{CH}_3\text{CH}=\text{CHCHO} + \text{H}_2\text{O}$$
Crotonaldehyde

Ketones also undergo the aldol condensation in the same manner as aldehydes. Thus acetone gives diacetone alcohol in the presence of barium hydroxide.

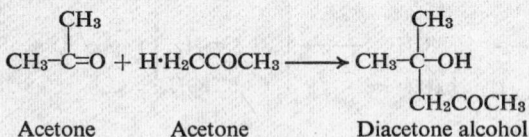

$$\overset{\text{CH}_3}{\underset{}{\text{CH}_3\text{-C}=\text{O}}} + \text{H·H}_2\text{CCOCH}_3 \longrightarrow \overset{\text{CH}_3}{\underset{\text{CH}_2\text{COCH}_3}{\text{CH}_3\text{-C-OH}}}$$

    Acetone   Acetone   Diacetone alcohol

Diacetone alcohol when heated eliminates a molecule of water, giving an unsaturated ketone known as mesityl oxide.

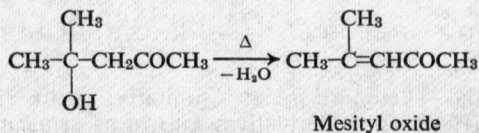

$$\overset{\text{CH}_3}{\underset{\text{OH}}{\text{CH}_3\text{-C-CH}_2\text{COCH}_3}} \overset{\Delta}{\underset{-\text{H}_2\text{O}}{\longrightarrow}} \overset{\text{CH}_3}{\underset{}{\text{CH}_3\text{-C}=\text{CHCOCH}_3}}$$

                                   Mesityl oxide

Mesityl oxide can be condensed with another molecule of acetone to give phorone.

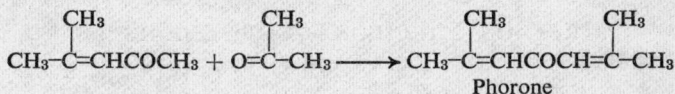

$$\underset{CH_3}{\overset{CH_3}{CH_3-\overset{|}{C}=CHCOCH_3}} + O=\overset{CH_3}{\overset{|}{C}-CH_3} \longrightarrow \underset{Phorone}{\overset{CH_3}{CH_3-\overset{|}{C}=CHCOCH=\overset{|}{C}-CH_3}}$$

Diacetone alcohol finds wide industrial application as a solvent.

(IX) **Polymerization. Aldehydes**, especially **formaldehyde** and **acetaldehyde**, polymerize in the presence of acid catalysts, formaldehyde giving **trioxane** and acetaldehyde giving **paraldehyde**. In both these polymerized forms the aldehyde function is absent, and hence an ether-like structure is assigned to each of them.

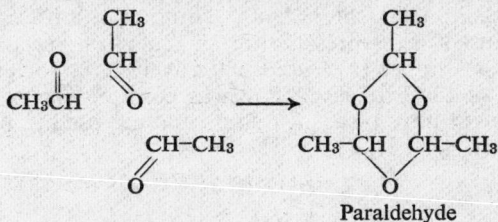

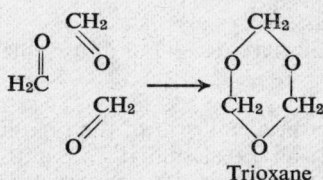

Paraldehyde

Trioxane

Evaporation of an aqueous solution of formaldehyde furnishes a solid polymer of formaldehyde, known as **paraformaldehyde**. Formaldehyde exists probably as the hydrate $CH_2(OH)_2$ in aqueous solution, and paraformaldehyde is evidently formed by elimination of water, as shown below between successive molecules of this hydrate.

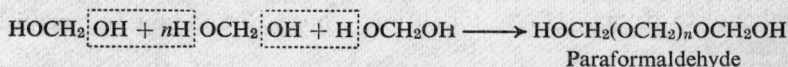

$$HOCH_2\overline{OH} + nH\overline{OCH_2}\overline{OH} + H\overline{OCH_2OH} \longrightarrow HOCH_2(OCH_2)_nOCH_2OH$$
$$\text{Paraformaldehyde}$$

The above three polymers are decomposed into the parent units under suitable conditions.

(X) **Acetals and Ketals.** Under the influence of acids, an alcohol reacts with a molecule of an aldehyde to give first a **hemiacetal** which again combines with another molecule of alcohol to give an acetal. For example:

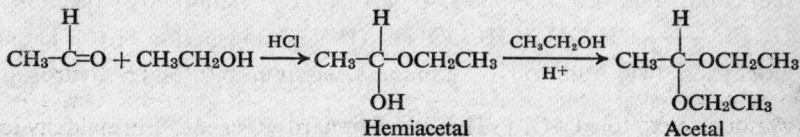

$$\underset{}{\overset{H}{CH_3-\overset{|}{C}=O}} + CH_3CH_2OH \xrightarrow{HCl} \underset{\underset{Hemiacetal}{OH}}{\overset{H}{CH_3-\overset{|}{\underset{|}{C}}-OCH_2CH_3}} \xrightarrow[H^+]{CH_3CH_2OH} \underset{\underset{Acetal}{OCH_2CH_3}}{\overset{H}{CH_3-\overset{|}{\underset{|}{C}}-OCH_2CH_3}}$$

In a similar way but less readily ketones furnish ketals.

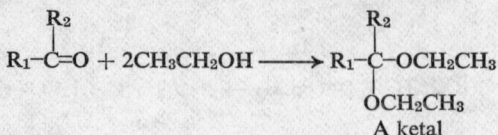

$$R_1-\underset{\underset{\displaystyle }{|}}{\overset{\overset{\displaystyle R_2}{|}}{C}}{=}O + 2CH_3CH_2OH \longrightarrow R_1-\underset{\underset{\displaystyle OCH_2CH_3}{|}}{\overset{\overset{\displaystyle R_2}{|}}{C}}-OCH_2CH_3$$

A ketal

Both acetals and ketals are easily hydrolysed to the parent carbonyl compounds by acids. They are, however, stable to alkalis and oxidizing agents and serve as intermediates wherein the carbonyl group is protected during reactions at other parts of the molecule.

Formaldehyde and acetaldehyde are the simplest aldehydes. Formaldehyde is used for the preparation of Bakelite plastics. It is an irritating gas soluble in water. A 40% solution of it in water, known as Formalin, is used in the preservation of biological specimens. Acetone is the simplest ketone. It is used extensively as a solvent and in the manufacture of the explosive known as cordite. The ketones of high molecular weight have pleasant odours and are used in perfumes.

SUMMARY

Aldehydes are of the type $R\overset{\overset{\displaystyle H}{|}}{C}{=}O$ and are obtained by oxidation of primary alcohols. Ketones conform to the type $R_1\overset{\overset{\displaystyle R_2}{|}}{C}{=}O$ and are oxidation products of secondary alcohols. Preparation: (1) Oxidation of primary or secondary alcohols with potassium dichromate and sulphuric acid. (2) Catalytic dehydrogenation of an alcohol vapour. (3) Rosenmund reduction of an acid chloride to an aldehyde using palladium on barium sulphate catalyst. (4) Decomposition by heating of calcium or magnesium salts of acids: e.g. calcium acetate to acetone. (5) Oxo reaction. (6) Special methods: (1) Hydration of acetylene to acetaldehyde in the presence of mercury catalyst. (2) Chlorination of ethanol to chloral. (3) Acetone from propene (important commercial method).

Properties and Reactions: (I) Tollen's test (oxidation with silver oxide) positive for aldehydes and negative for ketones. Same with Fehling's test. (II) Reduction of aldehydes to primary alcohols and ketones to secondary alcohols. Reduction carried out either catalytically or chemically. Use of lithium aluminium hydride in ether suspension for reduction. Meerwein–Ponndorf–Verley reduction of ketones to alcohols with aluminium isopropoxide in isopropyl alcohol. Pinacol reduction, that is, acetone → pinacol. (III) Cannizzaro reaction: $HCHO \xrightarrow[\text{NaOH}]{\text{Conc.}} CH_3OH + HCOONa$. (IV) Halogenation to α-halo-aldehydes. (V) Addition of ammonia, sodium bisulphite, hydrogen cyanide, etc., to $-\overset{\overset{\displaystyle }{|}}{C}{=}O$. (VI) With Grignard Reagent: Formaldehyde yields primary alcohol, other aldehydes yield secondary alcohols, ketones yield tertiary alcohols. (VII) Condensation products of carbonyl compounds with hydroxylamine, phenylhydrazine, semicarbazide to give respectively oximes, phenylhydrazones, and semicarbazones which are useful for characterization purposes. (VIII) Aldol condensation: Aldol from acetaldehyde and diacetone alcohol from acetone. (IX) Polymerization of formaldehyde to trioxane and paraformaldehyde; similarly paraldehyde from acetaldehyde. (X) Formation of acetals and ketals by reaction of aldehyde or ketone with alcohol in presence of mineral acids.

Problem Set No. 5

1. Write structural formulae for all the 5-carbon aldehydes and ketones with their I.U.C. names.
2. How are the following conversions achieved?

    (a) Acetone to 2-chloropropane.
    (b) Ethanol to acetone.
    (c) Mesityl oxide from acetic acid.
    (d) Propanal to butanal.

3. A compound A of the formula C_4H_8O gives a compound B when treated with hydroxylamine and does not reduce Tollen's or Fehling's solution. Write down structures for A and B.
4. Give a plausible synthesis of butene-2 from butanone-2.
5. How can one prepare: (I) $CH_3CH_2CH(OC_3H_7)_2$ from propanol-1? (II) $CH_3CHOHCH_2CH_3$ from CH_3CHO? (III) $CH_3CH_2CH_2OH$ from CH_3CH_2OH?

CHAPTER SEVEN

CARBOXYLIC ACIDS AND THEIR DERIVATIVES

The functional group of organic acids is the carboxyl group, $-\overset{\overset{\displaystyle O}{\|}}{C}-OH$, wherein a carbonyl carbon atom is linked to a hydroxyl group. It might be mentioned at the outset that in organic acids the carbonyl function is masked and the hydroxyl function predominates. Acids which contain one carboxyl group are called monobasic acids, those containing two, dibasic, and so on.

FATTY ACIDS

Fatty acids have the general formula $R-\overset{\overset{\displaystyle O}{\|}}{C}-OH$, in which R is as usual an alkyl radical. The higher members of this series of acids occur in nature in the combined form of esters of glycerol (fats), and hence all the members of this family are called fatty acids.

PREPARATION: (1) **By Oxidation.** Primary alcohols and aldehydes, as previously mentioned, give carboxylic acids on oxidation. For example, *n*-**heptaldehyde**, a product of thermal decomposition of castor oil, is oxidised in good yield to *n*-heptylic acid.

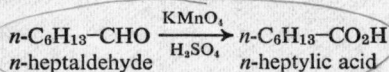

$$n\text{-C}_6\text{H}_{13}-\text{CHO} \xrightarrow[\text{H}_2\text{SO}_4]{\text{KMnO}_4} n\text{-C}_6\text{H}_{13}-\text{CO}_2\text{H}$$
$$\text{\textit{n}-heptaldehyde} \qquad\qquad \text{\textit{n}-heptylic acid}$$

Olefinic compounds and ketones also furnish acids on oxidation, but these acids have a smaller number of carbon atoms.

(2) **From Nitriles** ($R-C\equiv N$). The hydrolysis of alkyl cyanides or nitriles with acids or alkalis furnishes carboxylic acids. The nitriles are obtained by refluxing alkyl halides with alcoholic sodium or potassium cyanide.

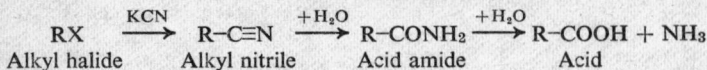

$$\underset{\text{Alkyl halide}}{\text{RX}} \xrightarrow{\text{KCN}} \underset{\text{Alkyl nitrile}}{\text{R}-\text{C}\equiv\text{N}} \xrightarrow{+\text{H}_2\text{O}} \underset{\text{Acid amide}}{\text{R}-\text{CONH}_2} \xrightarrow{+\text{H}_2\text{O}} \underset{\text{Acid}}{\text{R}-\text{COOH} + \text{NH}_3}$$

The initial product of hydrolysis is an amide which is further hydrolysed to the acid. Glutaric acid (a dibasic acid), for instance, is prepared as follows:

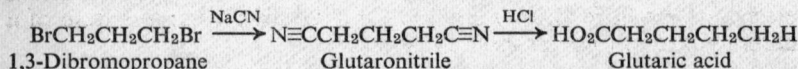

$$\underset{\text{1,3-Dibromopropane}}{\text{BrCH}_2\text{CH}_2\text{CH}_2\text{Br}} \xrightarrow{\text{NaCN}} \underset{\text{Glutaronitrile}}{\text{N}\equiv\text{CCH}_2\text{CH}_2\text{CH}_2\text{C}\equiv\text{N}} \xrightarrow{\text{HCl}} \underset{\text{Glutaric acid}}{\text{HO}_2\text{CCH}_2\text{CH}_2\text{CH}_2\text{CH}_2\text{H}}$$

The nitrile method is particularly applicable to primary halides. With secondary and tertiary halides the yields of nitriles (and therefore of acids) are less satisfactory due to side reactions.

(3) **Grignard Synthesis.** Alkyl magnesium halides react with carbon dioxide to give magnesium halide derivatives, which on hydrolysis give acid:

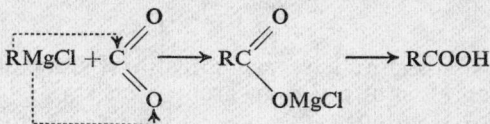

$$RMgCl + \overset{O}{\underset{O}{\overset{\|}{C}}} \longrightarrow RC\overset{O}{\underset{OMgCl}{\overset{\|}{\diagup}}} \longrightarrow RCOOH$$

The reaction can be carried out by pouring the Grignard solution on dry ice (solid CO_2) or bubbling CO_2 into the solution.

NOMENCLATURE: The common names for the first six members of the series of straight-chain saturated monocarboxylic acids are:

formic acid (HCOOH)
acetic acid (CH_3COOH)
propionic acid (CH_3CH_2COOH)
butyric acid ($CH_3CH_2CH_2COOH$)
valeric acid ($CH_3CH_2CH_2CH_2COOH$)
caproic acid ($CH_3CH_2CH_2CH_2CH_2COOH$)

Several of these names have been in use for a long period of time and usually refer to the natural source of the acid. The I.U.C. name for an acid has the ending *-oic* and is derived in the same way as that of an aldehyde, that is, from the name of the hydrocarbon corresponding to the longest chain containing the –COOH group. Names of several acids are given here for purposes of illustration.

H·COOH	Methanoic acid
CH_3COOH	Ethanoic acid
C_3H_7COOH	Butanoic acid
$CH_3CH_2\underset{\underset{CH_3}{\|}}{C}H–COOH$	2-Methylbutanoic acid
$CH_3\underset{\underset{CH_3}{\|}}{C}H–\underset{\underset{CH_2CH_3}{\|}}{C}H–CH_2COOH$	3-Ethyl-4-methylpentanoic acid

Very often the above system is ignored and substituted acids are named using Greek letters to indicate the position of the substituents, the letter α referring to the carbon atom to which the –COOH group is attached and not to the carboxylic carbon atom itself. Thus,

$$Br–CH_2–CH_2–CH_2–COOH$$

is γ-bromobutyric acid, $HOCH_2CH_2COOH$ is β-hydroxypropionic acid, and so on.

PROPERTIES: The oxygen atom is electronegative, that is it attracts electrons. We have already seen that this causes the carbon atom of a carbonyl group to be positively charged:

$$\overset{O^{\delta-}}{\underset{\underset{C}{|}}{\overset{\|}{C}}}{}^{\delta+}$$

Similarly, the oxygen atom in a hydroxyl group attracts the electrons in its bond to hydrogen and to the rest of the molecule:

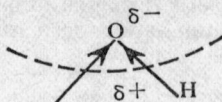

$$\overset{\delta -}{O}$$

In the carboxyl group we have such a hydroxyl group attached to the positive carbon atom of a carbon group, which will attract electrons from the hydroxyl group:

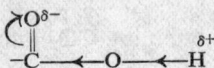

$$-C \longleftarrow O \longleftarrow H$$

This causes the hydroxylic hydrogen atom to have a larger positive charge in carboxylic acids than in alcohols.

In carboxylic acids, as will be seen from the formula, there is excellent scope for the formation of intermolecular hydrogen bonds (cf. page 57), and hence they are associated, as shown, to give cyclic dimers:

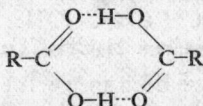

$$R-C \overset{O \cdots H-O}{\underset{O-H \cdots O}{}} C-R$$

The boiling points are consequently higher than those of comparable unassociated liquids. Though a carboxyl group is made up of a carbonyl and a hydroxyl group, we find that the typical properties of carbonyl compounds and alcohols are altered in the acids. The acids do not give the addition reactions characteristic of the carbonyl compound, and the carboxylic acids are stronger acids than alcohols, which are generally neutral. The carbonyl group by association with the hydroxyl group has rendered the −OH group more acidic; that is, it has increased the possibility for the hydrogen nucleus to be detached and carried away by a molecule of a suitable solvent, for example, water:

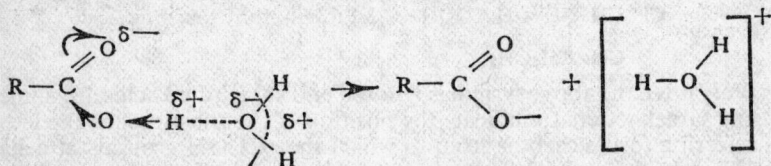

$$R-C \longrightarrow R-C + \left[H-O \overset{H}{\underset{H}{}} \right]^{+}$$

It will also be noticed that the acids undergo most of the reactions characteristic of alcohols.

(I) **Acid Properties.** They react with metals giving hydrogen, and are neutralized by alkali to give salt and water.

$$RCOOH + NaOH \longrightarrow RCOONa + H_2O$$

Carboxylic acids are stronger acids than carbonic acid: they decompose carbonates,

$$2R{\cdot}COOH + Na_2CO_3 \longrightarrow 2R{\cdot}COONa + H_2O + CO_2$$

They are weaker than the mineral acids, and so are displaced by them, for example, acetic acid is displaced from acetates on warming with concentrated sulphuric acid:

$$CH_3 \cdot COONa + H_2SO_4 \longrightarrow CH_3 \cdot COOH + NaHSO_4$$

giving the characteristic smell of vinegar.

(II) **Replacement of the Hydroxyl Group** leads to the formation of esters, acid chlorides, and amides. Esters, as we have already seen, are obtained by heating carboxylic acids with alcohols in the presence of inorganic acids used as catalysts.

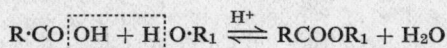

$$R \cdot CO \boxed{OH + H} O \cdot R_1 \overset{H^+}{\rightleftharpoons} RCOOR_1 + H_2O$$

Acid chlorides are prepared by treating acids with phosphorus penta-chloride (PCl_5) or better still with thionyl chloride:

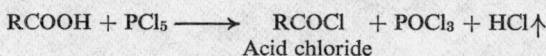

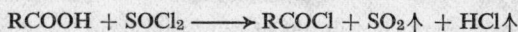

$$RCOOH + PCl_5 \longrightarrow \quad RCOCl \ + POCl_3 + HCl\uparrow$$
Acid chloride

$$RCOOH + SOCl_2 \longrightarrow RCOCl + SO_2\uparrow + HCl\uparrow$$

(III) **Properties of the Whole Carbonyl Group.** When the ammonium salt of an organic acid is heated, water is eliminated and an amide is obtained.

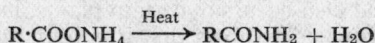

$$R \cdot COONH_4 \overset{Heat}{\longrightarrow} RCONH_2 + H_2O$$

As has already been mentioned, the acids do not take part in reactions characteristic of the carbonyl compounds—such as oxime or phenyl-hydrazone formation. Acids are more difficult to reduce than aldehydes, ketones, and even esters. The only satisfactory method of reducing acids or, better still, salts of acids is to use lithium aluminium hydride, pre-viously described.

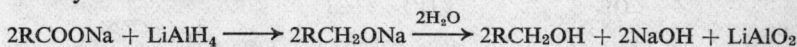

$$2RCOONa + LiAlH_4 \longrightarrow 2RCH_2ONa \overset{2H_2O}{\longrightarrow} 2RCH_2OH + 2NaOH + LiAlO_2$$

Carboxylic acids undergo decarboxylation (that is, loss of CO_2) when the sodium salt of the acid is heated with sodium hydroxide.

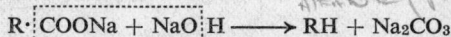

$$R \cdot \boxed{COONa + NaO} H \longrightarrow RH + Na_2CO_3$$

Pyrolysis of calcium salts, barium salts, and salts of other acids to give ketones has been previously discussed, as has the behaviour of ammo-nium salts upon heating. The nature of the metal in the salt apparently has profound influence on the course of decomposition of the salts. The silver salts of acids when heated with chlorine or bromine give alkyl halides.

$$RCOOAg + Cl_2 \longrightarrow RCl + CO_2 + AgCl$$

This is known as the HUNSDIECKER reaction, and the alkyl halides thus obtained are useful starting materials in synthesis.

(IV) **Properties of the Alkyl Group.** Acids may be halogenated directly at the α carbon atom. For example, acetic acid can be chlorinated in successive stages to mono-, di-, or tri-chloroacetic acids in the presence of catalytic amounts of iodine.

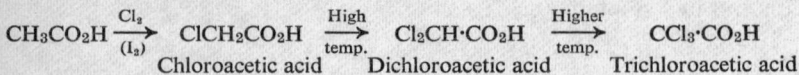

$$CH_3CO_2H \xrightarrow[(I_2)]{Cl_2} ClCH_2CO_2H \xrightarrow[\text{temp.}]{\text{High}} Cl_2CH \cdot CO_2H \xrightarrow[\text{temp.}]{\text{Higher}} CCl_3 \cdot CO_2H$$

Chloroacetic acid   Dichloroacetic acid   Trichloroacetic acid

Phosphorus trichloride and tribromide are also used as catalysts for the α halogenation of acids.

FORMIC ACID

In industry formic acid is produced by the action of carbon monoxide on sodium hydroxide at moderate temperatures and pressures. On acidification with hydrochloric acid, free formic acid is obtained.

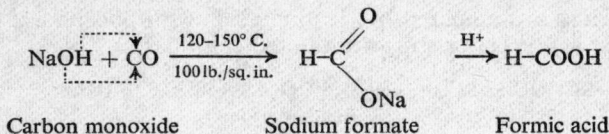

$$NaOH + CO \xrightarrow[\text{100 lb./sq. in.}]{120-150^\circ C.} H-C\underset{ONa}{\overset{O}{<}} \xrightarrow{H^+} H-COOH$$

Carbon monoxide   Sodium formate   Formic acid

PROPERTIES: Formic acid is a liquid of pungent odour, readily soluble in water and with a boiling point of 100·5° C. It is so named because it is a constituent of certain ants (Latin *formica* meaning ant). When an ant stings a person it injects a small quantity of formic acid which produces a blister. Formic acid also occurs in certain plants, including the nettle. The structure of formic acid shows that it is both an aldehyde and an acid. Like aldehydes, it reduces Fehling's solution and gives a silver mirror with Tollen's reagent, being easily oxidised to carbon dioxide and water.

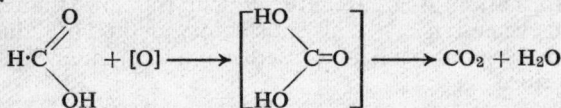

$$H \cdot C \underset{OH}{\overset{O}{<}} + [O] \longrightarrow \begin{bmatrix} HO \\ & C=O \\ HO \end{bmatrix} \longrightarrow CO_2 + H_2O$$

Unlike other carboxylic acids, it is readily decomposed by warm, concentrated sulphuric acid, with the liberation of carbon monoxide. In the laboratory, formic acid is useful in several syntheses. In industry it finds application for the tanning of hides and for the coagulation of rubber latex.

ACETIC ACID

Formerly, the pyroligneous acid from the destructive distillation of wood served as a source of acetic acid. Another source is vinegar obtained by the bacterial oxidation of wine waste which contains small amounts of ethanol. The oxidation is achieved by atmospheric oxygen in the presence of *Bacterium acetii*. Large quantities of acetic acid are now produced cheaply from acetylene. In the presence of mercuric sulphate, acetylene adds on a molecule of water to give acetaldehyde, which is then catalytically oxidised by air to acetic acid:

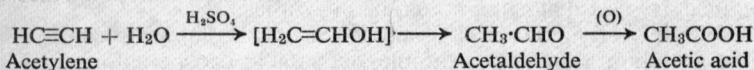

$$HC \equiv CH + H_2O \xrightarrow{H_2SO_4} [H_2C=CHOH] \longrightarrow CH_3 \cdot CHO \xrightarrow{(O)} CH_3COOH$$

Acetylene   Acetaldehyde   Acetic acid

Acetic acid is a pungent liquid miscible with water. It is one of the commonest acids and has been known from antiquity. It is said that Cleopatra, the Egyptian queen, recommended a drink of pearls dissolved in vinegar as an aid to beauty. Pure acetic acid is called glacial acetic acid, since on cooling it freezes to an ice-like block melting at 16·8° C. In industry it is used for the preparation of perfumes, dyes, pharmaceuticals, and plastics.

DERIVATIVES OF CARBOXYLIC ACIDS

(A) **Acid Chlorides.** These are obtained from acids by the replacement of the $-OH$ group using phosphorus trichloride or pentachloride or thionyl chloride. Thus glacial acetic acid gives acetyl chloride according to the equations:

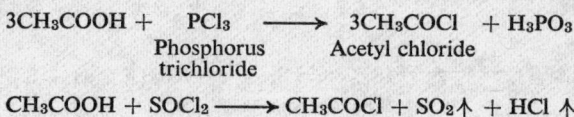

$$3CH_3COOH + PCl_3 \longrightarrow 3CH_3COCl + H_3PO_3$$

Phosphorus   Acetyl chloride
trichloride

$$CH_3COOH + SOCl_2 \longrightarrow CH_3COCl + SO_2\uparrow + HCl\uparrow$$

The use of thionyl chloride is advantageous since the products of the reaction other than the acid chloride are gases and leave the reaction vessel on heating. The group $R-C=O$ is sometimes called an acyl group and hence the acid chlorides are also referred to as acyl chlorides.

PROPERTIES: The acid chlorides are non-associated (hydrogen bonding is not possible), and hence their boiling points are less than those of the corresponding acids. Thus the boiling point of acetic acid is 118° C. and that of acetyl chloride is 51° C. The acid chlorides are generally pungent-smelling liquids that affect the eye and also fume in air due to hydrolysis and liberation of hydrogen chloride. They are useful as acylating agents, that is, they are used to replace an active hydrogen atom by the group $R-C=O$. For instance, they react readily with water, alcohol, and ammonia to give the corresponding acid, ester, and amide respectively:

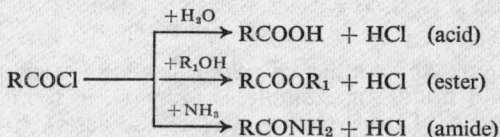

$$
RCOCl
\begin{cases}
\xrightarrow{+H_2O} RCOOH + HCl & \text{(acid)} \\
\xrightarrow{+R_1OH} RCOOR_1 + HCl & \text{(ester)} \\
\xrightarrow{+NH_3} RCONH_2 + HCl & \text{(amide)}
\end{cases}
$$

In all the above reactions hydrogen chloride is the other product, and a base like sodium hydroxide is usually utilized to neutralize it as it is formed.

Acid chlorides serve as important intermediates in chemical synthesis. Their reaction with alkyl (or aryl) cadmium halides is the basis for the preparation of a variety of ketones. A Grignard reagent, RMgX, in ether solution, when mixed with anhydrous cadmium chloride, affords the corresponding RCdX:

$$RMgX + CdCl_2 \longrightarrow RCdCl + MgXCl$$

Acid chlorides react with the above cadmium derivatives to give ketones.

$$R_1COCl + RCdCl \longrightarrow R_1COR + CdCl_2$$

The same reaction also takes place with Grignard reagents, but it is difficult to stop at the stage of ketone which undergoes further conversion to a tertiary alcohol. The above cadmium derivatives, though similar to Grignard reagents, are less reactive and add much less readily to the initially formed ketone.

Acid chlorides are also important intermediates in a method often used for preparing the next higher homologue of an acid. The reaction used to accomplish this is known as the ARNDT-EISTERT reaction. It consists of treating an acid chloride with diazomethane to obtain a diazoketone which on hydrolysis loses nitrogen and rearranges to give the higher carboxylic acid:

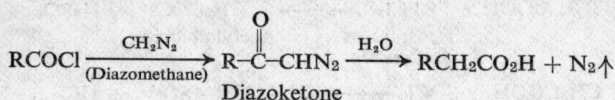

$$RCOCl \xrightarrow[\text{(Diazomethane)}]{CH_2N_2} R-\overset{\overset{\displaystyle O}{\|}}{C}-CHN_2 \xrightarrow{H_2O} RCH_2CO_2H + N_2\uparrow$$
Diazoketone

The conversion of the diazoketone to the acid actually involves the migration of the group R from the carbon atom of the carbonyl group to the carbon atom linked to the nitrogen atom in the diazoketone, and hence is an instance of molecular rearrangement.

(B) **Acid Anhydrides.** These may be looked upon as being derived from two molecules of an acid by the removal of one molecule of water, though they cannot be prepared in this way directly.

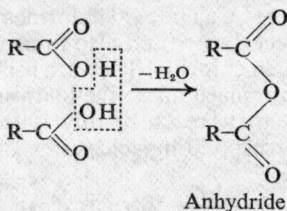

Anhydride

They can be prepared from the acid chloride and the sodium salt of the acid. Acetic anhydride, for example, is prepared in the laboratory by distilling a mixture of fused sodium acetate and acetyl chloride.

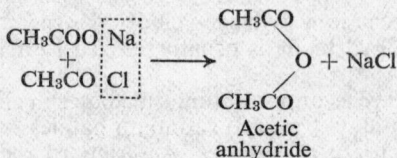

Acetic
anhydride

In industry acetone is thermally decomposed to give a reactive intermediate called ketene, which is then treated with acetic acid to give acetic anhydride.

$$\text{H} \vdots \underset{\substack{|\\ \text{CH}_2\text{-C=O}}}{\overset{\text{CH}_3 \vdots}{}} \xrightarrow{700-750°} \text{CH}_2\text{=C=O} + \text{CH}_4$$
Ketene

$$\text{CH}_3\text{COOH} + \text{CH}_2\text{=C=O} \longrightarrow \left[\begin{array}{c} \text{CH}_2\text{=C-OH} \\ | \\ \text{O} \\ | \\ \text{CH}_3\text{C=O} \end{array} \right] \longrightarrow \begin{array}{c} \text{CH}_3\text{C=O} \\ \backslash \\ \text{O} \\ / \\ \text{CH}_3\text{C=O} \end{array}$$

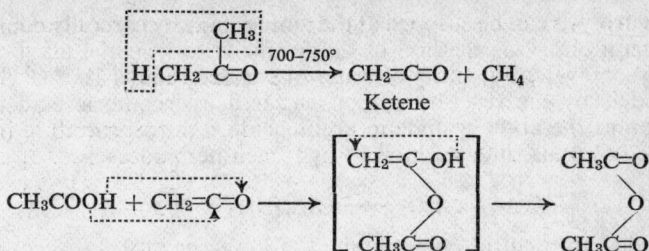

Acetic anhydride is a pungent-smelling liquid that is hydrolysed by warm water. It reacts with compounds containing an active hydrogen atom in exactly the same manner as acetyl chloride, giving rise to the acetylated products.

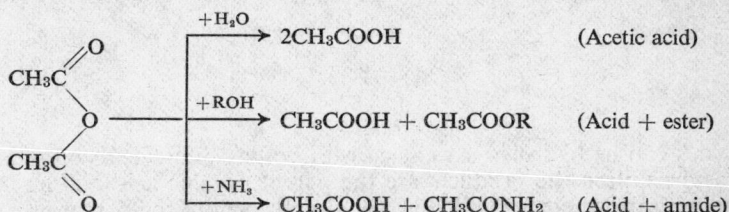

$$\begin{array}{c} \text{CH}_3\text{C} \overset{\displaystyle O}{\diagup} \\ \diagdown \\ \text{O} \\ \diagup \\ \text{CH}_3\text{C} \\ \diagdown_{\displaystyle O} \end{array} \left\{ \begin{array}{ll} \xrightarrow{+\text{H}_2\text{O}} 2\text{CH}_3\text{COOH} & \text{(Acetic acid)} \\ \\ \xrightarrow{+\text{ROH}} \text{CH}_3\text{COOH} + \text{CH}_3\text{COOR} & \text{(Acid + ester)} \\ \\ \xrightarrow{+\text{NH}_3} \text{CH}_3\text{COOH} + \text{CH}_3\text{CONH}_2 & \text{(Acid + amide)} \end{array} \right.$$

The acid anhydrides and acid chlorides are widely used for acylation of such compounds as amines, alcohols, and phenols. It should be mentioned that the anhydride is less vigorous in its reaction, and therefore more convenient to use for acylation than the acid chlorides.

(C) **Acid Amides.** An amide is a derivative of an acid, in which the hydroxyl group has been replaced by the amino group, and has the general formula $RCONH_2$. Monosubstituted amides, $RCONHR_1$, and disubstituted amides, $RCONR_1R_2$, are also known. Acid chlorides react with ammonia, primary amine, or secondary amine to give the unsubstituted, the monosubstituted, and disubstituted amides respectively.

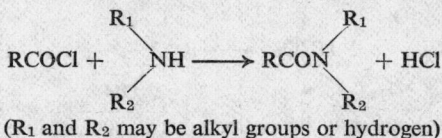

$$\text{RCOCl} + \underset{\underset{\text{R}_2}{|}}{\overset{\overset{\text{R}_1}{|}}{\text{NH}}} \longrightarrow \text{RCON} \underset{\text{R}_2}{\overset{\text{R}_1}{}} + \text{HCl}$$
(R_1 and R_2 may be alkyl groups or hydrogen)

The acylation can also be effected by replacing the acid chlorides with the acid anhydrides.

Amides may also be made in good yield by dehydrating ammonium salts of acids.

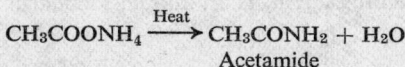

$$\text{CH}_3\text{COONH}_4 \xrightarrow{\text{Heat}} \text{CH}_3\text{CONH}_2 + \text{H}_2\text{O}$$
Acetamide

The amides, as has been already noted, are intermediates in the hydrolysis of nitriles to acids.

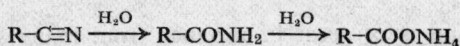

$$\text{R-C}\equiv\text{N} \xrightarrow{\text{H}_2\text{O}} \text{R-CONH}_2 \xrightarrow{\text{H}_2\text{O}} \text{R-COONH}_4$$

The hydrolysis can be stopped at the amide stage by carefully controlling the conditions. One method of doing this is to heat the alkyl cyanide with an equivalent quantity of water in a sealed tube at 180° C. Another method is by the reaction of concentrated hydrochloric acid upon a solution of the alkyl cyanide in acetic acid. A third procedure involves the use of an alkaline solution of hydrogen peroxide.

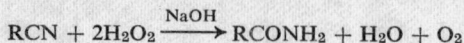

$$RCN + 2H_2O_2 \xrightarrow{\text{NaOH}} RCONH_2 + H_2O + O_2$$

Amides are neutral compounds, the lower members (except formamide) being crystalline solids soluble in water. Acetamide is associated by hydrogen bonding, as follows:

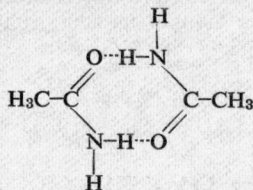

Amides can be hydrolysed to acids with boiling dilute acids or alkalis In acid solution the products are the parent organic acid and an ammonium salt, whereas in alkaline solution the salt of the parent organic acid and free ammonia are obtained.

$$RCONH_2 + H_3O^+ \longrightarrow RCOOH + NH_4^+$$
$$RCONH_2 + OH^- \longrightarrow RCOO^- + NH_3$$

When the amides are dehydrated with phosphorus pentoxide nitriles are obtained.

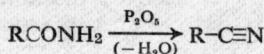

$$RCONH_2 \xrightarrow[(-H_2O)]{P_2O_5} R-C{\equiv}N$$

Nitrous acid reacts with amides to give the corresponding acid and elementary nitrogen.

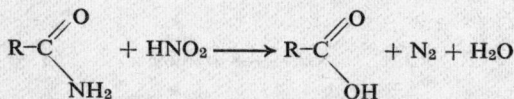

The nitrogen evolved may be collected and its volume measured. This reaction gives a method of estimating the nitrogen content of an amide.

For their degradation to an amine containing one fewer carbon atoms than the original amide (see page 130).

Amides are used in a wide variety of ways. Nylon is a polyamide. Sucaryl, the sweetening agent, is an amide, saccharin being an imide.

Urea. Urea ($H_2N-\overset{\overset{\displaystyle O}{\|}}{C}-NH_2$) is the diamide of carbonic acid ($HO-\overset{\overset{\displaystyle O}{\|}}{C}-OH$) and is of great importance from many standpoints. It occurs in the urine of mammals and also in other liquids of animal origin. A human adult excretes about 30 g. of urea in 24 hours. It will be recalled that urea was the first organic substance to be synthesized in the laboratory by Wohler in 1828. Wohler's synthesis consisted of the evaporation of an aqueous solution of ammonium cyanate, which isomerizes to give urea.

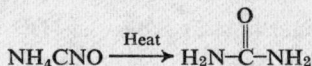

$$NH_4CNO \xrightarrow{\text{Heat}} H_2N{-}\overset{\overset{\displaystyle O}{\|}}{C}{-}NH_2$$

Urea may also be obtained by treating ammonia with phosgene.

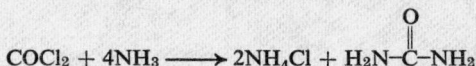

$$COCl_2 + 4NH_3 \longrightarrow 2NH_4Cl + H_2N{-}\overset{\overset{\displaystyle O}{\|}}{C}{-}NH_2$$

Industrially, urea is made by heating a mixture of ammonia and carbon dioxide under pressure.

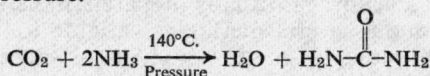

$$CO_2 + 2NH_3 \xrightarrow[\text{Pressure}]{140°C.} H_2O + H_2N{-}\overset{\overset{\displaystyle O}{\|}}{C}{-}NH_2$$

USES OF UREA: Urea is a valuable fertilizer because of its high nitrogen content (46·7%) and ease of hydrolysis.

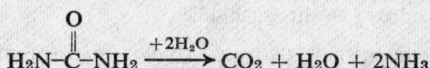

$$H_2N{-}\overset{\overset{\displaystyle O}{\|}}{C}{-}NH_2 \xrightarrow{+2H_2O} CO_2 + H_2O + 2NH_3$$

Urea and several of its derivatives are used to treat infections because of their bacteriostatic activity. The so-called ammoniated dentifrices con-contain urea, which serves as a source of ammonia. Urea is also used in the manufacture of barbiturates, the well-known hypnotics. These barbiturates are prepared by condensing a suitably substituted malonic ester with urea in the presence of sodium ethoxide. Veronal, for instance, is prepared as follows:

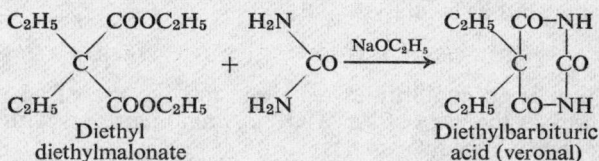

Diethyl diethylmalonate + Diethylbarbituric acid (veronal), with NaOC$_2$H$_5$

Since veronal itself is relatively insoluble in water, it is taken in the form of its sodium derivative which is water soluble.

(D) **Esters.** PREPARATION: (1) The method of esterification of alcohols with carboxylic acids in presence of inorganic acids, used as catalysts, was discussed in Chapter 5. In this method, by means of studies using isotopic tracer elements, the hydroxyl eliminated during esterification of primary or secondary alcohols has been shown to come from the acid rather than the alcohol. For example, when methanol containing O^{18} isotope (i.e. CH$_3$O^{18}H) is esterified with benzoic acid containing the ordinary O^{16} isotope (i.e. $C_6H_5CO_2H$) in the presence of hydrogen chloride the water formed contains only the ordinary O^{16} isotope:

$$C_6H_5C{\overset{O}{\underset{OH}{}}} + HO^{18}{-}CH_3 \xrightarrow{\ H^+\ } C_6H_5{-}C{\overset{O}{\underset{O^{18}CH_3}{}}} + H_2O$$

In the case of esters of tertiary alcohols, the –OH eliminated appears to originate from the alcohol used.

(2) An alternative route, especially for methyl esters, is used to treat a carboxylic acid with an ether solution of diazomethane.

$$R{-}C{\overset{O}{\underset{OH}{}}} + \underset{\text{Diazomethane}}{CH_2N_2} \longrightarrow R{-}C{\overset{O}{\underset{OCH_3}{}}} + N_2\uparrow$$

Diazomethane is a yellow gas usually prepared in the form of an ether solution. One among several methods available for its preparation consists of decomposing N-methyl-N-nitroso-N-nitroguanidine with potassium hydroxide:

$$\underset{\text{N-methyl-N-nitroso-N-nitroguanidine}}{CH_3N(NO)\overset{\overset{NH}{\|}}{C}NHNO_2} \xrightarrow{\ KOH\ } CH_2N_2$$

PROPERTIES: Esters can be hydrolysed by water in the presence of alkali or acid.

$$RCOOR_1 + H_2O \xrightarrow{\ H^+\ } RCOOH + R_1OH$$

$$RCOOR_1 + OH^- \longrightarrow RCOO^- + R_1OH$$

Alkaline hydrolysis of esters is also referred to as saponification (Greek *Sapon* for soap), since soaps are prepared by alkaline hydrolysis of fats and oils (see below). The alkoxy group of an ester of a primary or secondary alcohol may be exchanged readily for that of another alcohol using an inorganic acid as a catalyst.

$$RCOOR_1 + R_2OH \underset{}{\overset{H^+}{\rightleftharpoons}} RCOOR_2 + R_1OH$$

The reaction is an equilibrium reaction, and the use of an excess of R_2OH increases the yield of $R{-}COOR_2$. Ammonia reacts with esters at room temperature to give amides.

$$RCOOR_1 + NH_3 \longrightarrow RCONH_2 + R_1OH$$

The ester group can be reduced to the primary alcoholic group by sodium and alcohol or, better still, by lithium aluminium hydride.

$$4RCOOR_1 \xrightarrow{\ +2AlLiH_4\ } \underset{[+\ LiAl(OR_1)_4]}{LiAl(OCH_2R)_4} \xrightarrow{\ +4H_2O\ } \underset{[+\ LiOH + Al(OH)_3]}{4RCH_2OH}$$

Esters are generally used as solvents, especially for cellulose nitrate, in the formation of lacquers. The esters also find use in perfumes because of their pleasant smell.

Fats and Oils. By far the most important esters are the naturally occurring fats and oils. They are the esters of higher fatty acids and glycerol, and hence are also called glycerides. They have the following general formula, in which R_1, R_2, and R_3 are saturated alkyl groups like $C_{15}H_{31}$, $C_{17}H_{35}$, or higher unsaturated groups:

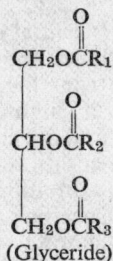

$$\begin{array}{c} O \\ \| \\ CH_2OCR_1 \\ | \\ O \\ \| \\ CHOCR_2 \\ | \\ O \\ \| \\ CH_2OCR_3 \end{array}$$
(Glyceride)

Chemically, there is no distinction between a fat and an oil. Those glycerides which are solids or semi-solids at room temperature are called fats, and those which are liquids are called oils. Vegetable fats are generally obtained from the fruits and seeds of plants, and are extracted by either cold pressing, hot pressing, or by solvent extraction. Olive oil, cottonseed oil, and peanut oil obtained by cold pressing are the most expensive edible oils. Animal fats are recovered by heating fatty tissues to a high temperature (dry rendering) or by treating them with steam and separating the liberated fats.

The glycerides are classified into simple and mixed glycerides. Simple glycerides are those in which all the acyl components are the same ($R_1=R_2=R_3$ in the formula above). In mixed glycerides the acyl radicals are different. The saturated acids occurring in fats and oils always contain an even number of carbon atoms, most often C_{12}, C_{14}, C_{16}, or C_{18}. These acids are lauric acid ($C_{11}H_{23}CO_2H$), myristic acid ($C_{13}H_{27}CO_2H$), palmitic acid ($C_{15}H_{31}CO_2H$), and stearic acid ($C_{17}H_{35}CO_2H$). The unsaturated acids are usually the C_{18} acids, oleic acid, linoleic acid, and linolenic acid with one, two, and three double bonds respectively.

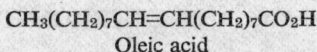

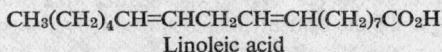

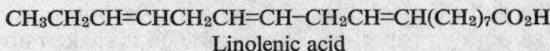

$$CH_3(CH_2)_7CH=CH(CH_2)_7CO_2H$$
Oleic acid

$$CH_3(CH_2)_4CH=CHCH_2CH=CH(CH_2)_7CO_2H$$
Linoleic acid

$$CH_3CH_2CH=CHCH_2CH=CH-CH_2CH=CH(CH_2)_7CO_2H$$
Linolenic acid

In a few cases, like chaulmoogra oil, used in the treatment of leprosy, the acids encountered are cyclic acids. The unsaturated glycerides usually have lower melting points than the corresponding saturated glycerides. The unsaturated glycerides can be partially or completely hydrogenated in presence of nickel catalysts. Hydrogenation raises the melting point of the product, and hence the process is referred to as one of hardening. Vegetable unsaturated oils, like cottonseed, peanut, or soyabean oil, are thus hydrogenated to give clean cooking greases like Trex. Hydrogenation also improves the keeping qualities of lard. Drying oils, like linseed and tung oils, are mixed glycerides of unsaturated acids. Their use in paints and varnishes in turn rests on the oxidizability of the unsaturated linkages by oxygen whereby a tough film is formed which protects the painted surface.

About 20–50% of the calorific intake of man consists of fats. The combustion of 1 g. of fat produces about 9,500 calories, which, when compared with 4,400 calories from 1 g. of protein and 3,961 calories from 1 g. of cane sugar, shows its high calorific value. The fats are the richest in energy of man's food and are stored in the body to be oxidized when needed. Fats and oils, in addition to the uses already mentioned, are used for soap making and in lubricants, lacquers, and medical preparations. Their use for any specific purpose depends on a number of physiochemical characteristics, among which are the following:

I. *Acid Value.* This is a measure of the degree of hydrolysis of the fat or its rancidity. The acid value is the number of milligrams of KOH required to neutralize the fatty acid in 1 g. of the fat.

II. *Saponification Number.* The saponification number is the number of milligrams of KOH required to saponify completely 1 g. of fat.

III. *Reichert–Meissl Number.* This gives an idea of the volatile water-soluble acid constituents of the fat. It is the number of millilitres of decinormal KOH required to neutralize the volatile water-soluble acids liberated in the hydrolysis of 5 g. of the fat under specified conditions.

IV. *Iodine Value.* It is, by definition, the number of grams of iodine absorbed by 100 g. of the fat or oil. The iodine value is a measure of the degree of unsaturation present in the fat and determines its classification as a drying or non-drying oil.

The waxes which are used in candlemaking, leather finishes, and in cosmetics (lipsticks) are also esters of long-chain acids—not of glycerol but of alcohols of high molecular weight. They are also constituents of secretions of certain insects and occur on the leaves of various plants. Apparently they serve to protect plants from attacks by bacteria or fungi and loss of moisture.

SOAPS AND DETERGENTS

Soaps are metallic salts (especially of sodium and potassium) of fatty acids containing 8–18 carbon atoms, obtained by the saponification of fats with alkali. The structural feature of a soap is that its salt end is water soluble and oil insoluble, and the hydrocarbon portion has just the reverse solubility properties. When some oil is shaken up vigorously with water a dispersed mixture known as an emulsion is obtained. On standing, the emulsion breaks up and the oil and water separate into two layers. If, however, soap is added, a permanent emulsion is formed. The use of soap to clean greasy surfaces is based on this ability to form an emulsion which may be washed off with water. In addition to soap, several other synthetic detergents are used. The sulphated oils are obtained by heating unsaturated fats with sulphuric acid. The double bonds add sulphuric acid to give the hydrogen sulphate of the hydroxy acid, and neutralization with sodium hydroxide gives the sodium salt. Sulphated castor oil is called Turkey Red Oil. The number and variety of synthetic detergents has increased enormously in recent years. Another group of detergents (e.g. trade name Dreft) are sulphates of higher alcohols (lauryl, myristyl, and palmityl) with the formula RCH_2OSO_3Na. The most widely used synthetic detergents are the sodium salts of alkylated aromatic sulphonic acids.

Many fats occur in nature along with fat-like substances in which one of the acid groups is replaced by the acid salt of phosphoric acid and an organic base. They are called **Phosphatides** or **Phospholipids**. Phosphatides are found in the lecithin of egg yolk and the cephalins of brain tissue. They have the general formula:

$$CH_2OCOR$$
$$CHOCOR$$
$$CH_2O-P=O$$
$$\text{with } O^- \text{ and } OCH_2CH_2\overset{+}{N}(CH_3)_3$$

DIBASIC ACIDS

Dibasic acids contain two carboxyl groups in the molecule. The structure and names of the first five members of the homologous series are given below:

COOH	COOH	COOH	COOH	COOH
COOH	CH_2	$(CH_2)_2$	$(CH_2)_3$	$(CH_2)_4$
	COOH	COOH	COOH	CO_2H
Oxalic acid	Malonic acid	Succinic acid	Glutaric acid	Adipic acid

The dibasic acids may be prepared by an extension of the general methods of preparation applied to the monobasic acids. Ethylene dibromide, for example, is converted into the dinitrile by reaction with potassium cyanide and then hydrolysed to **succinic** acid.

$$\begin{array}{c} CH_2Br \\ CH_2Br \end{array} + KCN \longrightarrow \begin{array}{c} CH_2C{\equiv}N \\ CH_2C{\equiv}N \end{array} \longrightarrow \begin{array}{c} CH_2CO_2H \\ CH_2CO_2H \end{array}$$

Ethylene dibromide · · · Succinonitrile · · · Succinic acid

Special methods are employed for the preparation of specific dicarboxylic acids. The commercial method for preparation of oxalic acid involves the heating of sodium formate under reduced pressure.

$$2H \cdot C \overset{O}{\underset{ONa}{\big\langle}} \xrightarrow{\text{Heat}} H_2 + (COONa)_2 \xrightarrow{H^+} \begin{array}{c} COOH \\ COOH \end{array}$$

Sodium formate · · · Sodium oxalate · · · Oxalic acid

Succinic acid is obtained in industry by reduction of maleic acid (see later), which is available by air oxidation of benzene in presence of vanadium pentoxide.

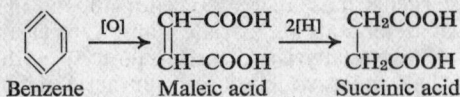

$$\text{Benzene} \xrightarrow{[O]} \begin{array}{l}\text{CH–COOH}\\ \|\\ \text{CH–COOH}\end{array} \xrightarrow{2[H]} \begin{array}{l}\text{CH}_2\text{COOH}\\ |\\ \text{CH}_2\text{COOH}\end{array}$$

Benzene   Maleic acid   Succinic acid

Adipic acid, used on a large scale for the manufacture of nylon, is made by the oxidation of *cyclo*-hexanol.

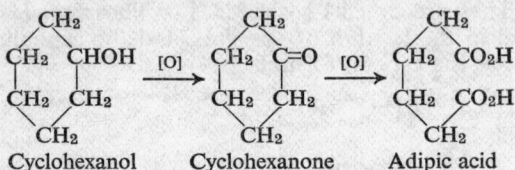

Cyclohexanol   Cyclohexanone   Adipic acid

PROPERTIES: All the dibasic acids are crystalline solids. At room temperature the lower members are soluble in water. The dicarboxylic acids form salts, esters, amides, acid chlorides, and other derivatives just like monocarboxylic acids. Either or both of the two carboxyls may be converted to the above derivatives. It is possible to convert one carboxyl group to the ester group and the other, for example, to the acid chloride group. Such mixed compounds are useful in organic synthesis. The dicarboxylic acids, like all polyfunctional compounds, have certain characteristic properties depending upon the relative position of the two carboxyl groups. The carboxyl group is an electron-attracting group, and therefore the presence of one carboxyl close to another increases the ease of ionization of the first hydrogen ion. This effect rapidly decreases as the carboxyl groups are increasingly separated. Thus oxalic acid is one of the strongest of organic acids whose first dissociation constant (K_1) is $6 \cdot 5 \times 10^{-2}$. For malonic acid K_1 is $1 \cdot 4 \times 10^{-3}$, but when two or more methylene groups intervene, the two carboxyl groups have little or no effect on each other. Thus succinic acid $(K_1 = 6 \cdot 4 \times 10^{-5})$ is only slightly stronger than acetic acid $(K = 1 \cdot 8 \times 10^{-5})$.

Oxalic acid on heating loses carbon dioxide to give formic acid, which in turn decomposes to give carbon monoxide and water.

$$\begin{array}{l}\text{COOH}\\ \text{COOH}\end{array} \xrightarrow[\text{Heat}]{-CO_2} \begin{array}{l}\text{H}\\ \text{COOH}\end{array} \xrightarrow{\text{Heat}} CO + H_2O$$

Malonic acid and substituted malonic acids when heated above the melting point lose carbon dioxide to give monocarboxylic acids.

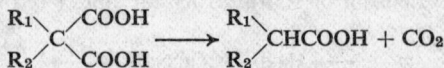

$$\begin{array}{l}R_1\\ \quad\diagdown\\ \quad\quad C\\ \quad\diagup\\ R_2\end{array}\begin{array}{l}\diagup\text{COOH}\\ \\ \diagdown\text{COOH}\end{array} \longrightarrow \begin{array}{l}R_1\\ \quad\diagdown\\ \quad\quad\text{CHCOOH} + CO_2\\ \quad\diagup\\ R_2\end{array}$$

Succinic and glutaric acids lose water on heating, preferably in the presence of a dehydrating agent like acetic anhydride, to give the stable five- and six-membered cyclic anhydrides.

$$CH_2CO_2H \atop CH_2CO_2H \xrightarrow{-H_2O} CH_2-C{\diagup O \atop \diagdown O} \atop CH_2-C{\diagup \atop \diagdown O}$$

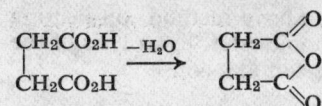

Succinic acid   Succinic anhydride   Glutaric acid   Glutaric anhydride

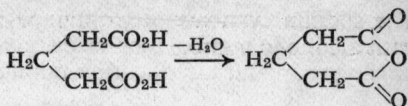

The ammonium salts of succinic and glutaric acids on heating easily lose ammonia giving amides, and then the imides.

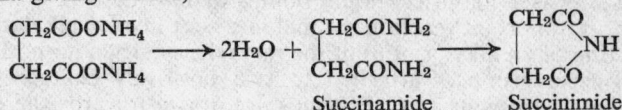

$$CH_2COONH_4 \atop CH_2COONH_4 \longrightarrow 2H_2O + {CH_2CONH_2 \atop CH_2CONH_2} \longrightarrow {CH_2CO \atop CH_2CO}{\diagdown \atop \diagup}NH$$

Succinamide   Succinimide

When bromine is added to an ice-cold alkaline solution of succinimide, *N*-bromosuccinimide is precipitated in excellent yield.

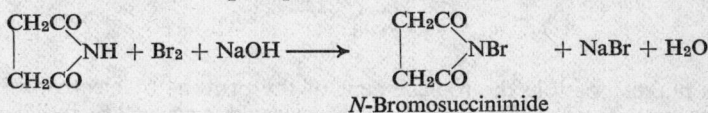

$${CH_2CO \atop CH_2CO} NH + Br_2 + NaOH \longrightarrow {CH_2CO \atop CH_2CO} NBr + NaBr + H_2O$$

N-Bromosuccinimide

This is a very valuable reagent and, in boiling carbon tetrachloride as a solvent, it brominates unsaturated and aromatic compounds on the α carbon relative to the double bond or ring thus:

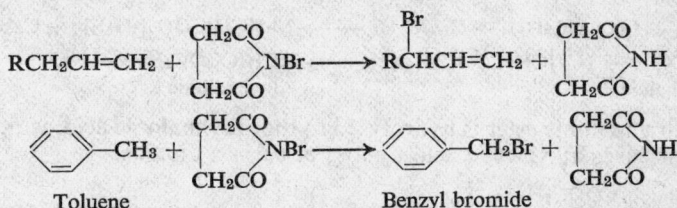

$$RCH_2CH=CH_2 + {CH_2CO \atop CH_2CO}NBr \longrightarrow RC\overset{Br}{H}CH=CH_2 + {CH_2CO \atop CH_2CO}NH$$

$$\langle\rangle-CH_3 + {CH_2CO \atop CH_2CO}NBr \longrightarrow \langle\rangle-CH_2Br + {CH_2CO \atop CH_2CO}NH$$

Toluene   Benzyl bromide

Bromination with *N*-bromosuccinimide leaves the double bond intact; hence its special importance. Adipic and higher acids on heating with dehydrating agents give linear polymeric anhydrides. When adipic acid is heated with barium hydroxide *cyclo*pentanone is obtained via the barium salt initially formed (see Chapter Six).

MALONIC ESTER

The reactions of diethylmalonate, commonly called malonic ester, have sufficient importance to merit special treatment. It is extensively used in organic synthesis for the preparation of a variety of compounds. The methylene group $-CH_2-$, hemmed in between two carbethoxy groups, exhibits pronounced reactivity and is referred to as an active methylene group. Malonic acid is prepared from chloroacetic acid as follows:

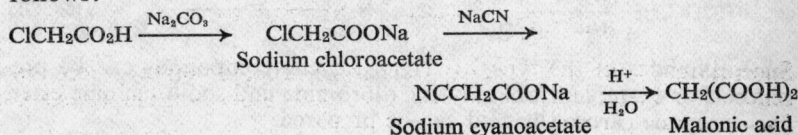

$$ClCH_2CO_2H \xrightarrow{Na_2CO_3} ClCH_2COONa \xrightarrow{NaCN}$$

Sodium chloroacetate

$$NCCH_2COONa \xrightarrow[H_2O]{H^+} CH_2(COOH)_2$$

Sodium cyanoacetate   Malonic acid

The sodium cyanoacetate, obtained by the above method, on heating with alcoholic hydrogen chloride, gives malonic ester, hydrolysis of the nitrile group and esterification taking place simultaneously.

$$\text{NCCH}_2\text{COONa} \xrightarrow[\text{H}^+]{\text{C}_2\text{H}_5\text{OH}} \text{CH}_2(\text{COOC}_2\text{H}_5)_2$$

Sodium cyanoacetate   Diethylmalonate or malonic ester

Malonic ester is a colourless liquid boiling at 199° C.

The most important reactions of malonic ester involve **alkylation** (i.e. introduction of an alkyl group) of the methylene carbon atom. Because the methylene group is attached to two positively charged carbon atoms, the electrons in its C–H bonds are drawn towards the carbon.

This makes possible the replacement of the protons by electropositive metal ions. The hydrogen atoms of the methylene group are sufficiently acidic to permit ready formation of a sodium derivative by the action of metallic sodium or sodium ethoxide. This derivative reacts with an alkyl halide to give an alkyl substituted malonic ester.

$$\text{CH}_2(\text{COOC}_2\text{H}_5)_2 + \text{NaOC}_2\text{H}_5 \longrightarrow \text{Na}^+[\text{CH}(\text{COOC}_2\text{H}_5)_2]^- + \text{C}_2\text{H}_5\text{OH}$$

$$\text{RX} + \text{Na}^+[\text{CH}(\text{COOC}_2\text{H}_5)_2]^- \longrightarrow \text{RCH}(\text{COOC}_2\text{H}_5)_2 + \text{NaX}$$

Alkyl halide   Alkylmalonic ester

The alkylmalonic ester is hydrolysed to the alkylmalonic acid, which on heating gives alkylacetic acid by loss of carbon dioxide.

$$\text{RCH}\begin{array}{l}\diagup\text{COOC}_2\text{H}_5 \\ \diagdown\text{COOC}_2\text{H}_5\end{array} \xrightarrow{\text{Hydrol.}} \text{RCH}\begin{array}{l}\diagup\text{COOH} \\ \diagdown\text{COOH}\end{array} \xrightarrow{\text{Heat}} \text{RCH}_2\text{COOH} + \text{CO}_2$$

For example, *n*-valeric acid can be prepared by the sequence below.

$$\text{CH}_2(\text{COOC}_2\text{H}_5)_2 \xrightarrow[\text{CH}_3\text{CH}_2\text{CH}_2\text{Br}]{\text{NaOC}_2\text{H}_5} \text{CH}_3\text{CH}_2\text{CH}_2\text{CH}(\text{COOC}_2\text{H}_5)_2 \xrightarrow{\text{KOH}}$$

$$\text{CH}_3\text{CH}_2\text{CH}_2\text{CH}(\text{COOH})_2 \xrightarrow[\text{Heat}]{\text{H}_2\text{SO}_4} \text{CH}_3(\text{CH}_2)_3\text{COOH}$$

n-Valeric acid

The net result is the conversion of a halide RX to the acid R–CH$_2$CO$_2$H.

A monosubstituted alkylmalonic ester still has a hydrogen atom replaceable by sodium and can be alkylated in a similar way to give a dialkylmalonic ester which could be hydrolysed and decarboxylated to give a dialkylacetic acid:

$$\text{RCH}(\text{COOC}_2\text{H}_5)_2 \xrightarrow[\text{NaOC}_2\text{H}_5]{\text{R}_1\text{X}} \begin{array}{l}\text{R}\diagdown \\ \text{R}_1\diagup\end{array}\text{C}(\text{COOC}_2\text{H}_5)_2 \xrightarrow{\text{(i) H}_2\text{O (ii) } -\text{CO}_2} \begin{array}{l}\text{R}\diagdown \\ \text{R}_1\diagup\end{array}\text{CHCOOH}$$

With dihalides of the type X(CH$_2$)$_n$X, cyclic compounds can be prepared. For example, from ethylene dibromide and sodio malonic ester, cyclopropane carboxylic acid can be prepared.

$$\begin{matrix} \text{CH}_2\text{Br} \\ | \\ \text{CH}_2\text{Br} \end{matrix} + \text{CH}_2(\text{COOC}_2\text{H}_5)_2 \xrightarrow{2\text{C}_2\text{H}_5\text{ONa}} \begin{matrix} \text{H}_2\text{C} \\ \\ \text{H}_2\text{C} \end{matrix} \Big\rangle \text{C}(\text{COOC}_2\text{H}_5)_2 \longrightarrow \begin{matrix} \text{H}_2\text{C} \\ \\ \text{H}_2\text{C} \end{matrix} \Big\rangle \text{CHCOOH}$$

Ethylene   Malonic ester   Cyclopropane   Cyclopropane
dibromide   malonic ester   carboxylic acid

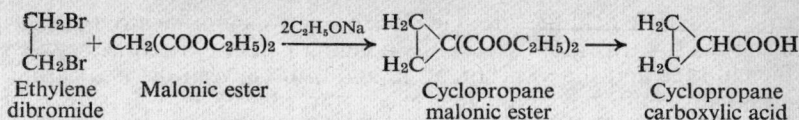

Similarly, trimethylene and pentamethylene dibromides give rise to cyclobutane, cyclopentane, and cyclohexane derivatives.

Ethyl malonate adds on α,β-unsaturated esters and other carbonyl compounds in the presence of basic catalysts like sodium ethoxide to give products which can be hydrolysed and decarboxylated to give β-substituted glutaric acids.

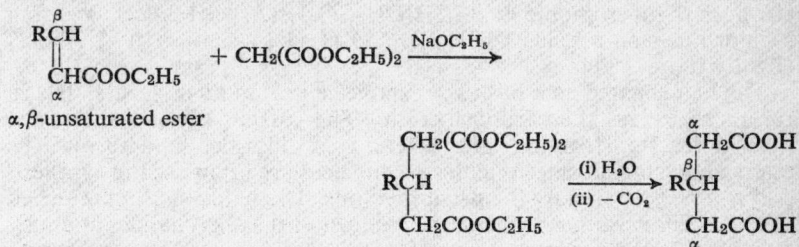

$$\begin{matrix} \overset{\beta}{\text{R}}\text{CH} \\ \| \\ \text{CHCOOC}_2\text{H}_5 \\ _\alpha \end{matrix} + \text{CH}_2(\text{COOC}_2\text{H}_5)_2 \xrightarrow{\text{NaOC}_2\text{H}_5}$$

α,β-unsaturated ester

$$\begin{matrix} \text{CH}_2(\text{COOC}_2\text{H}_5)_2 \\ | \\ \text{RCH} \\ | \\ \text{CH}_2\text{COOC}_2\text{H}_5 \end{matrix} \xrightarrow[\text{(ii)} -\text{CO}_2]{\text{(i) H}_2\text{O}} \begin{matrix} \overset{\alpha}{\text{CH}_2}\text{COOH} \\ \overset{\beta}{|} \\ \text{RCH} \\ | \\ \text{CH}_2\text{COOH} \\ _\alpha \end{matrix}$$

Ethyl malonate also condenses with aldehydes in the presence of amines to give α,β-unsaturated diesters which can be converted to α,β-unsaturated acids.

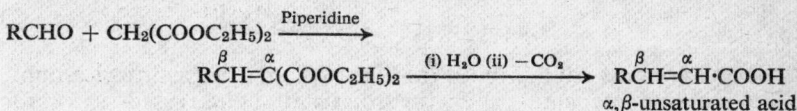

$$\text{RCHO} + \text{CH}_2(\text{COOC}_2\text{H}_5)_2 \xrightarrow{\text{Piperidine}}$$

$$\overset{\beta}{\text{R}}\text{CH}=\overset{\alpha}{\text{C}}(\text{COOC}_2\text{H}_5)_2 \xrightarrow{\text{(i) H}_2\text{O (ii)} -\text{CO}_2} \overset{\beta}{\text{R}}\text{CH}=\overset{\alpha}{\text{CH}}\cdot\text{COOH}$$

α,β-unsaturated acid

HALOGENATED ACIDS

Halogenation of acids in the presence of catalysts, as previously mentioned, gives α-haloacids. A β-halogen acid is often prepared by addition of hydrogen halide to an unsaturated acid, the halogen adding to the carbon farthest removed from the carboxyl group.

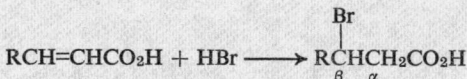

$$\text{RCH}=\text{CHCO}_2\text{H} + \text{HBr} \longrightarrow \overset{\begin{matrix}\text{Br}\\|\end{matrix}}{\text{R}}\underset{\beta}{\text{C}}\text{H}\underset{\alpha}{\text{CH}_2}\text{CO}_2\text{H}$$

The halogenated acids are much stronger acids than the unsubstituted acids. Thus monochloroacetic acid (K_a, $1\cdot4 \times 10^{-3}$) is stronger than acetic acid (K_a, $1\cdot8 \times 10^{-5}$); dichloroacetic acid (K_a, 5×10^{-2}) is still stronger, while trichloroacetic acid (K_a, $1\cdot3 \times 10^{-1}$) is as strong. This increase in strength of a halogenated acid is due to the electron-attracting nature or electronegativity of the halogen atom.

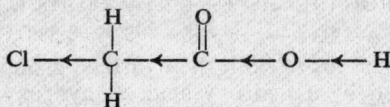

$$\text{Cl} \longleftarrow \overset{\overset{\textstyle\text{H}}{|}}{\underset{\underset{\textstyle\text{H}}{|}}{\text{C}}} \longleftarrow \overset{\overset{\textstyle\text{O}}{\|}}{\text{C}} \longleftarrow \text{O} \longleftarrow \text{H}$$

In chloroacetic acid the strongly electronegative chlorine atom attracts electrons towards itself. Hence the bonding pairs of electrons between the chlorine and the α carbon atom and also between the two carbon atoms are shifted by inductive effect towards the left, as shown in formula above. The carboxyl carbon thus becomes somewhat deficient in electrons and causes the further displacements indicated, resulting in the easy ionization of hydrogen as proton. For the same reason, dichloroacetic and trichloroacetic acids are still stronger.

The above inductive effect of a halogen atom decreases rapidly with increasing distance from the hydrogen atom of the −OH group. α-Chloropropionic acid, $CH_3CHClCO_2H$ ($K_a = 1.6 + 10^{-3}$), for example, is stronger than propionic acid, $CH_3CH_2CO_2H$ ($K_a = 1.34 \times 10^{-5}$), but β-chloropropionic acid, $ClCH_2CH_2CO_2H$ ($K_a = 8.4 \times 10^{-5}$) is only slightly more acidic.

The halogenated acid has two reactive centres in its molecule, that is, the halogen and the carboxyl group. The carboxyl group shows the normal reaction forming salts, esters, acid chlorides, and so on. The esters and acid chlorides of chloroacetic acid are often used in synthesis and can be prepared by the usual methods. The halogen atom attached to an α carbon atom reacts like the halogen of the alkyl halides and may be replaced by the cyano, hydroxyl or ethoxyl groups. The preparation of cyanoacetic acid from chloroacetic acid is an instance in point. The halogen attached to the β carbon atom is much less reactive and relatively more difficult to replace by other groups.

HYDROXY ACIDS

Hydroxy acids contain both the carboxyl and the hydroxyl groups. They can be prepared (1) by hydrolysis of halogenated acids, for example:

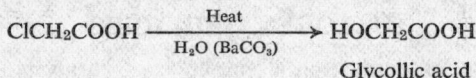

$$ClCH_2COOH \xrightarrow[\text{H}_2\text{O (BaCO}_3\text{)}]{\text{Heat}} HOCH_2COOH$$

Glycollic acid

and (2) by hydrolysis of hydroxynitriles (cyanohydrins) obtained by the action of hydrogen cyanide on aldehydes and ketones. Thus acetaldehyde cyanohydrin is readily hydrolysed to the hydroxy acid, lactic acid.

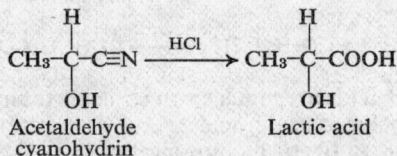

$$CH_3{-}\overset{\displaystyle H}{\underset{\displaystyle OH}{C}}{-}C{\equiv}N \xrightarrow{\text{HCl}} CH_3{-}\overset{\displaystyle H}{\underset{\displaystyle OH}{C}}{-}COOH$$

Acetaldehyde   Lactic acid
cyanohydrin

β-Hydroxy acids are often obtained by oxidation of aldols, for example:

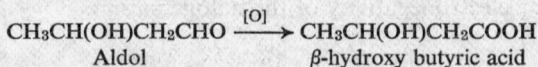

$$CH_3CH(OH)CH_2CHO \xrightarrow{\text{[O]}} CH_3CH(OH)CH_2COOH$$

Aldol   β-hydroxy butyric acid

The chemical reactions of both the hydroxyl group and the carboxyl group are exhibited by a hydroxy acid. A hydroxy acid forms salts, esters, anhydrides, etc., in the same manner as a monocarboxylic acid. The hydroxy group, as in alcohols, may be esterified, or converted into an ether group $-OR$ or exchanged for a halogen atom. The hydroxy acids, with the $-OH$ group in different positions with respect to the carboxyl group, differ in their behaviour on heating alone or with dehydrating agents. An α-hydroxy acid (e.g. lactic acid) on heating forms a cyclic ester known as a lactide.

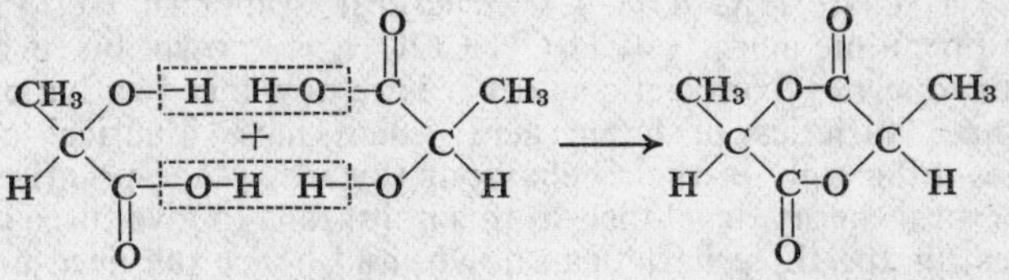

Lactic acid   Lactide

A β-hydroxy acid on heating loses a molecule of water giving an unsaturated acid.

$$HOCH_2CH_2COOH \xrightarrow{\text{Heat}} H_2O + CH_2{=}CHCOOH$$

Acrylic acid

The γ- and δ-hydroxy acids form ring compounds called lactones with 5 and 6 atoms respectively by esterification internally between the hydroxy and the carboxyl groups.

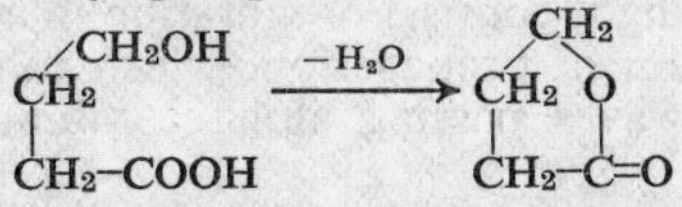

γ-Hydroxybutyric γ-Butyrolactone
acid

Since lactones are internal esters, the ring structure is cleaved by treatment with alkali with the formation of the alkali salt of the parent hydroxy acid.

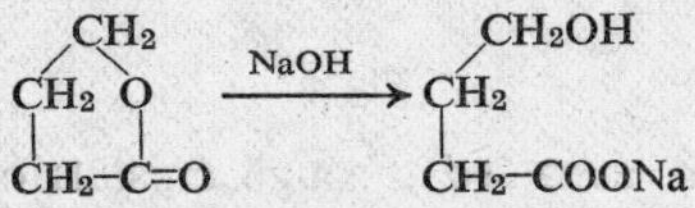

LACTIC ACID

Lactic acid occurs in sour milk; hence its name. It can be prepared from acetaldehyde cyanohydrin by hydrolysis. The commercial method involves fermentation of cane sugar or starch by lactic acid bacteria. Large quantities of lactic acid are used in dyeing wool and in removing lime from hides previous to tanning. Several salts of lactic acid (lactates) are used in food products and in medicine. Ethyl lactate is used in the lacquer industry as a solvent for nitrocellulose.

OPTICAL ISOMERISM

Lactic acid is a classical example of a type of isomerism called optical isomerism. Our study of organic compounds to this point has shown that the chemical properties of a molecule are a function of its structure. Therefore molecules with different properties should have different structures. Three distinct varieties of lactic acid are known to exist. One variety is obtained by the fermentation of lactose in milk, another is derived from an extract of meat and muscle, and the third one is secured synthetically in the laboratory. All three varieties have exactly the same chemical properties, and hence must be assigned the same formula $CH_3CH(OH)COOH$. (A structural isomer of lactic acid, β-hydroxy propionic acid, $HOCH_2CH_2COOH$, is known, but it has very different properties from lactic acid.) We are now faced with a problem. These three varieties of lactic acid which have identical chemical properties differ only in their behaviour towards plane-polarized light. The structural theory developed to this point fails to give three different structures for the three varieties known, and hence the need arises for the extension or modification of the accepted views in order to accommodate this new phenomenon. It is necessary to re-examine critically our ideas on the disposition of the valences of the carbon atom to see if reconciliation can be effected by considering three-dimensional models instead of the planar models we have been using until now.

Before passing on to a consideration of LE BEL and VAN'T HOFF's theory of stereoisomerism, it is necessary to understand the way in which the three varieties of lactic acid differ. It was noted that they differ in their behaviour towards 'plane-polarized light'. What is plane-polarized light? Ordinary light consists of vibration in all directions at right angles to the direction of propagation, which can be compared conveniently with the waves passing along a vibrating stretched rope.

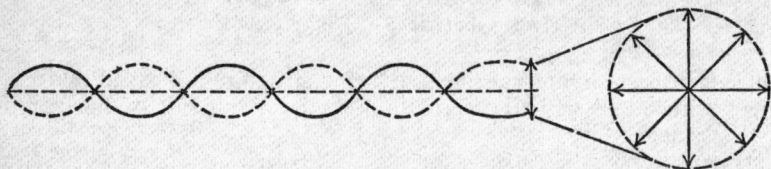

Fig. 6

When ordinary light is passed through crystals of Iceland spar (the mineral calcite, $CaCO_3$, found in Iceland) or a Nicol prism (a device made by bisecting a rhombohedron of Iceland spar and cementing the parts with Canada balsam) it vibrates only in one plane and is said to be plane-polarized somewhat like the effect of passing our vibrating rope through a narrow slit: only vibrations in the same plane as the slit can pass through. When a ray or polarized light is passed through a solution of meat-extract lactic acid the plane of polarization is rotated through a certain angle to the right. On repeating the experiments with the other varieties of lactic acid, it is found that sour-milk lactic acid does not turn the plane of polarization at all and the third variety rotates the plane of polarization to an equal angle to the left, as does the meat-extract variety to the right. When substances in the liquid or dissolved state have the ability to rotate the plane of polarization of polarized light they are said to be optically active. If they rotate the plane of polarization to the right they are called dextrorotatory (*d*). If they rotate it to the left they are called laevorotatory (*l*). On reflection, it may seem as though much fuss is being made about such a trivial difference in the properties of the three varieties of lactic acid which are almost identical. However, as we

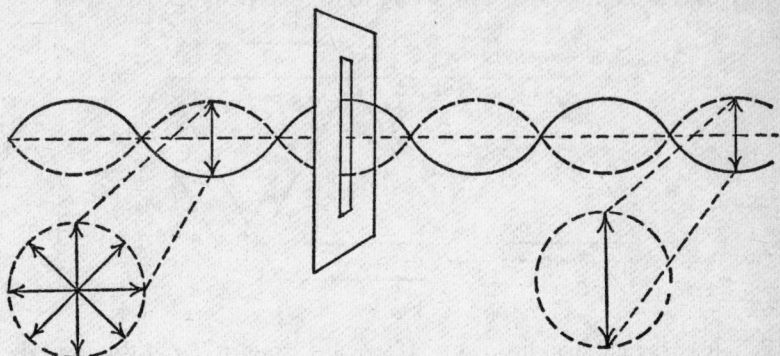

Fig. 7

shall see, the theory of optical isomerism, developed to explain this difference in the direction of rotation of the varieties of lactic acid, has had far-reaching results on the general structural theory in organic chemistry.

We shall now see how the lactic acid problem was solved. In 1848 LOUIS PASTEUR, who was working with the salts of tartaric acid, suggested that optical activity was associated with an asymmetric (not symmetric) arrangement of the atoms in space of the molecule. In 1873 WISLICENUS said, 'If it is once granted that molecules could be structurally identical, and yet possess different properties, it can only be explained on the ground that the difference is due to a different arrangement of the atoms in space.' The two great chemists had an inkling of the truth and suggested that we should consider the arrangements of the atoms in space, that is, think of three-dimensional models instead of the flat planar structures we have been satisfied with so far.

This germ of truth, contained in this suggestion of Pasteur and Wislicenus, developed into a most comprehensive and elegant theory of stereoisomerism at the hands of Le Bel and Van't Hoff. Working independently and almost simultaneously, they solved the lactic acid problem by formulating the tetrahedral arrangement of the valences of the carbon atom, which marked a milestone in the development of structural organic chemistry (cf. page 24). We shall examine the theory in some detail. We are in the habit of thinking of the flat picture of the

methane molecule, H–C–H, with the four valences of the carbon atom

pointing north, south, east, and west all in one plane. Le Bel and Van't Hoff suggested that the four valences of the carbon atom do not lie in a

plane, but are directed towards the corners of a **tetrahedron** at the centre
of which the carbon atom is situated. A **tetrahedron** is a symmetrical
solid figure with four faces, each of which is an equilateral triangle.

We can easily make a tetrahedral model by cutting a sheet of stiff
paper of the shape drawn below and folding it as indicated:

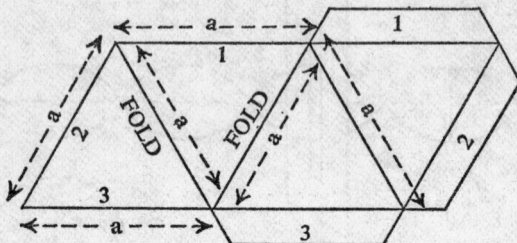

Fig. 8

The flaps are held by a little gum. If we place four different things at the
corners it will be found that two different arrangements are possible. If
we attach four different groups or atoms to the valences of the carbon
atom in the tetrahedral model such a molecule will be asymmetric, that
is, it cannot be divided into identical halves. A symmetrical object can be
divided into two identical halves. Further, the image of a symmetrical
object as seen in a mirror is identical with the object, while the mirror
image of an asymmetric object is not identical with the object. Let us
think in terms of a conical flask or a funnel:

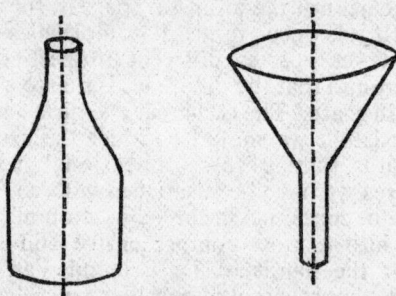

Fig. 9

They are symmetrical objects and can be divided into identical halves.
The image of a conical flask seen in a mirror is identical with the
object. Now if we take an asymmetric object such as a hand or foot we
would be unable (if we tried) to divide it into two identical parts by any
plane. Further, the image of the right hand seen in a mirror is the left
hand, and the two are not superimposable on each other, so that each
part of one does not actually coincide with the corresponding part of the
other.

This brings us to the essential characteristic of an asymmetric object,
that is, that no plane of symmetry can be drawn through it, and the

object and the image are not superimposable. Familiar asymmetric objects are a hand, a corkscrew, a foot, and so on. We can now add one more item to the list of asymmetric objects. A molecule in which at least one carbon atom is attached to four different groups or atoms is asymmetric, because the molecule cannot be divided into identical halves by a plane of symmetry, and the mirror image and the object are not superimposable.

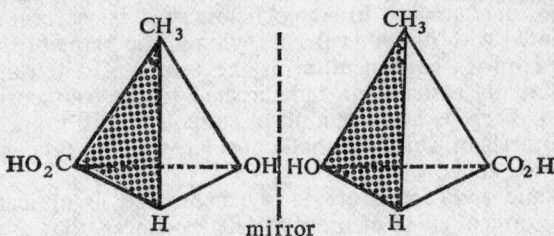

Fig. 10

In the case of lactic acid, $CH_3\overset{*}{C}H(OH)COOH$, the asterisked carbon atom has four different groups—a methyl, a hydrogen atom, a hydroxyl group, and a carboxyl group—attached to it. The molecular asymmetry and the two forms can be represented as shown in Fig. 8. Each form is related to the other as an object and its mirror image. These two molecules are not superimposable, for if we place the methyl group on methyl and the H on H, then the $-OH$ group falls on $-COOH$ and the $-COOH$ on the $-OH$.

It has been found that optical activity and molecular asymmetry go together. A compound which contains at least one asymmetric carbon atom is found to be optically active. An asymmetric carbon atom can be defined as an atom which is attached to four different groups or atoms. If only three different groups are attached to a carbon atom the molecule is no longer asymmetric and optical activity disappears. For example, lactic acid is optically active and contains an asymmetric carbon atom. In the lactic acid molecule, if we substitute one hydrogen atom for the hydroxyl group, the resulting molecule is propionic acid $CH_3\overset{*}{C}H_2COOH$. Here the four groups attached to the asterisked carbon atom are not all different; only three are different. If, in the above tetrahedral models of the two forms of lactic acid, the OH is replaced by H a plane of symmetry can be drawn through the CH_3, the $-COOH$, and the central carbon atom, bisecting the H to OH edge of the tetrahedron. The molecule is symmetrical with respect to this plane, and hence is superimposable on its mirror image. In the case of lactic acid we have two molecules; one is the object, and the other the image. That is, one variety is dextro or *d*-lactic acid and the other is laevo or *l*-lactic acid. The letters *d* and *l* are used by convention to indicate rotation to the right $(+)$ or rotation to the left $(-)$. The third variety is an equimolecular mixture of the two varieties and is called the racemic or *dl*-variety. It is optically inactive by external compensation and can be resolved by suitable means into the *d* and *l* isomers. In the *d* and *l* forms the relationship of any particular group to the rest of the molecule is identical, and hence the physical and chemical properties of the d and l varieties are identical. This type of isomerism, depending upon the different arrangements in space, is called **stereoisomerism**, meaning space isomerism. The word stereo is quite well known to us in such words as stereoscope, which is an instrument for seeing special kinds of pictures and which is capable of transforming a flat landscape into a three-dimensional reality. In stereochemistry a flat carbon atom, thus far represented with its four valences lying in the plane of the paper, is transformed into a carbon atom at the centre of a tetrahedron, the valences spatially distributed and directed to the four corners of the tetrahedron. There is another kind of stereoisomerism, known as geometrical isomerism, which we shall consider when we take up the study of unsaturated acids.

The racemic or dl-form, previously mentioned, is optically inactive because the optical effect of the d molecules cancels that of the l molecules. Its chemical properties are identical with those of the d and l forms. However, in physical properties such as melting point, solubility, etc., there are differences.

The optically active compounds so far considered have only one asymmetric carbon atom. Compounds with more than one asymmetric carbon atom are also encountered often. For example, the addition of bromine to cinnamic acid gives dibromocinnamic acid, in which the two carbon atoms marked with an asterisk are asymmetric.

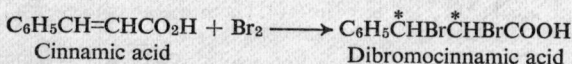

$$C_6H_5CH{=}CHCO_2H + Br_2 \longrightarrow C_6H_5\overset{*}{C}HBr\overset{*}{C}HBrCOOH$$
Cinnamic acid   Dibromocinnamic acid

In a case where the two asymmetric atoms are not identical each can contribute a dextro or laevo rotation to the molecule as a whole. If these contributions are designated $+a$ or $-a$ for one atom and $+b$ or $-b$ for the other atom the following four optically active forms become possible:

(1)	(2)	(3)	(4)
$+a$	$-a$	$+a$	$-a$
$+b$	$-b$	$-b$	$+b$

Of these (1) and (2) are mirror images and can combine to give a racemic form. So can (3) and (4) to give a second racemic form. In general, a molecule having n dissimilar asymmetric carbon atoms can exist in 2_n optically active forms. This rule does not apply to molecules having similar asymmetric atoms like tartaric acid, where the two atoms marked are similar.

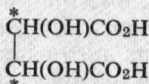

$$\overset{*}{C}H(OH)CO_2H$$
$$\underset{*}{C}H(OH)CO_2H$$

Using the same designations as before and remembering that $b = a$, the possible forms are:

(1)	(2)	(3)
$+a$	$-a$	$+a$
$+a$	$-a$	$-a$
Racemic form		Meso form

The first two forms are mirror images of each other and together form a racemic variety. The third form, in which the rotations cancel each other internally, is referred to as the meso form. Unlike the *dl* form, the meso form cannot be resolved because of this internal compensation. The two inactive forms have different physical properties. The different tartaric acids can be represented by the perspective tetrahedral formulae:

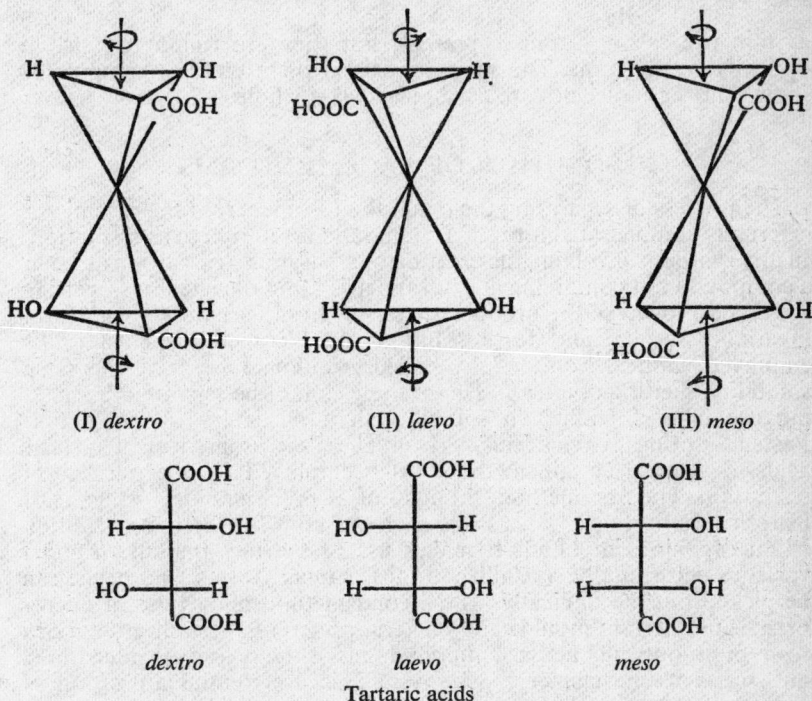

(I) *dextro*   (II) *laevo*   (III) *meso*

dextro   laevo   meso

Tartaric acids

In these representations the two tetrahedrons of the two carbon atoms are joined together and the different groups attached to the free corners. Furthermore, the solid line connecting the −H and −OH groups represents one edge of each tetrahedron above or in front of the plane of the paper whereas the dotted line indicates the edge of the tetrahedron, which will be away from view in a solid model. In (I) the groups H, OH, and CO_2H are arranged clockwise in both tetrahedrons and counterclockwise in (II). In (III) they are affixed clockwise in the upper half and counterclockwise in the lower half. The same models are represented, more simply, by the planar projections shown beneath each tetrahedral formula above. These are obtained by the imaginary flattening of the tetrahedrons both from the front and the rear so that the groups lie on the plane of the paper.

The existence of mesotartaric acid shows that the presence of asymmetric carbon atoms alone is not a sufficient condition for optical activity. In spite of the presence of two asymmetric carbon atoms, the molecule of mesotartaric acid as a whole has a plane of symmetry and is therefore optically inactive. A more fundamental condition for optical activity is molecular asymmetry. Thus, in the case of allene derivatives of the formula $\begin{smallmatrix} R_1 & & R_1 \\ & \diagdown & \\ & C=C=C & \\ & \diagup & \\ R_2 & & R_2 \end{smallmatrix}$ and certain derivatives of diphenyl, no asymmetric carbon atom is present, but they are found to exist in optically active forms. The space models of these types of compounds show that the molecules are asymmetric as a whole.

RESOLUTION OF RACEMIC FORMS

The process of separating the d and the l isomer from the racemic or externally compensated form of a compound is referred to as resolution. In any synthesis involving the creation of an asymmetric carbon atom in a compound not containing any asymmetric atom the chances of getting the d and l forms of the product are equal. In other words, the dl form invariably results, and for resolution suitable methods have to be employed. Optical isomers (also called **enantiomorphs**) have the same solubility, melting point, etc. Hence these cannot be separated by making use of a difference in solubility in a particular solvent. Louis Pasteur, during his classical work on the stereoisomerism of tartaric acids, developed certain methods often employed for the purpose of resolution. The first method, though not of much practical value, is of historical importance, since Pasteur employed it for the first resolution of an organic compound. It makes use of the fact that the d and l varieties occasionally crystallize in mirror image forms, and hence can be picked out mechanically. The second method makes use of micro-organisms. Certain moulds and bacteria preferentially utilize the d or l form of an optically active compound, and if the process is interrupted an excess of one isomer is left over. This preferential utilization of optical isomers by micro-organisms is due to enzymes present in them. This biochemical method has the drawback that one of the isomers is lost.

The best method of resolution is the chemical method. This method consists in making a racemic mixture react chemically with another optically active substance. For instance, in order to resolve a racemic sample of lactic acid it is neutralized with an optically active base, say an l-base. As a result of the reaction, two salts are formed thus:

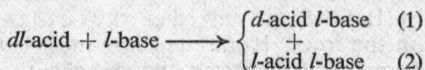

$$dl\text{-acid} + l\text{-base} \longrightarrow \begin{cases} d\text{-acid } l\text{-base} & (1) \\ \quad + \\ l\text{-acid } l\text{-base} & (2) \end{cases}$$

The two salts obtained (1) and (2) are neither identical nor are they related as object and mirror image. The mirror image of d-acid l-base (1) will be the l-acid d-base and not (2). Optical isomers of the type of (1) and (2) are referred to as diastereoisomers. Such diastereoisomers (unlike enantiomorphs) differ in such properties as solubility, melting point, etc. Hence the mixture of (1) and (2) formed in the above reaction can be easily separated by fractional crystallization from a suitable solvent. The active acids can then be obtained from the separated salts by acidification with mineral acids. The usual bases used for resolution of racemic acids are *l*-strychnine, *l*-brucine, and *l*-quinine. Similarly, a racemic base can be separated by salt formation with an optically active acid. *d*-Tartaric acid, which is easily available, can be used for the purpose.

During certain chemical transformations it was found that a *d*-compound was converted into an *l*-compound, and vice versa. This change was investigated by PAUL WALDEN, and hence is known as Walden inversion. The interconversions of active forms of malic acid and chlorosuccinic acid serve as examples:

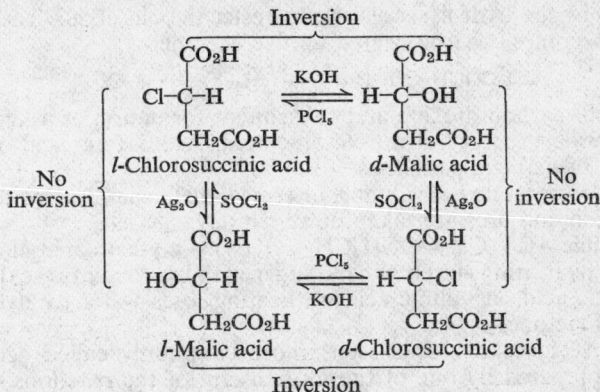

The Walden inversion is the result of a displacement reaction (analogous to the displacement of iodine in methyl iodide by the ^{-}OH group from an alkali) in which the entering group in an asymmetric molecule takes a place opposite to the group which it displaces.

ALDEHYDO AND KETONIC ACIDS

These compounds may be regarded as oxidation products of the hydroxy acids and contain a carbonyl group in addition to a carboxyl group. As might be expected, they display the properties of both acids and ketones or aldehydes.

Glyoxylic acid, $\begin{matrix} CHO \\ | \\ COOH \end{matrix}$, is the simplest aldehydo acid. Glyoxylic acid occurs in many unripe fruits. It may be obtained in the laboratory by heating dichloroacetic acid with water. It exists in the form of a hydrate, $\begin{matrix} CH(OH)_2 \\ | \\ COOH \end{matrix}$.

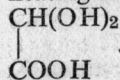

The ketonic acids are more important and conveniently studied by considering one example each of α, β, and γ ketonic acids. (The Greek letters α, β, and γ refer to the position of the keto group with respect to the carboxyl group.)

Pyruvic acid, $CH_3COCOOH$, the simplest of keto acids, is prepared by heating tartaric acid with potassium hydrogen sulphate:

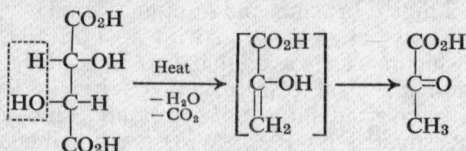

$$\begin{array}{c} CO_2H \\ | \\ H-C-OH \\ | \\ HO-C-H \\ | \\ CO_2H \end{array} \xrightarrow[-CO_2]{\substack{Heat \\ -H_2O}} \left[\begin{array}{c} CO_2H \\ | \\ C-OH \\ \| \\ CH_2 \end{array}\right] \longrightarrow \begin{array}{c} CO_2H \\ | \\ C=O \\ | \\ CH_3 \end{array}$$

Pyruvic acid plays a vital role in several physiological processes.

Acetoacetic Acid (CH_3COCH_2COOH) is a β-ketonic acid. It can be obtained by the hydrolysis of its ethyl ester in cold alkali. The acid on heating becomes decarboxylated to give acetone.

$$CH_3COCH_2COOH \longrightarrow CH_3COCH_3 + CO_2\uparrow$$

The loss of carbon dioxide and subsequent formation of a ketone is a characteristic property of a β-ketonic acid. Acetoacetic acid occurs in the urine of diabetic patients.

Ethyl acetoacetate is a valuable intermediate in chemical synthesis. Its preparation and uses are taken up separately.

Laevulinic Acid ($CH_3COCH_2CH_2COOH$) is a γ-keto acid and can be obtained by heating starch or cane sugar with hydrochloric acid. Unlike acetoacetic acid, laevulinic acid on heating loses water to give an unsaturated lactone.

Ethyl Acetoacetate. This ester, more commonly called acetoacetic ester, is prepared by one of the most useful of the reactions of esters known as CLAISEN condensation. In the presence of a strong base such as sodium ethoxide, the hydrogen atom of one molecule of an ester forms an alcohol with the alkoxy group of another ester molecule. Thus, in the case of ethyl acetate we get:

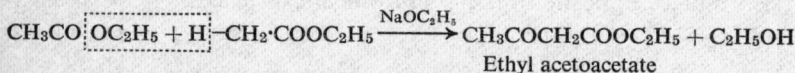

$$CH_3CO\underline{|OC_2H_5 + H|}-CH_2 \cdot COOC_2H_5 \xrightarrow{NaOC_2H_5} CH_3COCH_2COOC_2H_5 + C_2H_5OH$$

Ethyl acetoacetate

A high yield is obtained by distillation of ethyl alcohol as it is formed. Acetoacetic ester is one of the best known examples of **tautomerism**. Tautomerism is a spontaneous and reversible isomerization in which a hydrogen atom (really a proton) migrates in one direction and a covalent bond migrates in the opposite direction in a molecule. When three atoms X, Y, and Z are arranged so that there is a multiple bond between two of them and a hydrogen atom on the third, this isomerization may occur.

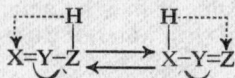

$$\underset{X=Y-Z}{\overset{\cdots H}{|}} \rightleftarrows \underset{X-Y=Z}{\overset{H\cdots}{|}}$$

The position of equilibrium depends upon the nature of the atoms X, Y, and Z. A situation is often encountered in which two atoms are carbon and the third oxygen. Acetoacetic ester is an instance in point.

$$CH_3C-CH_2C \overset{O}{\underset{OC_2H_5}{\Big\backslash}} \quad \rightleftharpoons \quad CH_3C=CHC \overset{O}{\underset{OC_2H_5}{\Big\backslash}}$$

Keto form   Enol form

Tautomeric system

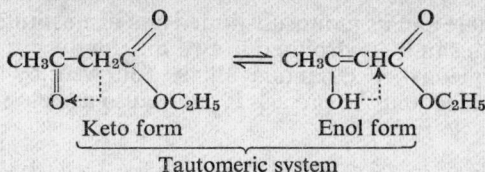

The hydroxyisomer is called an enol. The word *enol* is a combination of *en* to indicate the double bond C=C and *ol* to indicate the alcoholic group $-OH$. Other familiar examples of tautomerism are met in the case of nitroparaffins and amides.

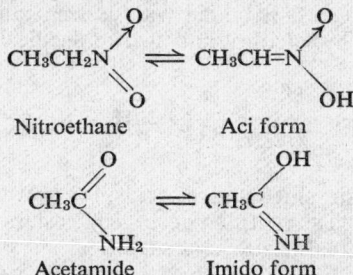

$$CH_3CH_2N \overset{O}{\underset{O}{\Big\backslash}} \quad \rightleftharpoons \quad CH_3CH=N \overset{O}{\underset{OH}{\Big\backslash}}$$

Nitroethane   Aci form

$$CH_3C \overset{O}{\underset{NH_2}{\Big\backslash}} \quad \rightleftharpoons \quad CH_3C \overset{OH}{\underset{NH}{\Big\backslash}}$$

Acetamide   Imido form

In the case of acetoacetic ester, the ketonic properties of the ester are exhibited by the formation of oxime, phenyl hydrazone, etc., whereas the enolic properties are evident in the addition of bromine and the formation of metallic salts. The copper derivative of the enolic form of acetoacetic ester is readily obtained when the ester is shaken with a solution of cupric acetate. The two forms which exist in equilibrium at room temperature are separated by dissolving the ester in hexane and cooling the solution to $-78°$ C. The pure ketonic compound separates in crystalline condition and is removed from the solution by filtration. The enolic form is obtained by suspending the sodium salt of the ester in hexane, cooling to $-78°$ C. and introducing gaseous hydrogen chloride. The precipitated sodium chloride is removed and the enol form obtained by evaporation of the solvent. Ordinarily acetoacetic ester exists predominantly in the keto form, roughly 7% being enolic.

Synthetic Uses of Acetoacetic Ester. Acetoacetic ester, like malonic ester previously discussed, contains an active methylene group. The active hydrogen atoms are replaceable by sodium to give sodio derivatives. As in the case of malonic ester, the sodio derivative of acetoacetic ester, formed by treatment of the ester with sodium or sodium ethoxide, can be alkylated by an alkyl halide thus:

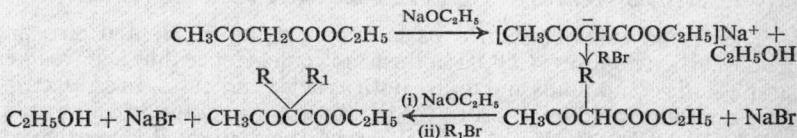

$$CH_3COCH_2COOC_2H_5 \xrightarrow{NaOC_2H_5} [CH_3COCHCOOC_2H_5]Na^+ + C_2H_5OH$$

$$C_2H_5OH + NaBr + CH_3CO\overset{R}{\underset{R_1}{C}}COOC_2H_5 \xleftarrow[\text{(ii) R}_1\text{Br}]{\text{(i) NaOC}_2\text{H}_5} CH_3CO\overset{R}{C}HCOOC_2H_5 + NaBr$$

The monoalkylated product still has an active hydrogen, and by repeating the alkylation the disubstituted product can be obtained.

Ethyl acetoacetate and its monosubstituted and disubstituted derivatives described above can be hydrolysed in two different ways. After boiling with dilute aqueous or ethanolic alkali, followed by acidification, ketones are formed and the process is referred to as ketonic hydrolysis.

$$CH_3 \cdot \underset{\underset{R_1}{|}}{\overset{\overset{O}{\|}}{C}} - \underset{\underset{R_1}{|}}{\overset{\overset{R}{|}}{C}} - \overset{\overset{O}{\|}}{C}_{OC_2H_5} + 2NaOH \xrightarrow[\substack{\text{Dilute aqueous} \\ \text{or ethanolic} \\ \text{alkali}}]{\text{Boil}} CH_3 \cdot \underset{\underset{R_1}{|}}{\overset{\overset{O}{\|}}{C}} - \overset{\overset{R}{|}}{C}H + Na_2CO_3 + C_2H_5OH$$

$$(R = \text{alkyl}, R_1 = H \text{ or alkyl})$$

This method is useful in the preparation of a variety of ketones.

When ethyl acetoacetate is boiled with concentrated ethanolic alkali, it is hydrolysed to ethyl alcohol and two molecules of sodium acetate.

$$CH_3 \cdot \overset{\overset{O}{\|}}{C} \vdots CH_2 - \overset{\overset{O}{\|}}{C} \overset{O}{\underset{\vdots OC_2H_5}{}} \longrightarrow 2CH_3 \cdot COONa + C_2H_5OH$$

$$Na\overset{\vdots}{O}H \quad Na\overset{\vdots}{O}H$$

This type of hydrolysis, since it gives acids, is known as acid hydrolysis. The mono- and dialkyl substituted acetoacetic esters furnished similarly mono- and disubstituted acetic acid salts.

$$CH_3 - \underset{\underset{R_1}{|}}{\overset{\overset{O}{\|}}{C}} - \underset{\underset{R_1}{|}}{\overset{\overset{R}{|}}{C}} - \overset{\overset{O}{\|}}{C}_{OC_2H_5} + 2NaOH \longrightarrow CH_3 \cdot COONa + \underset{\underset{R_1}{|}}{\overset{\overset{R}{|}}{H}C} - COONa + C_2H_5OH$$

Disubstituted
acetic acid salt

and thence

$$\underset{\underset{R_1}{|}}{\overset{\overset{R}{|}}{H}C}COONa \xrightarrow{H^+} H - \underset{\underset{R_1}{|}}{\overset{\overset{R}{|}}{C}} - COOH + Na^+$$

Disubstituted
acetic acid

Like malonic ester, acetoacetic ester reacts with ethylene dibromide to give a derivative of cyclopropane. With higher alkylene dibromides, ring compounds containing five or more carbon atoms can be obtained. In addition to the synthesis of ketones and acids mentioned above, acetoacetic ester has been used for the synthesis of a variety of other compounds, especially heterocyclic compounds.

AMINO ACIDS, POLYPEPTIDES, AND PROTEINS

The amino acids contain a basic amino group $-NH_2$ and also an acidic carboxyl group $-COOH$ in their molecule. Hence they behave like amphoteric compounds and ionize both as acids and bases. In an electric field an amino acid migrates to the cathode in acid solution and to the anode in basic solution. For each amino acid there is a particular hydrogen-ion concentration at which the acidic and basic ionizations are equal, and this particular pH value is called the isoelectric point.

The electrically neutral form of an amino acid is a zwitterion, that is, a dipolar ion carrying both a negative and a positive charge. The acid and the basic groups within the same molecule react with each other as follows:

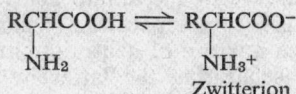

$$\underset{\underset{NH_2}{|}}{RCHCOOH} \rightleftharpoons \underset{\underset{NH_3^+}{|}}{RCHCOO^-}$$

Zwitterion

In the solid state these dipolar ions are arranged so as to display the maximum amount of interaction between the opposite charges on neighbouring molecules. The zwitterionic structure is in accord with the observed saltlike character of the amino acids, such as infusibility, low volatility, solubility in water, lack of solubility in organic solvents, and a high dipole moment.

The α-amino acids are the constituent units of proteins which form an essential part of all living tissues and so necessary a part of our diet. About two dozen α-amino acids have been isolated by hydrolysis of proteins. Certain of these amino acids cannot be synthesized by the animal organism and have to be supplied in the diet. Hence they are called essential amino acids. In their absence malnutrition results. The α-amino acids may be roughly classified into: (1) monoamino monocarboxylic acids; (2) diamino monocarboxylic acids; (3) monoamino dicarboxylic acids; and (4) amino acids containing an aromatic or heterocyclic ring.

Glycine or α-aminoacetic acid is the simplest amino acid. Alanine is α-aminopropionic acid, and is obtained from nearly all the proteins by hydrolysis. Serine is α-amino-β-hydroxypropionic acid and valine is α-aminoisovaleric acid.

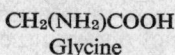

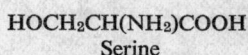

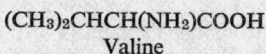

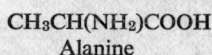

$CH_2(NH_2)COOH$
Glycine

$HOCH_2CH(NH_2)COOH$
Serine

$(CH_3)_2CHCH(NH_2)COOH$
Valine

$CH_3CH(NH_2)COOH$
Alanine

Among the diamino monocarboxylic acids, lysine (α,ε-diaminocaproic acid) may be mentioned. Monoamino dicarboxylic acids include aspartic acid (α-aminosuccinic acid) and glutamic acid (α-aminoglutaric acid). Phenylalanine and tyrosine are amino acids containing an aromatic nucleus, whereas proline and tryptophane contain heterocyclic rings.

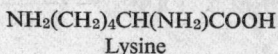

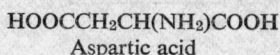

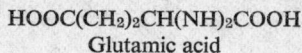

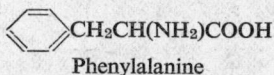

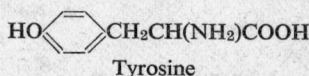

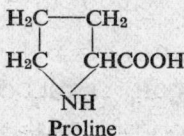

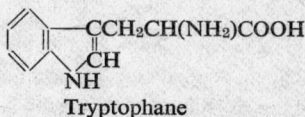

$NH_2(CH_2)_4CH(NH_2)COOH$
Lysine

$HOOCCH_2CH(NH_2)COOH$
Aspartic acid

$HOOC(CH_2)_2CH(NH)_2COOH$
Glutamic acid

Phenylalanine

Tyrosine

Proline

Tryptophane

Isolation of Amino Acids from Proteins. The usual method of hydrolysis of a protein consists in refluxing it with 20% hydrochloric acid (or 35% sulphuric acid). Invariably a mixture of several amino acids is formed. The saltlike nature of the amino acids makes it necessary to employ special methods for the separation of this mixture. One method consists of converting a mixture of neutral amino acids to their esters, $R-CH-CO_2CH_3$, and separating the latter by fractional distillation.
      |
    NH_2
Subsequent hydrolysis of the pure esters gives the corresponding amino acids. Another method involves preferential precipitation of a specific amino acid with reagents like mercuric sulphate, phosphotungstic acid, etc. Hydrolysis by means of enzymes is often useful for the preparation of certain amino acids sensitive to chemical treatment. Synthetic methods have also been worked out, but since these always produce racemic products, they must be used in conjunction with suitable enzymatic methods of resolution.

SYNTHESIS: The following are a few among a large number of synthetic methods available:

1. *Strecker Synthesis.* Cyanohydrins when treated with ammonia are converted into aminonitriles which when hydrolysed yield α-amino acids. For example:

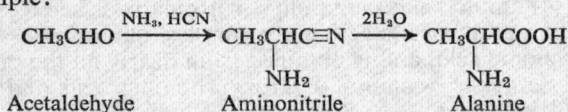

$$CH_3CHO \xrightarrow{NH_3, HCN} CH_3CHC\equiv N \xrightarrow{2H_2O} CH_3CHCOOH$$

                                     NH_2 NH_2

    Acetaldehyde   Aminonitrile   Alanine

In practice, the formation of cyanohydrin and its reaction with ammonia are combined in one step by the use of a mixture of potassium cyanide and ammonium chloride.

2. *The Phthalimide Synthesis.* An α-haloester, especially an α-halomalonate, is treated with potassium phthalimide to produce a monosubstituted malonate which is suitably alkylated and then converted to the desired amino acid by conventional procedures:

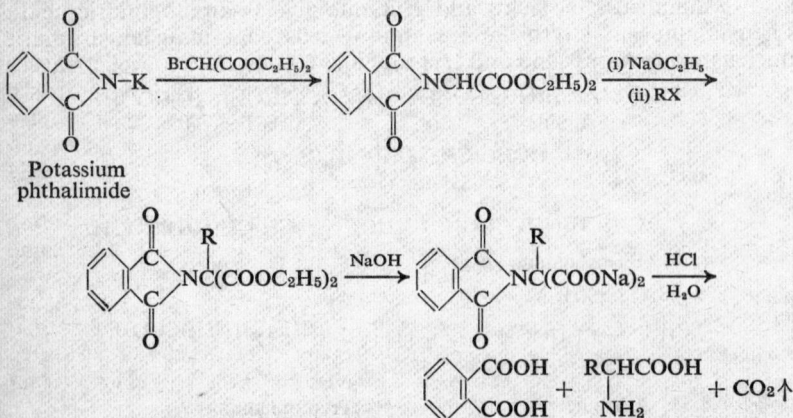

Potassium
phthalimide

3. *Synthesis Via Acylaminomalonates.* A modification of the above method consists of alkylation of an *n*-acyl derivative of aminomalonic ester. These *n*-acyl derivatives are prepared from malonic ester by the sequence of steps shown:

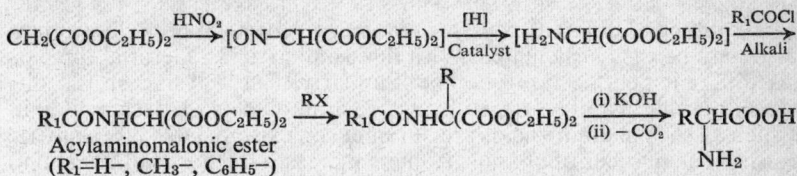

$$CH_2(COOC_2H_5)_2 \xrightarrow{HNO_2} [ON-CH(COOC_2H_5)_2] \xrightarrow[Catalyst]{[H]} [H_2NCH(COOC_2H_5)_2] \xrightarrow[Alkali]{R_1COCl}$$

$$R_1CONHCH(COOC_2H_5)_2 \xrightarrow{RX} R_1CONH\overset{R}{\underset{|}{C}}(COOC_2H_5)_2 \xrightarrow[(ii) -CO_2]{(i) KOH} R\overset{}{\underset{|}{C}}HCOOH \atop NH_2$$

Acylaminomalonic ester
(R_1=H−, CH_3−, C_6H_5−)

The brackets indicate that the compounds bracketed are formed but not isolated. This general method has been successfully employed for preparing a variety of α-amino acids.

PROPERTIES: The α-amino acids give the general reactions characteristic of the amino and carboxyl functions. As acids they form esters, amides, acid chlorides, etc. As amines, they can be acylated and they also react with nitrous acid, as shown below:

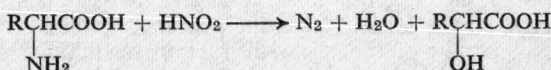

$$R\overset{}{\underset{|}{C}}HCOOH + HNO_2 \longrightarrow N_2 + H_2O + R\overset{}{\underset{|}{C}}HCOOH \atop OH$$
$$NH_2$$

This reaction is used in the analysis of proteins to estimate the amount of amino nitrogen present in them (Van Slyke analysis). Because of their amphoteric nature, they can form salts with both acids and bases.

The most important reaction of an amino acid is what is known as **peptide formation**. The carboxyl group of one molecule of an amino acid may react with the amino group of a neighbouring molecule, with the result that amino acids may form amides among themselves.

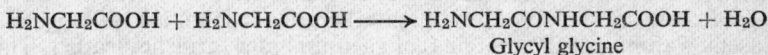

$$H_2NCH_2COOH + H_2NCH_2COOH \longrightarrow H_2NCH_2CONHCH_2COOH + H_2O$$
Glycyl glycine

Such intermolecular acid amides are known as **peptides**. Glycyl glycine is a dipeptide, since it is made up from two molecules of an amino acid and it may be hydrolysed easily at the peptide link −CO−NH− to return the two molecules of glycine. The reaction may be continued at either end of the dipeptide with the resulting formation of a tri, tetra, and a polypeptide. The proteins are essentially polypeptides derived from a number of different α-amino acids. The number of molecules of amino acids entering into the formula of one molecule of a protein ordinarily ranges from 300 to 1,000.

Proteins are important complex substances associated with living matter. They play a vital role in all the activities of the cell. The name protein is derived from the Greek *proteios*, meaning 'of first importance'. One of the three classes of food—the others being fats and carbohydrates—the importance of proteins can hardly be overemphasized. The fats and carbohydrates contain only the elements carbon, hydrogen, and oxygen, but proteins, in addition to these elements, also contain nitrogen, and sometimes sulphur and phosphorus.

A protein is essentially, as previously mentioned, a polypeptide of the type

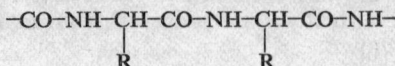

$$-CO-NH-\underset{\underset{R}{|}}{C}H-CO-NH-\underset{\underset{R}{|}}{C}H-CO-NH-$$

The number of different peptides that can be formed with a few amino acids can be easily calculated. With three amino acids, factorial three or six different peptides are possible. Similarly, factorial five or 120 different peptides can be formed from five amino acids. A decapeptide in which all the ten amino acids are different can exist in 3,628,800 isomeric forms. The number of possible isomers increases further if all the amino acids are optically active. The properties and biological functions of a protein are governed by the sequence and the type of amino acids in the chain. In building protein molecules, nature does not seem to function in a haphazard way, since the amino acids are found to recur in definite cycles.

In the light of the complex structure of the proteins any achievement in the field is greatly to be appreciated. The ordinary chemical analysis of a protein does not give an indication of the manner in which the amino acids are joined together nor does it give the spatial configuration of the proteins. In the initial stages the problem seemed hopeless. But with the development of the X-ray-diffraction technique, much order has been brought into what was once a chaotic realm. LINUS PAULING and his co-workers at the California Institute of Technology got an insight into the shape of the protein molecule, and SANGER and his associates (1954) at Cambridge reported for the first time the total chemical structure of a protein. VINCENT DU VIGNAUD and his associates at Cornell University even succeeded in synthesizing a natural protein (oxytocin), for the first time in chemical history.

USES: In addition to their use as food, proteins have several industrial uses. Plastic articles and fabric (lanital) are made from the milk protein Casein. Tanning of the protein (collagen) in hides furnishes leather. The use of gelatin and glue, which are products of hydrolysis of collagen, is also familiar. The usefulness of wool and silk fibres which are nothing but proteins may also be mentioned.

UNSATURATED ACIDS—GEOMETRICAL ISOMERISM

The unsaturated acids contain one or more double bonds in their molecule in addition to their carboxyl group. A study of the unsaturated acids will necessarily have to be prefaced by a consideration of a special type of stereoisomerism that is found to occur in unsaturated compounds (not necessarily acids) containing a double bond. This type of isomerism, known as geometrical isomerism, is also due to different spatial arrangement of the atoms within the molecule, but is quite different from the optical isomerism that has been discussed previously.

On the basis of experiment, it is inferred that in the case of a carbon–carbon single bond free rotation of the two carbon atoms along with their attached hydrogen or other atoms exists about the C–C axis. That is to say, in a compound like ethylene dichloride (I) rotation of each

CH_2Cl group about the C–C axis is possible. If free rotation is not possible, isomeric compounds of ethylene dichloride corresponding to the planar projection formulae (I) and (II) (in fact, an infinite number of isomers corresponding to different relative positions of the halogens) must exist.

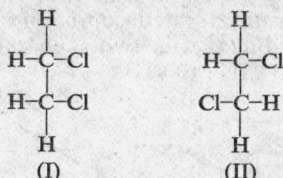

(I)   (II)

The two structures given above are different for in (I) the two chlorine atoms are near each other and on the same side of the C–C bond and in (II) the two chlorine atoms are much farther apart and lie on the opposite sides of the C–C axis. Actually only one variety of ethylene dichloride exists, and this must be accounted for. If each carbon atom along with its attached atoms is assumed to rotate around the C–C axis it will be found that when the lower half of (I) rotates with respect to the upper half around the C–C axis through 180°, molecule (I) becomes molecule (II). In other words, the groups attached to one carbon atom are not fixed in space relative to the groups attached to the other carbon atom. With the tetrahedral models (III) and (IV) corresponding to (I) and (II) the situation can perhaps be better understood.

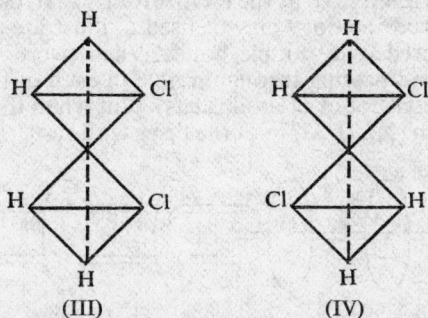

(III)   (IV)

We find that we can only account for the number and types of isomers formed by assuming that the second bond somehow hinders free rotation about the double bond and the groups attached to the doubly bound carbon atoms are permanently fixed in position. This will be discussed further in Chapter Eleven, but for the time being the following simple interpretation will suffice:

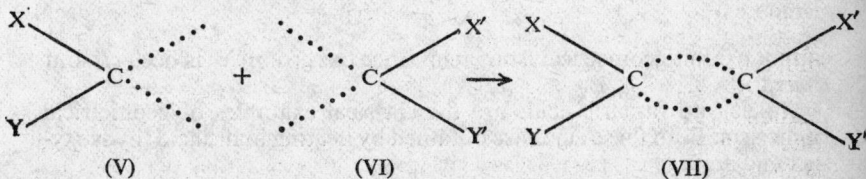

(V)   (VI)   (VII)

A saturated carbon atom may be represented by a model showing only the tetrahedrally directed linkages as in (V) and (VI). In this representation if the two solid lines are assumed to be in the plane of the paper, the dotted lines will lie in a plane at right angles to the plane of the paper, one of these dotted lines being above and the other below the plane of the paper. An unsaturated compound may be considered to originate by the dotted linkages in two such carbon atoms coalescing together to form either (VII) or (VIII),

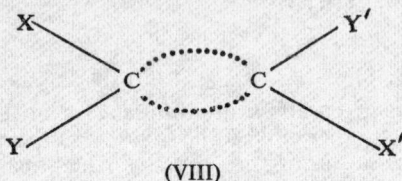

(VIII)

in which the groups X, Y, X', and Y' as well as the carbon atoms will lie in the plane of the paper and the double bond shown by the dotted lines in a plane at right angles to it. (VII) and (VIII) differ in the mutual relative distances of the groups attached to the carbon atoms, X and X', lying on the same side in (VII) and the same groups lying on opposite sides in (VIII). The double bond makes the structure rigid, and there is no possibility of free rotation of the carbon atoms with respect to each other. (VII) and (VIII) are then typical geometrical isomers. If X and X' are the same, (VII) is referred to as the *cis* form and (VIII) as the *trans* form.

A somewhat crude analogy may be used to illustrate the hindrance to free rotation caused by a double bond. When two discs of wood are supported on a rod passing through a hole in each of them, as in (IX), rotation of the discs about the rod is easy. But when there are two supporting rods, as in (X), rotation of the discs with respect to each other is not possible.

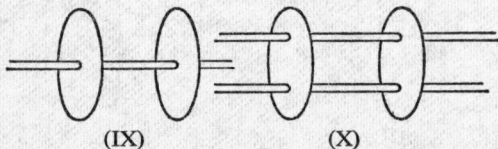

(IX)   (X)

Fig. 11

A little reflection will show that the two groups on either carbon atom in (VII) or (VIII) must be unlike or else geometrical isomerism will not be exhibited. Thus, for example, a compound represented by the formula

$$X\text{-}C\text{-}X$$
$$X\text{-}C\text{-}Y$$

cannot exhibit geometrical isomerism since the group 'Y' is both *cis* and *trans* to 'X'.

Maleic and fumaric acids are the classical examples of geometrical isomerism. Both these acids are obtained by heating malic acid (hydroxy-succinic acid).

$$H-C-COOH$$
$$\|$$
$$H-C-COOH$$
Maleic acid (*cis*)

$$\text{CH(OH)COOH} \xrightarrow{-H_2O}$$

$$\text{CH}_2\text{COOH}$$
Malic acid

$$H-C-COOH$$
$$\|$$
$$HOOC-C-H$$
Fumaric acid (*trans*)

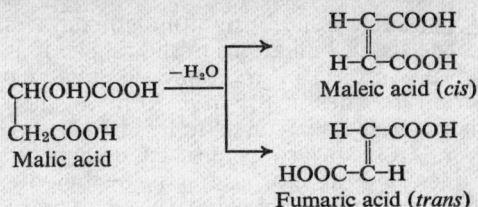

On the basis of the preceding discussion, a double bond and geometrical isomerism are as inseparably associated as cause and effect. So in the case of maleic and fumaric acids, if we remove the cause, namely the double bond (by saturation), the effects should disappear. Such is indeed the case. On reduction both maleic and fumaric acids give rise to one and the same succinic acid.

Unlike fumaric acid, maleic acid is easily converted into its anhydride by heating above its melting point, and should have the *cis*-structure in which the two carboxyl groups are on the same side of the double bond and could interact with each other to split off water and form the anhydride.

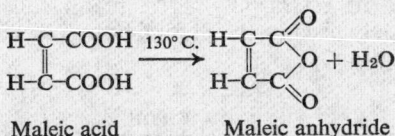

$$H-C-COOH \xrightarrow{130°C.} \quad + H_2O$$
$$\|$$
$$H-C-COOH$$

Maleic acid   Maleic anhydride

The configurations of a pair of geometrical isomers may be determined by relating them to compounds of known configuration or by physical measurements, especially the measurement of dipole moments. The *cis* isomer of a symmetrical compound has a definite dipole moment, while the *trans* isomer has zero moment. Thus *cis* dibromoethylene has a dipole moment of 1·35 units, while the *trans* isomer has zero moment. Usually the *trans* isomer is more stable than the *cis*. Hence fumaric acid is more stable than maleic acid. On heating maleic acid to 200° C., it is converted into fumaric acid. Maleic acid has the lower melting point, higher solubility in water, greater heat of combustion, and higher value for the first ionization constant.

Geometrical isomerism is not confined to unsaturated compounds alone. Saturated ring compounds also display geometrical isomerism. Thus, for example, in the case of 1,4-cyclohexane dicarboxylic acid, both *cis* and *trans* isomers are known.

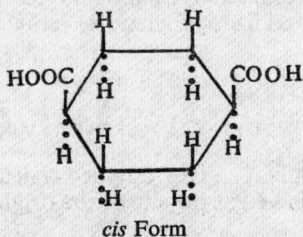

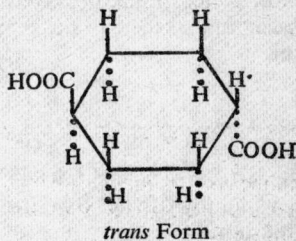

cis Form   *trans* Form

The ring, like the double bond in unsaturated compounds, makes the structure rigid. The six carbon atoms of the ring lie in a plane, and the two carboxyl groups project on the same side of this plane in the *cis* form and on opposite sides in the *trans* form. In the representations above, the carbon atoms are assumed to lie in a plane perpendicular to that of the paper. Atoms or groups projecting above the ring are indicated by solid lines, and those projecting below by dotted lines.

Acrylic acid is the simplest unsaturated acid. It can be prepared by the removal of hydrogen bromide from β-bromopropionic acid with alkali:

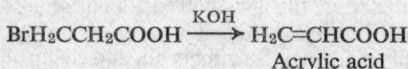

$$BrH_2CCH_2COOH \xrightarrow{KOH} H_2C{=}CHCOOH$$
Acrylic acid

It can also be obtained by oxidizing acrolein ($CH_2{=}CH{-}CHO$) or by hydrolysis of acrylonitrile ($CH_2{=}CH{-}CN$). Since acrylic acid has two hydrogen atoms attached to one of the doubly bound carbon atoms, it cannot exhibit geometrical isomerism. Acrylic acid gives the usual reactions of an acid in getting converted into salts, esters, amides, etc., and the double bond shows the usual olefinic properties.

Acrylic acid and its esters polymerize readily on standing. Methyl methacrylate is widely used in the manufacture of plastics. It is synthesized from acetone thus:

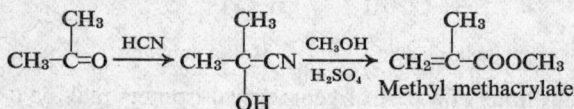

$$CH_3{-}\overset{CH_3}{\underset{}{C}}{=}O \xrightarrow{HCN} CH_3{-}\overset{CH_3}{\underset{OH}{C}}{-}CN \xrightarrow[H_2SO_4]{CH_3OH} CH_2{=}\overset{CH_3}{\underset{}{C}}{-}COOCH_3$$
Methyl methacrylate

Polymerized methyl methacrylate is a colourless transparent plastic sold under the name Perspex.

Oleic acid contains a double bond and can exhibit geometrical isomerism.

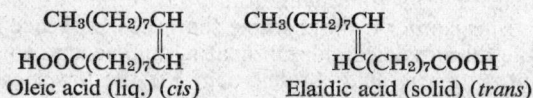

$$\begin{array}{ll} CH_3(CH_2)_7CH & CH_3(CH_2)_7CH \\ \| & \| \\ HOOC(CH_2)_7CH & HC(CH_2)_7COOH \end{array}$$
Oleic acid (liq.) (*cis*)   Elaidic acid (solid) (*trans*)

As has been seen, it is an important acid obtained by the hydrolysis of fats. Reduction of the *cis* oleic acid or the *trans* elaidic acid as is to be expected leads to stearic acid. Unsaturated acids containing additional double bonds, like linoleic and linolenic acids, were mentioned before. The position of double bonds in the many unsaturated acids occurring in fats is determined by oxidation experiments. Alkaline potassium permanganate, it will be remembered, oxidizes ethylene to ethylene glycol.

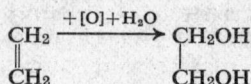

$$\overset{CH_2}{\underset{CH_2}{\|}} \xrightarrow{+[O]+H_2O} \overset{CH_2OH}{\underset{CH_2OH}{}}$$

An acid like oleic acid when treated similarly gives dihydroxystearic acid, which by further oxidation can be cleaved at the site of the original double bond.

$$CH_3(CH_2)_7CH=CH(CH_2)_7COOH \xrightarrow{KMnO_4 \text{ or } H_2O_2}$$

$$CH_3(CH_2)_7\underset{\underset{OH}{|}}{CH}-\underset{\underset{OH}{|}}{CH}(CH_2)_7COOH \xrightarrow{HIO_4}$$

9,10-Dihydroxystearic acid

$$CH_3(CH_2)_7CH=O \ + \ O=CH(CH_2)_7COOH$$

Pelargonic aldehyde   Azelaic half-aldehyde

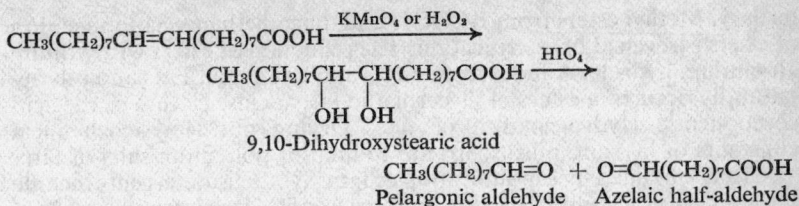

By identifying the fragments thus obtained, the double bond is deduced to be at the centre of the carbon chain.

SUMMARY

Monocarboxylic acids: RCOOH. Preparation: (1) Oxidation of RCH_2OH to RCOOH. (2) $RCN \xrightarrow[H_2O]{Alkali} RCOOH$. (3) $RX \xrightarrow{Mg} RMgX \xrightarrow[(ii) H^+]{(i) CO_2} RCOOH$. Properties and reactions: association of acids, acidity of acids. Conversion of acids to acid chlorides, esters, amides, etc. Reduction of acids with lithium aluminium hydride. Difference in behaviour on heating metal salts: $RCOONa \longrightarrow RH$; $(CH_3COO)_2 Ca \longrightarrow CH_3COCH_3$; $RCOONH_4 \longrightarrow RCONH_2$. Hunsdiecker reaction: $RCOOAg \xrightarrow{Br_2} RBr$. Halogenation of acids: acetic acid to α-halo-acids. Common acids include (i) formic acid. Prepared industrially from sodium hydroxide and carbon monoxide; $CO + NaOH \longrightarrow HCOONa$. Behaviour of formic acid as an acid and as an aldehyde. (ii) Acetic acid. Previously obtained from pyroligneous acid. Manufactured now from acetylene via acetaldehyde. Derivatives: (A) Acid chlorides are of the type RCOX. Lower b.p.s than acids. Preparation by treating acids with thionyl chloride, phosphorus pentachloride, etc. Use in acylation of hydroxy and amino compounds. Synthetic uses: (a) $R_1MgX \xrightarrow{CdCl_2} R_1CdX \xrightarrow{RCOCl} R_1COR$. (b) Arndt Eistert reaction to prepare higher homologues. $RCOCl \xrightarrow{CH_2N_2} RCOCHN_2 \xrightarrow[Rearrangement]{H_2O} RCH_2COOH$. (B) Acid anhydrides: Type formula $(RCO)_2O$; prepared from RCOONa and RCOCl. Acetic anhydride prepared in industry from acetone. Use of anhydrides in acylations. (C) Acid amides: type formula $RCONR_1R_2$ (R_1, R_2 are alkyl groups or hydrogen atoms); prepared from acid chlorides and amines of the type R_1R_2NH. Simple amides ($RCONH_2$) obtained also by heating ammonium salts and also by partial hydrolysis of nitriles: $RCN \longrightarrow RCONH_2$. Amides are neutral compounds easily hydrolysed to component parts. Dehydration of amides to nitriles; acids from amides by treatment with nitrous acids. Urea, a diamide of commercial importance. Preparation: (1) From ammonium cyanate. (2) Treating phosgene with ammonia. (3) Industrial method consists in reacting carbon dioxide with ammonia under pressure at 140°. Uses of urea as fertilizer and in the manufacture of barbiturates; Veronal, the hypnotic. (D) Esters prepared by esterification: $RCOOH + R'OH \longrightarrow RCOOR' + H_2O$. The hydroxyl eliminated during esterification is from the acid unless the alcohol used is tertiary. Methyl esters from RCOOH and diazomethane. Saponification of esters is reversal of esterification. The reduction of esters with lithium aluminium hydride. Amides from esters and ammonia. Fats and oils are naturally occurring esters of glycerol and fatty acids. Nature of the acid components. Hydrogenation of fats. Drying oils. Physicochemical constants of fats and oils. Soaps are sodium or potassium salts of fatty acids and obtained by saponification of fats. Synthetic detergents include sodium salts of sulphated oils, sulphates of higher alcohols and of aromatic sulphonic acids.

Dibasic Acids. Type formula $(CH_2)_n(COOH)_2$. Preparation: Oxalic acid by heating sodium formate; malonic acid by hydrolysis of cyanoacetic acid; succinic acid by hydrolysis of succinonitrile and glutaric acid by hydrolysis of glutaronitrile. Adipic acid (important in the manufacture of Nylon) is made from cyclohexanol. Properties and reactions: dibasic acids form derivatives characteristic of monocarboxylic acids. Effect of proximity of −COOH groups on the acid strength. Formation of cyclic anhydrides from succinic and glutaric acids and also cyclic imides. N-Bromosuccinimide and its usefulness. Diethyl malonate, its preparation and usefulness in organic synthesis. Active −CH₂− group easily alkylated with alkyl halides and product converted to alkylacetic acids. Cyclic compounds obtained by using dihalides. Condensation with aldehydes to give intermediates for the preparation of αβ-unsaturated acids.

Halogenated Acids. α and β halogen acids. Preparation: effect of halogen on the acidity of the acid. Reactions centre around the halogen atom and the −COOH group.

Hydroxy Acids. These contain a hydroxyl group in addition to −COOH. Preparation: (1) Hydrolysis of halo acids. (2) Hydrolysis of cyanohydrins. (3) Oxidation of β-hydroxy aldehydes to β-hydroxy acids. Chemical reactions characteristic of both −OH and −COOH groups. Difference in behaviour of the different hydroxy acids on heating. Lactones. Lactic acid: important commercially. Manufactured by fermentation process; also obtained by hydrolysis of acetaldehyde cyanohydrin.

Optical Isomerism. Lactic acid, a classical example which shows optical activity. Spatial nature of organic compounds. Van't Hoff–Le Bel concept of the tetrahedral carbon atom; asymmetric carbon atom; nature of *d*, *l*, *dl*, and *meso* forms. Dibromocinnamic acid, an example of a compound having two different asymmetric carbon atoms. Different forms of tartaric acid and their planar projection formulae. Resolution of racemic forms. Walden inversion.

Aldehydo and Ketonic Acids. These have −CHO or −C=O group in addition to −COOH. Pyruvic acid ($CH_3COCOOH$) prepared by heating tartaric acid with potassium bisulphate. Esters of acetoacetic acid important in the synthesis of organic compounds. Ethyl acetoacetate prepared by Claisen condensation. Keto-enol tautomerism of ethyl acetoacetate. Alkylation of ethyl acetoacetate with alkylhalides. Ketonic and acid hydrolysis of resulting products to give a variety of ketones and acids respectively.

α-**Amino Acids, Polypeptides, and Proteins.** Zwitterionic nature of amino acids. Their isolation by acid hydrolysis of proteins and chemical classification. Synthetic methods: (1) Strecker's cyanohydrin synthesis. (2) Phthalimide synthesis involving alkylation of potassium phthalimide with bromomalonic ester by alkylation of the product, hydrolysis, and decarboxylation. (3) Use of acylamino malonates instead of phthalimidomalonate in method (2). Properties and reactions: amino acids show reactions of both $-NH_2$ and $-COOH$ groups. Peptides and their formation. Proteins are polypeptides containing different α-amino acids. Recent synthetic achievements in the protein field. Uses of proteins.

Geometrical Isomerism. Shown by compounds of the type
$$\begin{matrix} X & X' \\ & C{=}C \\ Y & Y' \end{matrix}$$
and a result of rigid structure due to double bond. Meaning of *cis* and *trans* forms. Maleic and fumaric acids as examples of geometrical isomers. Methods for determination of *cis* and *trans* forms include study of ease of formation of cyclic compounds, dipole moment measurements, and other physicochemical constants. Geometrical isomerism in cyclic compounds.

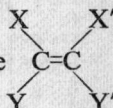

Unsaturated Acids. Acrylic acid $(CH_2{=}CH{-}COOH)$. Preparation: (1) Dehydrobromination of β-bromopropionic acid. (2) Oxidation of acrolein. (3) Hydrolysis of acrylonitrile. Methyl methacrylate
$$\overset{\displaystyle CH_3}{\underset{\displaystyle |}{}}$$
$(CH_2{=}C{-}COOCH_3)$. Used in the manufacture of plastics. Oleic acid and elaidic acids—their difference. Use of potassium permanganate—oxidation of unsaturated compounds to locate position of unsaturation.

Problem Set No. 6

1. How can *n*-butyric acid be prepared from: (*a*) 1-chloropropane; (*b*) ethyl propionate $(CH_3CH_2COOC_2H_5)$?
2. How would you distinguish between: (*a*) acetic acid and formic acid, and (*b*) chloroacetic acid and acetyl chloride?
3. An unknown compound may be *n*-hexane or hexanol-1 or *n*-hexanoic acid. Using only sodium bicarbonate and sodium, how can the sample be identified?
4. Indicate how to effect the following transformations:

   (*a*) Acetylene to acrylic acid.
   (*b*) Ethanol to acetic anhydride.
   (*c*) Ethanol to butanoic acid.
   (*d*) Propanoic acid to pyruvic acid.
   (*e*) Ethanol to pentanone-2.
   (*f*) Diethyl malonate to: (i) butanoic acid; (ii) 2-methylpentanoic acid; (iii) 2-pentenoic acid (assume that necessary halide intermediates are available for the syntheses mentioned in (*f*) and (*g*).
   (*g*) Ethyl acetoacetate to: (i) butanoic acid; (ii) pentanone-2; (iii) 5-oxohexanoic acid.
   (*h*) Ethyl acetamidomalonate to α-alanine.

5. Indicate the product that will be obtained by: (*a*) the action of P_2O_5 on butyramide; (*b*) heating succinic acid; (*c*) action of acetyl chloride on diethylamine; (*d*) action of one mole of 1,4-dibromobutane on 2 moles of sodiumalonic ester; (*e*) if product obtained in (*d*) is hydrolysed and decarboxylated, what is the product? (*f*) action of sodium ethoxide on 2 moles of ethyl propionate; (*g*) heating 4-hydroxy-butanoic acid; (*h*) action of *n*-bromosuccinimide on heptene-1; (*i*) the action of sodium hydroxide on ethyl propionate.

6. A certain oil may be a mineral oil, i.e. a petroleum fraction or a vegetable oil. How would you decide between the two possibilities?

7. Give the structural formulae of the optically active pentanols.

8. How many different forms are possible for the following structures?

   (*a*) $CH_3CHBrCOOH$
   (*b*) $CH_2OH-CH(OH)-CO_2H$
   (*c*) $CH_3-CH(OH)-CH(OH)-CH_3$
   (*d*) $CH_3-CH=CH-C_2H_5$
   (*e*) $CH_3-CH_2CH(OH)CH_2-CH_3$
   (*f*) $CH_3CH=CHCOOH$
   (*g*) $CH_3-CHBrCH(OH)-CH_3$

9. 0·2220 g. of a monocarboxylic acid requires 15 ml. of $N/5$ sodium hydroxide for neutralization. What product will be obtained by heating the sodium salt of the acid with sodium hydroxide?

10. A compound $C_6H_{14}O_6$ is acetylated with acetic anhydride and sodium acetate. The resultant product has a molecular weight of 434. How many hydroxyl groups are present in the original compound?

CHAPTER EIGHT

ORGANIC COMPOUNDS OF NITROGEN AND SULPHUR

AMINES

The amines are basic substances and may be pictured to be derived from ammonia NH_3 by the replacement of 1, 2, or 3 atoms of hydrogen by alkyl groups. Thus we have:

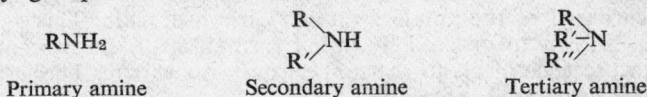

$$RNH_2 \qquad \begin{array}{c} R \\ \diagdown NH \\ R' \diagup \end{array} \qquad \begin{array}{c} R \\ R'{-}N \\ R'' \diagup \end{array}$$

Primary amine $\qquad$ Secondary amine $\qquad$ Tertiary amine

The characteristic group of a primary amine is $-NH_2$, that of a secondary amine $\diagdown NH$ and that of a tertiary amine $-N$. The word primary in the case of an amine refers to the fact that one carbon atom is directly attached to the nitrogen atom. Similarly, in the case of secondary amines two carbon atoms are directly attached to a nitrogen atom, and in tertiary amines three carbon atoms are directly attached to a nitrogen atom. Tertiary amines when treated with an organic halide RX give compounds similar to ammonium salts. Since they have four groups attached to the nitrogen atom; they are called quaternary ammonium salts.

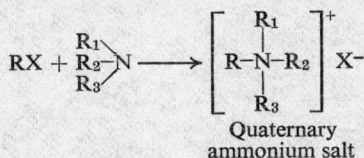

$$RX + \begin{array}{c} R_1 \\ R_2{-}N \\ R_3 \end{array} \longrightarrow \left[\begin{array}{c} R_1 \\ | \\ R{-}N{-}R_2 \\ | \\ R_3 \end{array} \right]^{+} X^{-}$$

Quaternary
ammonium salt

NOMENCLATURE: Simple amines are named by adding the ending 'amine' to the list of radicals attached to the nitrogen atom. These radicals are usually indicated by their common names rather than by their I.U.C. names. For example, CH_3NH_2 is methylamine, $(CH_3)_2CHNH_2$ is isopropylamine rather than 2-propylamine, $CH_3CH_2CH_2CH_2CH_2NH_2$ is n-amylamine rather than 1-pentylamine, $(C_2H_5)_2NH$ is diethylamine, etc.

PREPARATION OF AMINES: (1) **Alkylation of Ammonia**. Ammonia can be alkylated with an alkyl halide when a primary amine is formed as the initial product.

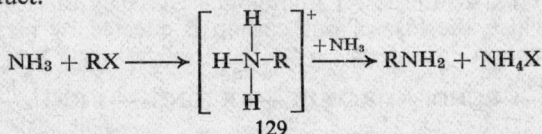

$$NH_3 + RX \longrightarrow \left[\begin{array}{c} H \\ | \\ H{-}N{-}R \\ | \\ H \end{array} \right]^{+} \xrightarrow{+NH_3} RNH_2 + NH_4X$$

129

This primary amine may react with excess alkyl halide to give a secondary amine and then a tertiary amine. The tertiary amine, in turn, may be converted to a quaternary ammonium salt. A mixture of products always results, and it is cumbersome to separate the products.

(2) **Hofmann Reaction.** When an amide is treated with a mixture of bromine and sodium hydroxide an amine with one carbon atom less than the amide is obtained. Thus, for example, acetamide gives methylamine by this reaction, known as the Hofmann reaction.

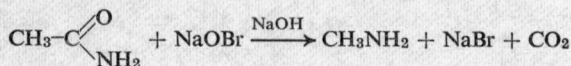

$$CH_3{-}C\overset{O}{\underset{NH_2}{\diagup}} + NaOBr \xrightarrow{NaOH} CH_3NH_2 + NaBr + CO_2$$

The sodium hypobromite, formed by reaction of bromine with sodium hydroxide in the cold, is the actual reagent which causes the elimination of the carbonyl of the amide group as carbon dioxide. This reaction has been shown to proceed by initial formation of the bromoamide which in presence of alkali rearranges to an isocyanate. The latter on hydrolysis furnishes the amines.

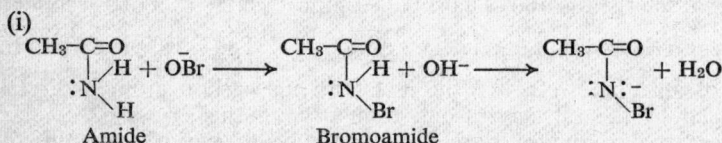

(i)

$$\underset{\text{Amide}}{CH_3{-}C{=}O\ \overset{H}{\underset{H}{:N}}} + \bar{O}Br \longrightarrow \underset{\text{Bromoamide}}{CH_3{-}C{=}O\ \overset{H}{\underset{Br}{:N}}} + OH^- \longrightarrow CH_3{-}C{=}O\ \overset{:-}{\underset{Br}{:N}} + H_2O$$

(ii) The alkyl group then migrates with its electron pair from the carbonyl carbon atom to the nitrogen atom. Simultaneously the bromine atom leaves the nitrogen atom, taking its bonding electron pair, as a bromide ion:

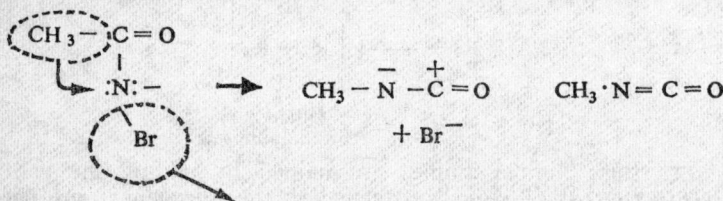

$$\overset{CH_3{-}\ C{=}O}{\underset{Br}{:N:{-}}} \longrightarrow CH_3 - \overset{-}{N} - \overset{+}{C} = O \qquad CH_3{\cdot}N{=}C{=}O$$
$$+ Br^-$$

This is followed by hydrolysis of the resulting isocyanate:

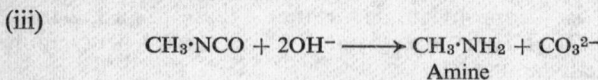

(iii)

$$CH_3{\cdot}NCO + 2OH^- \longrightarrow \underset{\text{Amine}}{CH_3{\cdot}NH_2} + CO_3^{2-}$$

The Hofmann reaction is of general applicability and provides a method for descending a homologous series. For instance, an alcohol $R{-}CH_2OH$ can be converted to the lower homologue ROH by the following sequence, in which the loss of one carbon is effected by means of the Hofmann reaction.

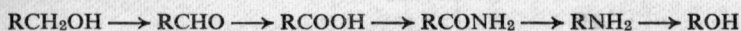

$$RCH_2OH \longrightarrow RCHO \longrightarrow RCOOH \longrightarrow RCONH_2 \longrightarrow RNH_2 \longrightarrow ROH$$

(3) **Reduction of Nitriles and Amides.** The versatile reagent lithium aluminium hydride has been found effective for reduction of nitriles and amides to amines.

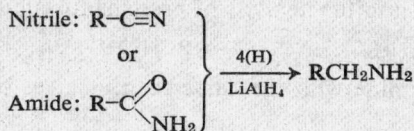

Nitrile: $R-C\equiv N$

or $\qquad \xrightarrow[\text{LiAlH}_4]{4(\text{H})} RCH_2NH_2$

Amide: $R-C \overset{O}{\underset{NH_2}{\diagup}}$

For example, caprylonitrile can be reduced in 90% yield to *n*-octyl-amine, $C_7H_{15}CN \longrightarrow C_7H_{15}CH_2NH_2$. In the case of nitriles the reduction can also be effected catalytically.

Commercially, methyl and ethyl amines are produced by heating the vapours of the corresponding alcohols with ammonia in the presence of a catalyst. There is an alternative method of manufacture of methyl amine, which is obtained as the hydrochloride by heating formaldehyde with ammonium chloride.

$$2CH_2O + NH_4Cl \longrightarrow CH_3NH_2 \cdot HCl + HCO_2H$$

PROPERTIES: The lower members of the family of amines are gases soluble in water. They have peculiar odours, somewhat fishy and ammoniacal. Primary and secondary amines can form hydrogen bonds of

the type $\quad N-H \cdots N-H$, and hence have higher boiling points than iso-

meric tertiary amines, which cannot form hydrogen bonds. Since nitrogen is less electronegative than oxygen, hydrogen bonding between two nitrogen atoms of two molecules of an amine occurs to a smaller extent than between two oxygen atoms of two molecules of an alcohol.

Chemically, the important property of the amines is their basicity. The nitrogen atom in an amine has an unshared pair of electrons which is available for combination with electron acceptors like H^+, BF_3, etc.

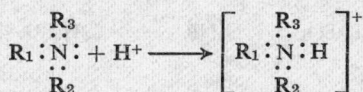

$$R_1 : \overset{R_3}{\underset{R_2}{N}} : + H^+ \longrightarrow \left[R_1 : \overset{R_3}{\underset{R_2}{N}} : H \right]^+$$

Amines are alkaline in aqueous solution mainly because of the abstraction of the proton of water and the resultant increase in the concentration of hydroxyl ions.

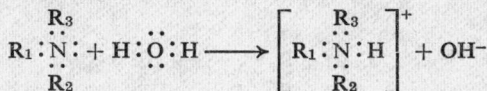

$$R_1 : \overset{R_3}{\underset{R_2}{N}} : + H : \overset{..}{\underset{..}{O}} : H \longrightarrow \left[R_1 : \overset{R_3}{\underset{R_2}{N}} : H \right]^+ + OH^-$$

Being basic, they form salts with acids like hydrochloric acid. The aliphatic amines, whether primary, secondary, or tertiary, are more basic than ammonia. This increased basicity is due to the electron-releasing nature of the alkyl substituents which make a lone pair of electrons on the nitrogen atom more readily available for combination with a proton.

The three classes of amines react differently with nitrous acid. A primary amine on treatment with nitrous acid gives the corresponding alcohol, and elementary nitrogen is liberated. This reaction, already referred to, is used in the analysis of proteins for amino nitrogen.

$$RCH_2NH_2 + HNO_2 \longrightarrow RCH_2OH + N_2\uparrow + H_2O$$

A secondary amine with the same reagent gives oily nitroso compounds

$$\begin{array}{c} R_1 \\ \diagdown \\ R_2 \diagup \end{array} NH + HNO_2 \longrightarrow \begin{array}{c} R_1 \\ \diagdown \\ R_2 \diagup \end{array} N-NO + H_2O$$

Nitroso compound

There is no characteristic reaction with nitrous acid in the case of a tertiary amine except salt formation.

Primary amines react with aldehydes to form the so-called Schiff's bases.

$$RNH_2 + O{=}\underset{\underset{H}{|}}{C}{\cdot}R_1 \longrightarrow RN{=}CHR_1 + H_2O$$

Schiff's base

Primary and secondary amines, since they contain replaceable hydrogen atoms, react with acid chlorides (or anhydrides) to give *N*-acyl derivatives:

$$RCOCl + \begin{array}{c} R_1 \\ \diagdown \\ R_2 \diagup \end{array} NH \longrightarrow RCON \begin{array}{c} \diagup R_1 \\ \diagdown R_2 \end{array} + HCl$$

Since tertiary amines do not have a replaceable hydrogen atom, they are unreactive towards the same reagents. The above acyl derivatives are usually high-melting crystalline compounds, often used in the characterization of amines. The reaction of amines with benzenesulphonyl chloride is often used to distinguish between and separate the different types of amines present in a mixture.

$$C_6H_5SO_2Cl + H_2NR \xrightarrow{-HCl} C_6H_5SO_2NR$$

Primary amine

(soluble in alkali)

$$C_6H_5SO_2Cl + HNR_2 \xrightarrow{-HCl} C_6H_5SO_2NR_2$$

Secondary amine

(insoluble in alkali)

$$C_6H_5SO_2Cl + \begin{array}{c} R_1 \\ R_2{-}N \\ R_3 \diagup \end{array} \longrightarrow \text{No reaction}$$

The benzenesulphonyl derivatives of primary amines are soluble in alkali, while those of secondary amines are insoluble. Tertiary amines, as usual, are indifferent towards the reagent. Furthermore, the benzenesulphonamides of both types are easily hydrolysed to give the parent amines.

A primary amine, when heated with chloroform and alkali, gives an isocyanide (carbylamine) with an intensely disagreeable and poisonous odour.

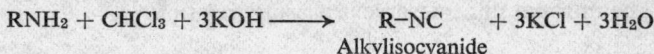

$$RNH_2 + CHCl_3 + 3KOH \longrightarrow \quad R-NC \quad + 3KCl + 3H_2O$$
$$\text{Alkylisocyanide}$$

This is a delicate test for primary amines and is referred to as Hofmann's carbylamine reaction.

When a quaternary ammonium salt is treated with moist silver oxide (which reacts as AgOH) tetra-alkyl ammonium hydroxides of the formula $\overset{+}{N}R_4\overset{-}{OH}$ are obtained. They are strong bases comparable to alkalis. When quarternary ammonium hydroxides of the type

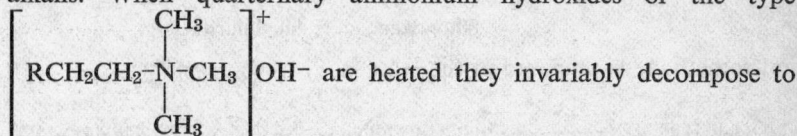

$$\left[RCH_2CH_2-\overset{\overset{\displaystyle CH_3}{|}}{\underset{\underset{\displaystyle CH_3}{|}}{N}}-CH_3 \right]^+ OH^- \text{ are heated they invariably decompose to}$$

give trimethyl amine, an olefin $RCH=CH_2$, and water. This reaction, known as Hofmann degradation, has served as the basis for determination of the structures of a number of naturally occurring alkaloids which contain one or more basic nitrogen atoms. Exhaustive methylation of such a compound gives a quaternary salt which with silver oxide gives the corresponding hydroxide. The latter, when heated, decomposes into a tertiary amine and a nitrogen-free fragment, namely, an olefin whose identification serves to throw light on the structure of the original alkaloid. For example:

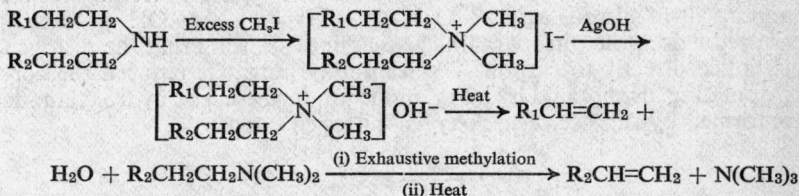

$$\underset{R_2CH_2CH_2}{\overset{R_1CH_2CH_2}{>}}NH \xrightarrow{\text{Excess } CH_3I} \left[\underset{R_2CH_2CH_2}{\overset{R_1CH_2CH_2}{>}}\overset{+}{N}\underset{CH_3}{\overset{CH_3}{<}}\right]I^- \xrightarrow{\text{AgOH}}$$

$$\left[\underset{R_2CH_2CH_2}{\overset{R_1CH_2CH_2}{>}}\overset{+}{N}\underset{CH_3}{\overset{CH_3}{<}}\right]OH^- \xrightarrow{\text{Heat}} R_1CH=CH_2 +$$

$$H_2O + R_2CH_2CH_2N(CH_3)_2 \xrightarrow[\text{(ii) Heat}]{\text{(i) Exhaustive methylation}} R_2CH=CH_2 + N(CH_3)_3$$

Any tertiary amine [$R_2CH_2CH_2N(CH_3)_2$ above] formed in the initial degradation can be subjected to the same process as the starting material and converted to a second olefin ($R_2CH=CH_2$) as outlined above. A tertiary amine on oxidation with aqueous hydrogen peroxide gives a stable amine oxide, as follows:

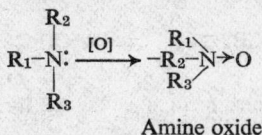

$$R_1-\overset{\overset{\displaystyle R_2}{|}}{\underset{\underset{\displaystyle R_3}{|}}{N}}: \xrightarrow{[O]} \underset{R_3}{\overset{R_1}{>}}R_2-N\rightarrow O$$

$$\text{Amine oxide}$$

Amines are used for producing amides used as detergents. High-molecular-weight amines are used as fungicides. Amines are also used in the manufacture of photographic developers, dyes, and textiles. Trimethylamine is used as a warning agent for leaks in gas pipes.

Other organic compounds of nitrogen, like the nitriles, amides, etc., have been discussed before, and they do not merit any special mention here. The nitroparaffins deserve a passing mention.

ALIPHATIC NITRO COMPOUNDS

When an alkyl halide is heated with silver nitrite a nitroparaffin is formed along with the isomeric alkyl nitrite.

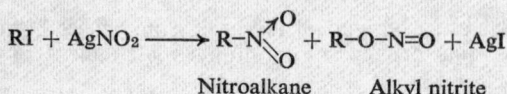

$$RI + AgNO_2 \longrightarrow R-N{\nearrow O \atop \searrow O} + R-O-N=O + AgI$$

Nitroalkane   Alkyl nitrite

Nitromethane is prepared in the laboratory as follows:

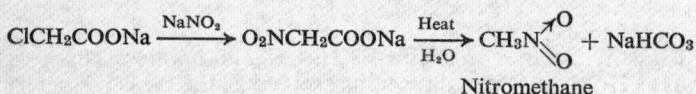

$$ClCH_2COONa \xrightarrow{NaNO_3} O_2NCH_2COONa \xrightarrow[H_2O]{Heat} CH_3N{\nearrow O \atop \searrow O} + NaHCO_3$$

Nitromethane

Industrially, alkanes are nitrated in the vapour phase to give nitro-paraffins.

The nitroparaffins are colourless liquids with an agreeable odour. In a nitroparaffin the nitrogen atom is linked directly to the carbon, while in the isomeric alkyl nitrite, which is an ester, the nitrogen atom is joined to carbon through an oxygen atom. In keeping with its structure an alkyl nitrite can be hydrolysed with alkali to give an alcohol (ROH) and a salt of nitrous acid. $R-O-N=O + NaOH \longrightarrow ROH + NaNO_2$ On reduction, the nitroparaffins give amines showing that the nitrogen is linked directly to carbon. When an alkyl nitrite is reduced the corresponding alcohol and hydroxylamine are formed. The hydroxylamine is formed by rupture of the oxygen–nitrogen bond.

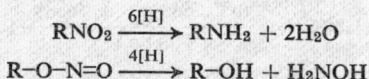

$$RNO_2 \xrightarrow{6[H]} RNH_2 + 2H_2O$$
$$R-O-N=O \xrightarrow{4[H]} R-OH + H_2NOH$$

A nitroparaffin containing at least one hydrogen atom attached to the carbon atom linked to the nitro group, though it is neutral, forms salts with sodium methoxide. The salt formation is attributed to the so-called 'aci' form, which is a tautomer of the nitro form.

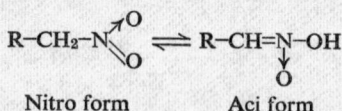

$$R-CH_2-N{\nearrow O \atop \searrow O} \rightleftharpoons R-CH=N-OH$$

Nitro form   Aci form

The nitroparaffins are used as solvents in the plastic industry and in the production of insecticides, detergents, explosives, and pharma-ceuticals.

SULPHUR COMPOUNDS

Of the large variety of organic sulphur compounds known, only two types, the mercaptans and the alkyl sulphides, will be dealt with here. A mercaptan (RSH) may be considered as derived from an alcohol (ROH) by substituting a sulphur atom for the oxygen atom. A mercaptan derived from ethanol is ethyl mercaptan (CH_3CH_2SH). A mercaptan derived from propanol is propyl mercaptan ($CH_3CH_2CH_2SH$) and so on. The name mercaptan refers to the fact that mercaptans as a rule form insoluble mercury salts.

Alkyl halides when heated with sodium hydrosulphide give mercaptans.

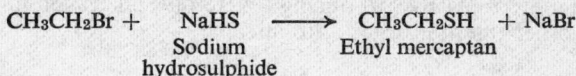

$$CH_3CH_2Br + \underset{\substack{\text{Sodium} \\ \text{hydrosulphide}}}{NaHS} \longrightarrow \underset{\text{Ethyl mercaptan}}{CH_3CH_2SH} + NaBr$$

The action of phosphorus pentasulphide on alcohols also gives mercaptans.

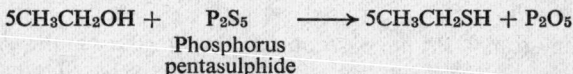

$$5CH_3CH_2OH + \underset{\substack{\text{Phosphorus} \\ \text{pentasulphide}}}{P_2S_5} \longrightarrow 5CH_3CH_2SH + P_2O_5$$

The mercaptans have an unpleasant odour. The odour of ethyl mercaptan is detectable even at a dilution of 1 part in 5×10^{10} parts of air. The mercaptans form salts readily with alkali.

$$CH_3CH_2SH + NaOH \longrightarrow CH_3CH_2SNa + H_2O$$

This is to be contrasted with the fact that the alcohols are neutral and do not form salts with alkali. The mercaptans are easily oxidized. Mild oxidation of ethyl mercaptan with a dilute solution of hydrogen peroxide gives diethyl disulphide.

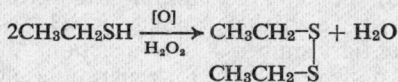

$$2CH_3CH_2SH \xrightarrow[H_2O_2]{[O]} \begin{array}{c} CH_3CH_2-S \\ | \\ CH_3CH_2-S \end{array} + H_2O$$

Stronger oxidizing agents like permanganate yield ethylsulphonic acid.

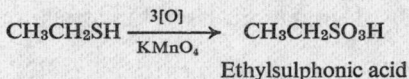

$$CH_3CH_2SH \xrightarrow[KMnO_4]{3[O]} \underset{\text{Ethylsulphonic acid}}{CH_3CH_2SO_3H}$$

Thiokol, a rubber substitute, is structurally related to disulphides.

The thio-ethers, or alkyl sulphides, are colourless liquids with the formula RSR. They are prepared by the action of alkyl halides on the sodium salt of a mercaptan. This method is easily recognized as an adaptation of Williamson's synthesis of ethers.

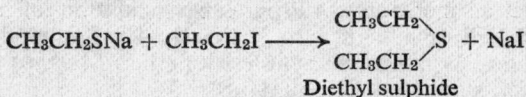

$$CH_3CH_2SNa + CH_3CH_2I \longrightarrow \underset{\text{Diethyl sulphide}}{\begin{array}{c} CH_3CH_2 \\ \\ CH_3CH_2 \end{array}\!\!\!\!>S} + NaI$$

Mustard gas, which was used in the First World War, is a thio-ether and is prepared by the reaction of ethylene with sulphur chloride.

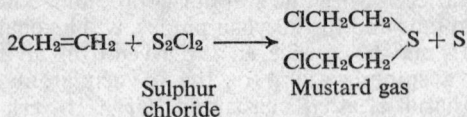

$$2CH_2{=}CH_2 + S_2Cl_2 \longrightarrow \begin{matrix} ClCH_2CH_2 \\ ClCH_2CH_2 \end{matrix}\!\!\Big\rangle S + S$$

Sulphur chloride   Mustard gas

It is so named since it has an odour like that of mustard in high concentrations. Mustard gas is actually not a gas but a high-boiling liquid. It is, however, so designated, since when dispersed as a cloud of fine droplets it remains in the air over a battleground for some time. It penetrates the skin easily, producing severe blisters. When inhaled, the heart and other internal organs are affected.

SUMMARY

Amines. General formula $R_1R_2R_3N$ (R_1, R_2, and R_3 = alkyl groups or hydrogen atoms). Primary, secondary, and tertiary amines are of the type $R{-}NH_2$, R_1R_2NH, and $R_1R_2R_3N$ respectively. Quaternary ammonium salts are of the type $[NR_1R_2R_3R_4]^+X^-$. Preparation: (1) Alkylation of ammonia. (2) Hofmann rearrangement involving an amide: $RCONH_2 \xrightarrow{NaOH+Br_2} RNH_2$. It is useful for descending a homologous series. (3) $LiAlH_4$ reduction of nitriles and amides. Commercially, methylamine and ethylamine are obtained from CH_3OH and C_2H_5OH respectively. Another process for CH_3NH_2 is by reaction of formaldehyde with ammonium chloride. Properties and reactions: Lower members are gases with peculiar odours. Basicity of amines due to lone pair of electrons on the nitrogen atom. Aliphatic amines more basic than ammonia. Reaction of aliphatic amines with nitrous acid serves to identify the class to which they belong: $RNH_2 \longrightarrow ROH\!:\!R_1R_2NH \longrightarrow R_1R_2NNO$; tertiary amines $\longrightarrow$ no reaction. Acylation of amines. Benzenesulphonyl derivatives used for separation and identification of different types of amines. Hofmann's carbylamine reaction of primary amines. $RNH_2 \longrightarrow R{-}NC$ (carbylamine). Exhaustive methylation of amines followed by degradation. Useful in structure determination.

Aliphatic Nitro Compounds. Type formula $RN\!\!\begin{smallmatrix}\nearrow O\\\searrow O\end{smallmatrix}$ Preparation:

(1) Reaction of RX with $AgNO_2$ (alkyl nitrite $R{-}O{-}NO$ side product). (2) Vapour-phase nitration of alkanes in industry. Nitro-paraffins are colourless, pleasant-smelling liquids. Tautomerism of nitro compounds.

Sulphur Compounds. Mercaptans (RSH) and sulphides ($R{-}S{-}R$). Preparation: (1) RX with NaHS. (2) ROH with P_2S_5. Salt formation characteristic of mercaptans. Progressive oxidation of mercaptans: $RSH \longrightarrow R{-}S{-}S{-}H$ and then to $R{-}SO_3H$. Alkyl sulphides prepared by adaptation of Williamson's synthesis of ether. Mustard gas, the war gas, is an alkyl sulphide $(ClCH_2CH_2)_2S$.

Problem Set No. 7

1. Indicate by equations the products obtained by treating: (*a*) *n*-propylamine with sodium hydroxide and chloroform; (*b*) propionamide with bromine and sodium hydroxide; (*c*) diethylamine with benzenesulphonyl chloride in the presence of sodium hydroxide; (*d*) ethylamine with nitrous acid; (*e*) triethylamine with ethyl iodide; (*f*) tetraethylammonium iodide with moist silver oxide; (*g*) propionamide with phosphorus pentoxide; (*h*) propyl iodide with silver nitrite; (*i*) *n*-butyramide with lithium aluminium hydride; (*j*) triethylamine with nitrous acid; (*k*) *l*-nitropropane with sodium methoxide.

2. How are the following transformations effected: (*a*) ethylamine from ethyl bromide; (*b*) ethylamine from methyl bromide; (*c*) Di-*n*-propyl sulphide from propanol-1?

3. Write down the products of Hofmann degradation of the following cyclic base:

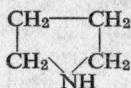

$$CH_2\text{———}CH_2$$
$$CH_2\diagdown\quad\diagup CH_2$$
$$NH$$

CHAPTER NINE

CARBOHYDRATES

The carbohydrates are an important class of compounds containing carbon, hydrogen, and oxygen. The last two elements, hydrogen and oxygen, are usually present in the same proportion as in water, namely two atoms of hydrogen to one of oxygen. They can be represented in general by the formula $C_x(H_2O)_y$, making it appear as though they are hydrates of carbon. The French called this family of compounds *hydrate de carbon*, and the English adopted the term carbohydrate. However, it is necessary to emphasize the fact that the chemical properties of these compounds show no similarity whatsoever to those of hydrates as a class. Further, many other compounds may be thought of which have no similarity in chemical properties to the carbohydrates but still can be represented by the molecular formula $C_x(H_2O)_y$. Examples are acetic acid CH_3COOH, for the present purpose written as $C_2(H_2O)_2$, and lactic acid $CH_3CHOHCOOH$, written as $C_3(H_2O)_3$. These substances also have no resemblance to the carbohydrates. Nevertheless, the name carbohydrates has come to stay and denotes a certain class of organic compounds occurring extensively in nature. From a structural point of view, this class comprises polyhydroxy aldehydes or polyhydroxy ketones or substances which give rise to these two types of compounds on hydrolysis.

The carbohydrates, fats, and proteins form the most important classes of food. The carbohydrates are not merely articles of food. They have found a wide variety of uses in industry. The cotton industry, the paper industry, the production of rayon, the fermentation industries, and so on bear witness to the usefulness of the carbohydrates. All these industries lean heavily on the availability of a large supply of carbohydrates. Nature, in her own munificent manner, has provided in growing plants an almost inexhaustible supply of carbohydrates. In order to be able to take advantage of such a large supply of carbohydrates from nature, and to fashion them for specific purposes and needs, an understanding of their properties and molecular structure is quite essential. Knowledge of the carbohydrates was meagre and fragmentary until about 1880. In the last decades of the nineteenth century, thanks mainly to the brilliant and pioneering investigations of the great German chemist, EMIL FISCHER, the structures of several carbohydrates were elucidated and their properties explained.

CLASSIFICATION: The carbohydrates usually have the termination *-ose*. Thus the words triose, tetrose, pentose, hexose, etc., denote the number of carbon atoms forming a straight chain in the carbohydrate. The carbohydrates may be aldehydic or ketonic, and therefore the appropriate prefix *aldo* or *keto* is also used. Therefore when we speak of an aldopentose we refer to an aldehydic carbohydrate containing five carbon atoms. Similarly, a ketohexose denotes a ketonic carbohydrate containing six carbon atoms.

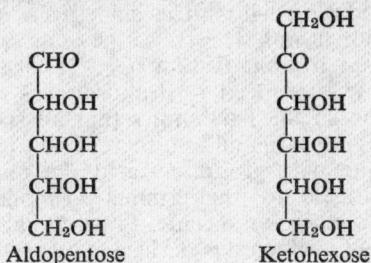

```
                              CH₂OH
                               |
        CHO                    CO
         |                     |
        CHOH                   CHOH
         |                     |
        CHOH                   CHOH
         |                     |
        CHOH                   CHOH
         |                     |
        CH₂OH                  CH₂OH
      Aldopentose            Ketohexose
```

These carbohydrates which cannot be hydrolysed to other carbohydrates of smaller carbon content are called monosaccharides or sugar units. In nature two or more such monosaccharide units are joined together by the loss of water molecules. Carbohydrates consisting of two monosaccharide units are called disaccharides, those containing three monosaccharide units trisaccharides, and those containing several units polysaccharides. In general, the monosaccharides and disaccharides are crystalline, water soluble, ether insoluble, and sweet to taste. The polysaccharides, on the other hand, are usually powdery or amorphous, insoluble in water and ether, and tasteless. The most common examples of polysaccharides are starch and cellulose.

MONOSACCHARIDES

Glucose is the most important of the monosaccharides. The elucidation of the structure of glucose constitutes a fascinating chapter in organic chemistry. It is illustrative of the methods and logic employed by the organic chemist in the determination of the structure of any unknown organic substance and will be dealt with in some detail.

What does a chemist do when he wants to find the structure of a compound? He first purifies it and then analyses it qualitatively and quantitatively for the elements in the compound to learn which elements are present and in what proportion. Having done that, he calculates its empirical formula, and after determining its molecular weight ascertains its molecular formula. The molecular formula of glucose is $C_6H_{12}O_6$.

The chemist then must ascertain the nature of the functional groups present in glucose. Glucose when acetylated with acetic anhydride gives a penta-acetyl derivative, indicating that five hydroxyl groups are present in the molecule. This knowledge coupled with the fact that two hydroxyl groups cannot be attached to the same carbon atom makes it imperative that of the six carbon atoms in the glucose molecule five carry one hydroxyl group each. Glucose has properties characteristic of a carbonyl group, as is evidenced by the fact that it reacts with hydroxylamine, phenylhydrazine, hydrogen cyanide, etc. It reduces Tollen's reagent and Fehling's solution, and in addition is oxidized by bromine water to give gluconic acid, a monocarboxylic acid, without loss of any carbon atom. This points to the fact that the carbonyl group in glucose is aldehydic in nature. The fact that it adds on only one mole of hydrogen cyanide shows that it contains only one aldehydic group. Gluconic acid on further oxidation with nitric acid gives a dicarboxylic acid without loss of carbon atoms. Therefore, the other group that gives rise to the second carboxyl group must be a $-CH_2OH$ or a primary alcoholic group. When glucose is reduced with phosphorus and hydriodic acid normal hexane is obtained. This shows that all six carbon atoms in glucose form a straight chain and that there is no forking or branching. The monovalent aldehyde group, as also the monovalent primary alcoholic group, must occupy the terminal positions of the six-carbon straight chain in the glucose molecule. From the above discussion the skeleton formula (I) for glucose is derived.

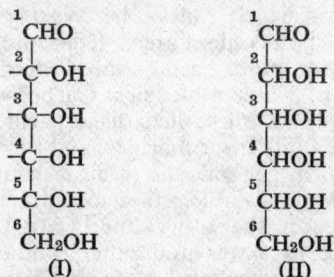

$$
\begin{array}{ll}
\overset{1}{C}HO & \overset{1}{C}HO \\
\overset{2}{-}C-OH & \overset{2}{C}HOH \\
\overset{3}{-}C-OH & \overset{3}{C}HOH \\
\overset{4}{-}C-OH & \overset{4}{C}HOH \\
\overset{5}{-}C-OH & \overset{5}{C}HOH \\
\overset{6}{C}H_2OH & \overset{6}{C}H_2OH \\
\quad (I) & \quad (II)
\end{array}
$$

This skeleton accounts for all but four hydrogen atoms in the molecular formula. To conform with valence rules the remaining four hydrogen atoms are attached to carbon atoms 2, 3, 4, and 5, each of which has one valence free. The picture of the glucose molecule as it emerges now is shown in (II) above.

The representation (II) is still far from complete. The final structure of the glucose molecule must also indicate the relative orientation (or configuration) of the four $-OH$ groups around the four central carbon atoms numbered 2, 3, 4, and 5. These four carbon atoms are dissimilar asymmetric carbon atoms, and hence glucose must exist in 2^4 or 16 optically isomeric forms. Three of these isomers (glucose, mannose, and glactose) occur in nature, and the other thirteen have been synthesized. Assigning the correct configuration to these sixteen isomers is no mean task. Even though it is beyond the scope of this book to deal in detail with work which preceded the establishment of the configuration of all the hexoses, the salient arguments used to fix the relative orientation of the $-OH$ groups in glucose may be briefly mentioned.

Before discussing the configuration of glucose it is necessary to refer to certain conventions used to represent configurations of asymmetric compounds. It has been established that there is absolutely no correlation between the sign of rotation and the absolute configuration of an optically active compound. For example, a dextro acid, its salts, esters, or other derivatives still containing the asymmetric carbon atom (or atoms) present in the parent acid need not all exhibit dextro rotation, even though they belong to the same stereochemical series. In order to emphasize relationships in absolute configurations between different optically active compounds, Emil Fischer introduced a certain convention whereby dextro (+) glyceraldehyde was selected as the standard to which the configuration of other optically active compounds could be referred. It was arbitrarily assigned the tetrahedral configuration (*a*), wherein the line connecting H and OH is one edge of the tetrahedron in front of or above the plane of the paper, and the dotted line connecting $-CHO$ and $-CH_2OH$ is the edge of the tetrahedron which will be hidden from view in a solid model.

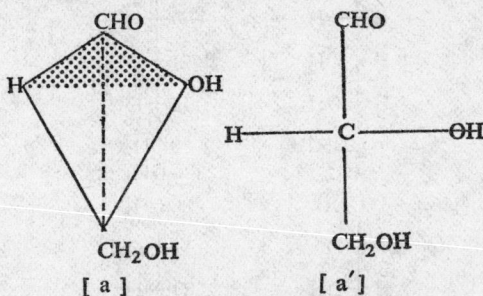

Fig. 12. D-(+)-Glyceraldehyde.

The same when projected on the plane of the paper appears as (*a'*). Similarly, L-glyceraldehyde has the tetrahedral configuration (*b*) and the projection formula (*b'*).

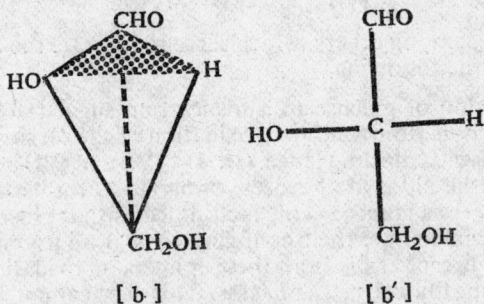

Fig. 13. L-(+)-Glyceraldehyde.

Optically active compounds that have been established to contain an asymmetric carbon atom with the same configuration as that of D-glyceraldehyde are represented as belonging to the same stereochemical series by the prefix D (pronounced 'dee' and not dextro). Similarly, configurations of compounds related to L-(—)-glyceraldehyde are indicated by the prefix L (pronounced 'ell' and not laevo). The sign of rotation, which, as said before, need not be the same for all the members of a stereochemical series, is indicated by the symbols + or —.

Experimentally, natural glucose was found to have the same configuration around C_5 as in D-(+)-glyceraldehyde, and hence is described as D-(+)-glucose. Of the sixteen theoretically possible isomers, only eight will have a configuration at C_5 corresponding to that of D-(+)-glyceraldehyde, i.e. belong to the D series of hexoses, while in the other eight the configuration at C_5 will be the same as in L-(−)-glyceraldehyde. Glucose then must have one of the following eight configurations that are possible for D-hexoses:

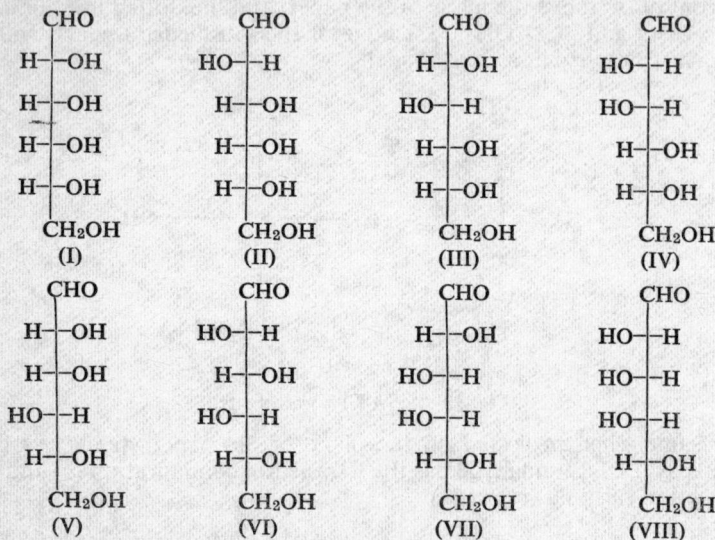

CHO	CHO	CHO	CHO
H——OH	HO——H	H——OH	HO——H
H——OH	H——OH	HO——H	HO——H
H——OH	H——OH	H——OH	H——OH
H——OH	H——OH	H——OH	H——OH
CH₂OH	CH₂OH	CH₂OH	CH₂OH
(I)	(II)	(III)	(IV)

CHO	CHO	CHO	CHO
H——OH	HO——H	H——OH	HO——H
H——OH	H——OH	HO——H	HO——H
HO——H	HO——H	HO——H	HO——H
H——OH	H——OH	H——OH	H——OH
CH₂OH	CH₂OH	CH₂OH	CH₂OH
(V)	(VI)	(VII)	(VIII)

The arguments, among others, which serve to indicate the configuration of glucose are the following:

1. Conversion of glucose to a five-carbon sugar (after loss of the aldehydo carbon atom) and then oxidation of the C_5 sugar furnish an optically active C_5 diacid. Hence, (I), (II), (V), and (VI) are eliminated.

2. One of the aldehyde hexoses, D-mannose, forms the same osazone (see later) as glucose. This fact indicates that glucose and mannose are 2-epimers, i.e. their configurations at all asymmetric atoms are the same except at C_2. Both these epimers on oxidation with nitric acid give optically active C_6-diacids. This fact cannot be accounted for by the structures (VII) and (VIII), which though epimeric at C_2 will give an optically active acid on oxidation only in the case of (VIII) and not in the case of (VII).

3. A transposition of the terminal −CHO and −CH₂OH groups in glucose by chemical means results in the formation of a new aldo hexose (L-gulose). Such a transposition of the groups in the case of structure (IV) does not result in the formation of a new sugar. This can be seen by rotating the configuration of the transposed structure in the plane of the paper. Hence glucose must have the configuration shown in (III), whereas structure (IV) must be assigned to D-mannose.

The configurations of the other naturally occurring hexoses D-(+)-galactose and D-(−)-fructose have been arrived at similarly and are represented by (VII) above and (IX) below respectively.

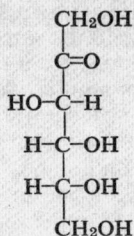

$$
\begin{array}{c}
CH_2OH \\
| \\
C=O \\
| \\
HO-C-H \\
| \\
H-C-OH \\
| \\
H-C-OH \\
| \\
CH_2OH
\end{array}
$$

(IX) D-(−)-Fructose

REACTIONS: Glucose on catalytic reduction or reduction with sodium amalgam gives D-sorbitol.

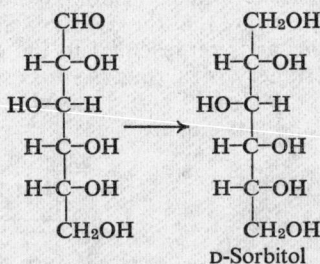

$$
\begin{array}{ccc}
CHO & & CH_2OH \\
| & & | \\
H-C-OH & & H-C-OH \\
| & & | \\
HO-C-H & \longrightarrow & HO-C-H \\
| & & | \\
H-C-OH & & H-C-OH \\
| & & | \\
H-C-OH & & H-C-OH \\
| & & | \\
CH_2OH & & CH_2OH
\end{array}
$$

D-Sorbitol

Whereas glucose reacts with hydroxylamine in the normal manner to give an oxime, its reaction with phenylhydrazine is somewhat unusual. Initially, the expected phenylhydrazone is formed. In the presence of excess phenylhydrazine two further molecules of the reagent are completely consumed to give the yellow crystalline glucosazone. The formation of the latter has been explained by assuming that after the phenylhydrazone formation oxidation of the second carbon of the chain to a carbonyl takes place, followed in turn by the usual reaction of a keto group.

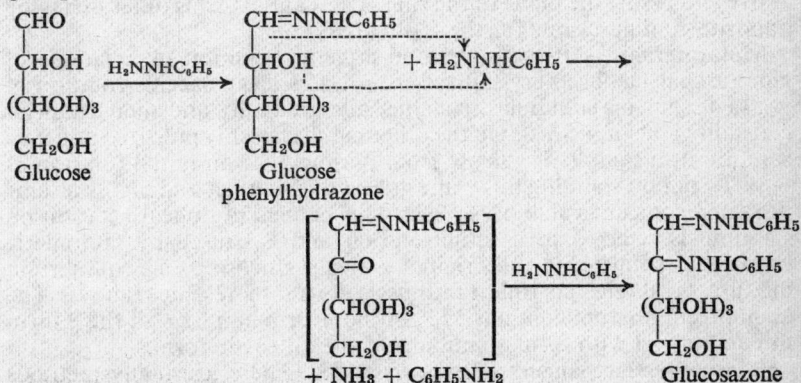

$$
\begin{array}{ccc}
CHO & & CH=NNHC_6H_5 \\
| & & | \\
CHOH & \xrightarrow{H_2NNHC_6H_5} & CHOH \qquad + H_2NNHC_6H_5 \longrightarrow \\
| & & | \\
(CHOH)_3 & & (CHOH)_3 \\
| & & | \\
CH_2OH & & CH_2OH \\
\text{Glucose} & & \text{Glucose} \\
& & \text{phenylhydrazone}
\end{array}
$$

$$
\begin{bmatrix}
CH=NNHC_6H_5 \\
| \\
C=O \\
| \\
(CHOH)_3 \\
| \\
CH_2OH
\end{bmatrix}
\xrightarrow{H_2NNHC_6H_5}
\begin{array}{c}
CH=NNHC_6H_5 \\
| \\
C=NNHC_6H_5 \\
| \\
(CHOH)_3 \\
| \\
CH_2OH \\
\text{Glucosazone}
\end{array}
$$

+ NH_3 + C_6H_5NH_2

The word osazone is derived from the terminations of the words hex*ose* and hydra*zone*. The osazones are all bright yellow crystalline compounds with unique crystalline forms and with definite melting points. They are very useful derivatives for characterization of sugars. Glucose and mannose give rise to the same osazone, indicating the same configuration at all carbon atoms except C_2. Fructose (IX) also gives the same phenylosazone, even though it is a ketone, because it has the same structure as glucose and mannose below C_2. In the case of glucose the aldehyde group first condenses with phenylhydrazine, and a new keto group is produced in the second stage, which in turn condenses with another molecule of phenylhydrazine to give the osazone. In the case of fructose a keto group first condenses with phenylhydrazine, then a new aldehyde group is produced (by the oxidation of the $-CH_2OH$ group), which in turn condenses with another molecule of phenyl-hydrazine.

Tautomerism of Glucose. Glucose exhibits tautomerism of a cyclic type made possible by the fact that the chain of carbon atoms in glucose is not actually straight, as is represented in the open chain form, but bends around the periphery of a ring. As a consequence, the $-CHO$ group is very close to the C_5-OH and reacts with it to give two stereo-isomers known as α- and β-D-glucose.

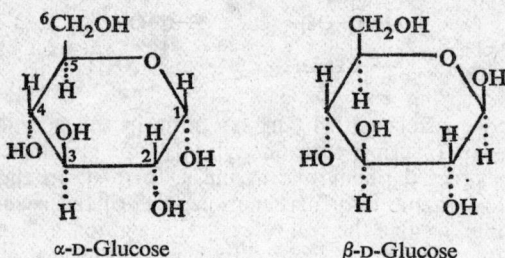

α-D-Glucose   β-D-Glucose

The rings are assumed to lie in a plane at right angles to that of the paper, with the thicker bonds nearer the reader and the substituents above and below the plane of the ring. The α and β forms differ in having opposite configurations for the C_1-hydroxyl.

Mutarotation. A freshly prepared aqueous solution of a sample of glucose that has been crystallized from water has a specific rotation of $+113\cdot4°$, but on standing this value falls gradually and then attains a constant value of $+52\cdot2°$. On the other hand, a freshly prepared aqueous solution of a sample of glucose from pyridine has an initial rotation of $+19\cdot7°$, but on standing this value gradually increases and finally attains the same constant value of $+52\cdot2°$. This change in optical rotation on standing is referred to as mutarotation and is caused by tautomeric changes in solution in the original α and β glucose to an equilibrium mixture of all the possible tautomeric forms represented above. The equilibrium mixture contains 37% of the α form and 63% of the β form in equilibrium with a small amount of the aldehyde form.

Interconversions among Monosaccharides. There are many methods known for effecting the conversion of one monosaccharide to another either by: (*a*) removing carbon atoms from a chain; (*b*) lengthening the chain by introducing additional carbon atoms; or (*c*) changing an aldose to a ketose.

(*a*) *Conversion of a Higher to a Lower Aldose*

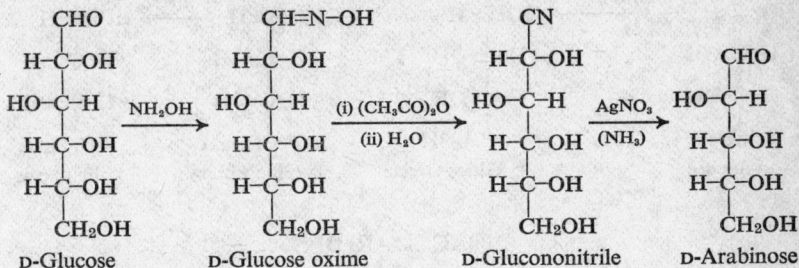

$$
\begin{array}{c}
\text{CHO} \\
\text{H-C-OH} \\
\text{HO-C-H} \\
\text{H-C-OH} \\
\text{H-C-OH} \\
\text{CH}_2\text{OH}
\end{array}
\xrightarrow{\text{NH}_2\text{OH}}
\begin{array}{c}
\text{CH=N-OH} \\
\text{H-C-OH} \\
\text{HO-C-H} \\
\text{H-C-OH} \\
\text{H-C-OH} \\
\text{CH}_2\text{OH}
\end{array}
\xrightarrow[\text{(ii) H}_2\text{O}]{\text{(i) (CH}_3\text{CO)}_2\text{O}}
\begin{array}{c}
\text{CN} \\
\text{H-C-OH} \\
\text{HO-C-H} \\
\text{H-C-OH} \\
\text{H-C-OH} \\
\text{CH}_2\text{OH}
\end{array}
\xrightarrow[\text{(NH}_3\text{)}]{\text{AgNO}_3}
\begin{array}{c}
\text{CHO} \\
\text{HO-C-H} \\
\text{H-C-OH} \\
\text{H-C-OH} \\
\text{CH}_2\text{OH}
\end{array}
$$

D-Glucose   D-Glucose oxime   D-Gluconotrile   D-Arabinose

The aldose is converted to its oxime, which, when dehydrated with acetic anhydride, gives a penta-acetyl nitrile. Hydrolysis removes the acetyl groups to give a nitrile with which ammoniacal silver nitrate loses hydrogen cyanide to give the next lower aldose.

(*b*) *Conversion of a Lower to a Higher Aldose*

An aldose is converted to its cyanohydrin, which on hydrolysis gives an acid. The latter is converted to its γ-lactone, which on reduction with sodium amalgam gives the higher aldose.

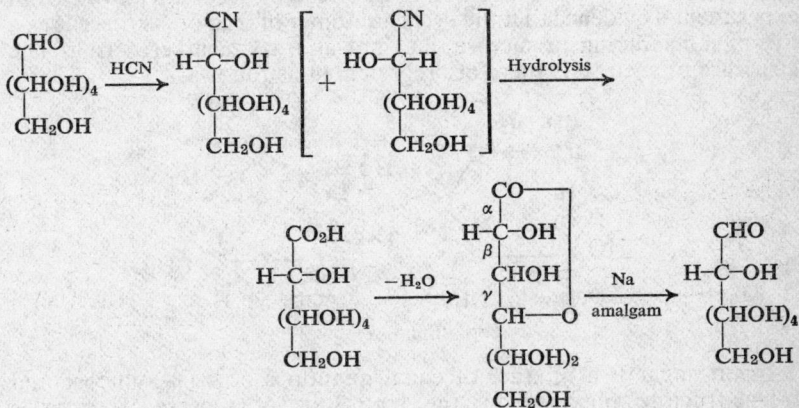

$$
\begin{array}{c}
\text{CHO} \\
\text{(CHOH)}_4 \\
\text{CH}_2\text{OH}
\end{array}
\xrightarrow{\text{HCN}}
\left[
\begin{array}{c}
\text{CN} \\
\text{H-C-OH} \\
\text{(CHOH)}_4 \\
\text{CH}_2\text{OH}
\end{array}
+
\begin{array}{c}
\text{CN} \\
\text{HO-C-H} \\
\text{(CHOH)}_4 \\
\text{CH}_2\text{OH}
\end{array}
\right]
\xrightarrow{\text{Hydrolysis}}
$$

$$
\begin{array}{c}
\text{CO}_2\text{H} \\
\text{H-C-OH} \\
\text{(CHOH)}_4 \\
\text{CH}_2\text{OH}
\end{array}
\xrightarrow{-\text{H}_2\text{O}}
\begin{array}{c}
\text{CO} \\
\alpha \\
\text{H-C-OH} \\
\beta \\
\text{CHOH} \\
\gamma \\
\text{CH---O} \\
\text{(CHOH)}_2 \\
\text{CH}_2\text{OH}
\end{array}
\xrightarrow[\text{amalgam}]{\text{Na}}
\begin{array}{c}
\text{CHO} \\
\text{H-C-OH} \\
\text{(CHOH)}_4 \\
\text{CH}_2\text{OH}
\end{array}
$$

In the addition of hydrogen cyanide, two stereoisomers may be formed, differing in the configuration around the new centre of asymmetry. Each of the cyanohydrins will, when subjected to the above sequence of reactions, furnish a different higher aldose. These two higher aldoses are 2-epimers of each other.

(*c*) The osazone of an aldohexose is hydrolysed to give a compound called **osone**, which on reduction with zinc and hydrochloric acid gives a ketose. For example, D-(+)-glucose is converted to D-(−)-fructose by the following sequence of reactions:

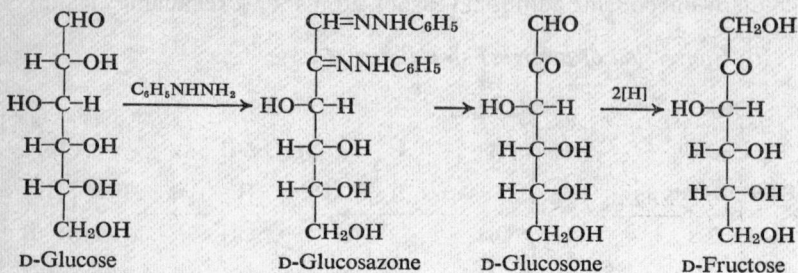

D-Glucose   D-Glucosazone   D-Glucosone   D-Fructose

DISACCHARIDES

Sucrose or cane sugar is the commonest of the disaccharides. It has the molecular formula $C_{12}H_{22}O_{11}$. It does not form a phenylhydrazone or exhibit carbonyl properties. On hydrolysis with dilute acids, it yields one molecule of D-(+)-glucose and one molecule of D-(−)-fructose. It is not a reducing sugar. It has neither the aldehydic nor the ketonic group among its two components and shows no mutarotation. Hence the potential aldehyde group (CHOH) in position 1 in the cyclic form of glucose and the potential keto group (C−OH) of fructose in position 2 of the cyclic form must have interacted with the elimination of a molecule of water and formation of an oxide bridge. It is known from experimental evidence that the cyclic tautomer of fructose is present as a five-membered ring in sucrose, and not as a six-membered ring. The structure of sucrose is therefore represented as follows:

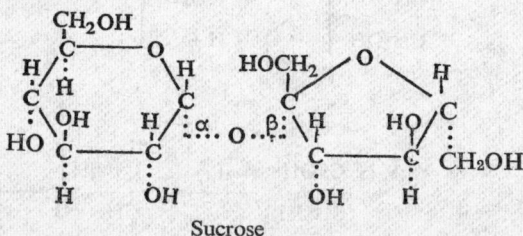

Sucrose

Invert sugar is a mixture of equal quantities of D-(+)-glucose and D-(−)-fructose obtained by the hydrolysis of sucrose. Sucrose is dextrorotatory (+66·4°), and the rotation becomes negative during hydrolysis, because the negative rotation of fructose is greater than the positive rotation of glucose.

Maltose is a disaccharide obtained by the partial hydrolysis of starch. On hydrolysis, maltose gives two molecules of glucose. It is a reducing sugar and contains a potential aldehyde group. Structurally it is glucose-4-α-glucoside. A glycoside is an ether obtained by combination of a sugar with a hydroxyl compound by loss of water. The specific names glucoside, galactoside, etc., indicate that the sugar unit is glucose, galactose, etc., respectively.

Lactose or milk sugar is obtained from the whey of milk by evaporation. On hydrolysis, it gives glucose and galactose (C_4 epimer of glucose). Lactose is a reducing sugar and forms an osazone. Therefore, at least one carbonyl group is present, and this has been found to be in the glucose half of the molecule. Actually, lactose has been established to be glucose-4'-β-galactoside. It is used in baby foods and in pharmaceutical preparations.

POLYSACCHARIDES

The disaccharides consist of two monosaccharides united through an oxygen atom as in sucrose. The polysaccharides contain several such units combined to form high-molecular-weight substances. The monosaccharides may be pentoses, as in xylan of wood, hexoses, as in starch or cellulose, or a mixture of pentoses or hexoses, as in wood gums. The polysaccharides are not soluble in water, though starch forms a dispersed colloidal solution.

Starch is found in almost all plants. It is stored in all grains and tubers as a future food supply for the germinating seeds. Starch granules from different sources vary considerably in size, but are characteristic of the plants they originate from. The granules of potato starch are coarse, whereas those of rice are fine. When starch granules are heated with water a colloidal solution is formed in which part of the starch is in solution. The water-soluble portion, called amylose, has a molecular weight of 50,000–200,000 and gives a blue colouration with iodine. The water-insoluble portion, called amylopectin, amounts to 80–90% of the starch granules, has a molecular weight in the range 50,000–1,000,000, and gives a red colouration with iodine. Upon acid hydrolysis, starch (both amylase and amylopectin) yields, as the ultimate product, glucose. Since starch has no reducing properties, the potential aldehyde group of the glucose unit must be locked up in joining the sugar units. Actually the oxide linking in starch has been shown to be between C_1 of one glucose unit and C_4 of the next unit.

Dextran is a polysaccharide which differs from starch in that its oxide linkage is between C_1 and C_6 of successive glucose units. It is produced by fermenting sucrose to a polysaccharide of very high molecular weight, which is then partially hydrolysed to fragments in the molecular-weight range of 60,000–250,000. Dextran is used as a substitute for blood plasma, and has an advantage over plasma, because it can be preserved without refrigeration for a long period of time.

Inulin, another polysaccharide, is found in varying quantities in potatoes and Jerusalem artichokes. But most of it is produced from dahlia tubers. Inulin gives a faint yellow colouration with iodine. On hydrolysis, it gives exclusively *P*-(−)-fructose. Just as starch is a polysaccharide having repeating glucose units, inulin is a polysaccharide having repeating fructose units. Inulin is the chief source of fructose. Animals store carbohydrate in the form of glycogen, which has a molecular weight of about 2,000 and gives a red colouration with iodine. It hydrolyses very easily to D-glucose. The storage of this material in the liver serves as a readily available reserve supply of glucose to the blood stream.

Cellulose. Cellulose is the most widely distributed of the polysaccharides. It constitutes about 50% of wood and about 90% of cotton. Cellulose is the chief constituent of the cell wall of plants; in other words, it is the framework of the vegetable kingdom. The naturally occurring cellulose may be purified to free it from waxes, fats, and pectin-like substances. These impurities are removed by washing the cellulose with alcohol and ether and then treating with alkali. The absorbent cotton used in hospitals has been purified in this manner.

A series of prolonged investigations spread over several years led to the conclusion that its molecular structure consists of a long chain of glucose units linked together in the 1 and 4 positions by elimination of water. It is also known that the union of the sugar units is β-glucosidic. Many bacteria decompose cellulose, but man and other carnivorous animals are unable to digest it because they lack the enzymes necessary for its hydrolysis. The enzyme emulsin easily hydrolyses β-glucosides, and hence serves to detect the presence of β-glucosidic linkages in polysaccharides. Cellulose is not a uniform chemical individual, and can be divided into at least two groups, α and β. β-Cellulose dissolves in 17–18% sodium hydroxide solution, whereas α does not. α-Cellulose is the type that is used for the manufacture of rayons. Hydrolysis of either variety of cellulose is difficult, and the conversion of cellulose, like that of starch, goes through a number of intermediate stages, and immediately prior to the formation of glucose gives a reducing disaccharide, cellobiose. Cellobiose holds the same relationship to cellulose that maltose holds to starch. It must be noted that while maltose is a glucose-4-α-glucoside, cellobiose is a glucose-4-β-glucoside.

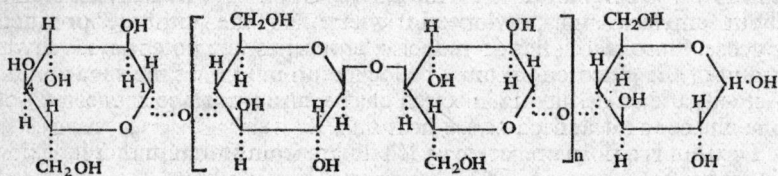

Cellulose

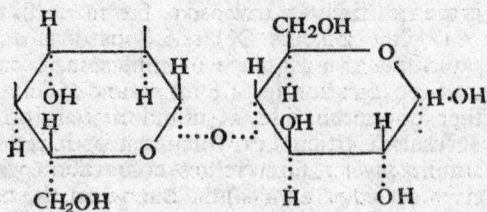

Cellobiose

Cellulose is the raw material for a variety of commercial products. Cellulose for paper manufacture is obtained from wood by alkali extraction. The lignin portion is dissolved out and cellulose is obtained. The lignin (in solution), which was at one time a waste product of paper manufacture, has found use in the manufacture of plastics as a result of recent research. It is also worthwhile to mention that one of the products of hydrolysis of lignin is vanillin, the substance used for giving flavour to ice-cream. Much of the vanillin used is obtained from lignin, a by-product in the paper industry.

The alcoholic structure of cellulose may be expected to furnish esters and ethers. Several ethers and esters of cellulose have been prepared, and they have many industrial applications. Among the esters, the nitrate, the acetate, the butyrate, and the xanthate are important. The methyl, ethyl, and benzyl ethers of cellulose have been prepared and all are useful materials. Cellulose dissolves in cupric ammonium hydroxide solution (Schweitzer's solution) and is precipitated on the addition of acid. When paper is immersed in moderately concentrated sulphuric acid for a few seconds and washed immediately with water and dilute ammonia it is transformed into parchment paper, which when dried becomes stiff and is semi-waterproof. Such sheets are used for wrapping food products.

Cellulose Nitrate (Nitrocellulose). Cellulose contains three hydroxyl groups per sugar unit, and hence can be esterified to different degrees. The actual reaction of cellulose with nitric acid does not proceed in a discrete manner to give mono-, di-, and trinitrates. The degree of nitration (esterification with nitric acid) is measured by the percentage of nitrogen in the resulting ester. Complete nitration to cellulose tri-nitrate corresponds to a nitrogen content of 14.6% in the ester; cellulose dinitrate has 11.11% of nitrogen, and in the case of mononitrate the percentage of nitrogen is 6.76. Nitrated cellulose containing more than 13% nitrogen is explosive and is used in smokeless powders as gun-cotton. Two high explosives, nitroglycerine and gun-cotton, are blended with other materials to obtain formulations that have desirable burning characteristics. Cordite, one of the better known of such formulations, contains 30% nitroglycerine, 65% gun cotton, and 5% Vaseline. A paste of these ingredients is incorporated with organic solvents (usually acetone) and extruded through dies in the form of perforated rods or cords (hence the name cordite). These are suitably cut and then freed of acetone.

Pyroxylin (containing about $11-12\%$ nitrogen) roughly corresponds to cellulose dinitrate. It is not explosive, but is highly inflammable. It is soluble in inorganic solvents. The solutions of pyroxylin in organic solvents are graded according to their thickness or viscosity and are widely used in the manufacture of plastics and quick-drying lacquers. Films made by evaporation of such a solution are brittle and, therefore, to eliminate this undesirable property a plasticizer is used. When camphor is used as the plasticizer we get the familiar celluloid. Pyroxylin does not dissolve in ether or alcohol. But when the two solvents are mixed together in the proportion of three parts of ether to one of alcohol the mixture dissolves pyroxylin and the resulting product is called collodion. The addition of castor oil and camphor to collodion gives flexible collodion. This collodion is nicknamed 'new-skin' and is used as a protective coating for minor cuts and skin abrasions. When spread on the skin the solvent quickly evaporates to leave a colourless transparent film.

Cellulose triacetate is made by the action of acetic acid and acetic anhydride on cotton linters in the presence of sulphuric acid. The excess of the acetylating agent is then removed by washing with water. The cellulose acetate is used in making films, lacquer, and plastic objects in the same manner as cellulose nitrate. A solution of cellulose acetate in acetone is used as a cement or seal in constructing model aeroplanes. Partial hydrolysis of cellulose triacetate to an intermediate stage between the triacetate and the diacetate gives a product which is soluble in acetone and which is of greater industrial utility than the triacetate. The acetate is much less inflammable than the nitrate. It transmits ultraviolet light and is employed in place of glass in the windows of greenhouses.

The cellulose acetate rayon, Celanese, is made by forcing out a viscous solution of cellulose acetate in acetone through fine openings of a tube, called a spinneret, downward in a rising current of warm air. Acetone evaporates off, and the cellulose acetate obtained in the form of long continuous fibres is wound on spools. In the manufacture of plastics the moulding qualities of the butyrate are superior to those of the acetate.

Cellulose Xanthate. Most of the rayon on the market is made by the viscose process. Sodium hydroxide solution and cellulose react with

$$\overset{\text{S}}{\underset{\|}{}}$$

carbon disulphide to give a xanthate, $R(OCSNa)_n$, if $R(OH)n$ is taken to represent cellulose. The solution of xanthate is a viscous colloidal solution called viscose. When viscose is treated with acid cellulose is regenerated. If the viscose is extruded through a spinneret into an acid solution fine rayon filaments are formed. It is to be emphasized that in this process no chemical change has occurred, and the glittering shimmering rayon is really the good old cellulose but with a new look. The physical properties of cellulose have been modified. When the viscose is extruded on to rollers in an acid medium films, called cellophane, are deposited. Cellophane is often used as a packing material.

Cellulose reacts with ethyl chloride or ethyl sulphate in the presence of alkali to give ethyl ethers corresponding to different degrees of etherification. These ethers are extensively used in the manufacture of plastics.

SUMMARY

Occurrence and importance of carbohydrates. Classification into mono-, di-, and polysaccharides. Aldoses are aldehydo and ketoses keto monosaccharides. Open structure of glucose based on: (i) presence of 5-OH groups as determined by acetylation. (ii) Formation of carbonyl derivatives. (iii) Reduction of Fehling's solution, etc. (iv) Reduction of glucose to *n*-hexane. (v) Oxidation to produce a dicarboxylic acid still containing all the carbon atoms intact. Optically active compounds of the same configuration do not all have same sign of rotation, and hence need for a standard to which the configurations could be referred. D-Glyceraldehyde as standard (H to left and $^-$OH to the right). C_5 in glucose has the same configuration as D-glyceraldehyde and hence D-glucose. Determination of the relative configuration of $^-$OH groups in glucose. Reactions: osazones formed by reaction between 3 moles of phenylhydrazine and 1 of aldose. Useful for identification purpose. Fructose, glucose, and mannose form same osazone showing carbon atoms other than C_1 and C_2 are the same. Oxide structure of glucose. Mutarotation of freshly prepared α and β forms. Interconversions among monosaccharides.

Disaccharides. Sucrose derived from one mole of D-(+)-glucose and D-(−)-fructose. Not a reducing sugar, and hence linking of C_1 of glucose moiety with C_2 of fructose moiety. Invert sugar. Maltose and lactose. Their structures. Glycosides are ethers obtained by combination of sugars with other $^-$OH containing compounds.

Polysaccharides. These contain many units of monosaccharides. Starch: can be separated into water-soluble amylose and water-insoluble amylopectin. Glucose obtained on drastic hydrolysis of starch and maltose on enzymatic hydrolysis. Hence starch is composed of glucose units. Nature of this linking. Dextran used as a substitute for blood plasma and inulin. Cellulose: widely distributed in nature and of great industrial importance. Obtained from wood by removing lignin by extraction with alkali. Partial hydrolysis of cellulose gives cellobiose, while complete hydrolysis gives glucose. Hence the presence of cellobiose units. Commercial products from cellulose. Cellulose trinitrate, an explosive. Cordite and its composition. Pyroxylin (cellulose dinitrate) used for making plastics and lacquers for drying, celluloid, etc. Cellulose triacetate used in making films, plastics, etc. Acetate rayon (Celanese) and viscose rayon, cellulose ethers.

Problem Set No. 8

1. What are the products obtained in the following reactions?
   (a) D-glucose is treated with bromine water. (b) D-glucose is treated with acetic anhydride and sodium acetate. (c) D-fructose is treated with excess of phenylhydrazine. (d) D-glucose is treated with hydrogen cyanide.
2. Give examples of: (I) Epimers; (II) Osone; (III) Mutarotation; (IV) an L-aldehydopentose; (V) a reducing sugar; (VI) a glucoside.
3. How are the following transformations carried out?

   (a) D-glucose to D-arabinose.
   (b) D-arabinose to D-glucose.
   (c) Sucrose to invert sugar
   (d) Starch to maltose.

CHAPTER TEN

ALICYCLIC COMPOUNDS

To this point we have studied compounds with a number of carbon atoms joined in a chain as, for example, C–C–C–C–C–. The chain can be transformed into a ring by bringing the end carbon atoms together. If we start with the compound 1,3-dichloropropane the end chlorine atoms can be eliminated by reaction with sodium as follows:

$$CH_2\Big\langle\begin{matrix}CH_2Cl\\CH_2Cl\end{matrix} + 2Na \longrightarrow CH_2\Big\langle\begin{matrix}CH_2\\ \\CH_2\end{matrix} + 2NaCl$$

This is a Wurtz-type reaction. Similarly, 1,4-dibromobutane can give rise to cyclobutane. These ring structures are nothing more than the paraffin straight-chain structures tied together at the ends of the chain. Therefore they are called **cycloparaffins**. During the process of tying up the end carbon atoms of a chain it becomes necessary to remove two hydrogen atoms, one from each end. Hence the general molecular formula for the cycloparaffins, C_nH_{2n}, is two hydrogen atoms less than that of the alkanes. This general formula is the same as the one for alkenes. They are thus structural isomers of the alkenes but have different chemical properties. Cycloparaffins and their derivatives are in fact more akin to open-chain saturated aliphatic compounds, and since they have cyclic structures, they are also referred to as alicyclic compounds.

The cycloparaffins are found in large quantities in Russian petroleum and in the petroleum obtained in California and other western states in the United States. The cycloparaffins occurring in petroleum are referred to in petroleum technology as the **naphthenes**. Before giving an account of the preparation and properties of the cycloparaffins, certain speculations of BAEYER on the stability and the ease of formation of ring systems referred to as Baeyer's strain theory may be dealt with. This theory has been modified by SACHSE and MOHR, and in the modified form the theory gives an excellent picture of the general behaviour of ring compounds with particular reference to their stability and ease of formation.

It has been stated that glutaric and succinic acids readily give cyclic anhydrides containing five and six members in the ring, while adipic and higher acids give rise to linear anhydrides. There are several other instances which suggest the peculiar stability of five- and six-membered rings. We shall first consider the exact significance of the term **straight-chain compound**. We shall imagine the building up of a straight-chain hydrocarbon pentane using five carbon atoms. We shall bear in mind the tetrahedral nature of the carbon valences. When two carbon atoms are united by a single bond one valence of each carbon atom is utilized

152

for the formation of the bond. The two-carbon system may be represented as

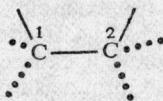

where one of the dotted bonds on each carbon atom is above and the other below the plane of the paper and the solid bonds lie in the plane of the paper. If six hydrogen atoms are attached to the six valences we get ethane. If one more carbon atom is joined to the above two-carbon system we get the pattern given below:

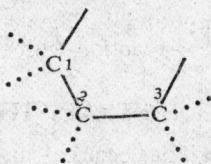

In this three-carbon propane skeleton we find the molecule developing quite an angle due to the tetrahedral disposition of the valences in space. A five-carbon skeleton is similarly found to assume quite a zigzag pattern.

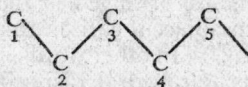

The straight chain does not seem to be so straight after all! The zigzag shape of a carbon chain follows as a result of the tetrahedral theory, and X-ray evidence has proved the essential correctness of this idea. The zigzag pattern of, for example, the pentane molecule is the normal or the ground state of the molecule. During a chemical reaction, when energy is supplied, the zigzag pattern by a single rotation of the bond between C_2 and C_3 can take up a winding pattern, as shown below:

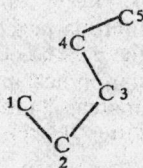

When the pentane molecule is more agitated, as when more energy is supplied in a reaction, rotation between C_3 and C_4 will give the ring pattern shown below:

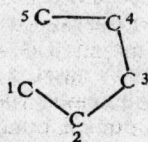

In this arrangement carbon atoms 1 and 5 are in close proximity. They seem set for direct linking and a consequent closure of the ring, given sufficient temptation. Thus, for example, 1,5-dichloropentane on the straight-chain model is $ClCH_2CH_2CH_2CH_2CH_2Cl$. In a chemical reaction, given sufficient stimulus, it can rearrange itself into the ring pattern and then react with sodium as shown:

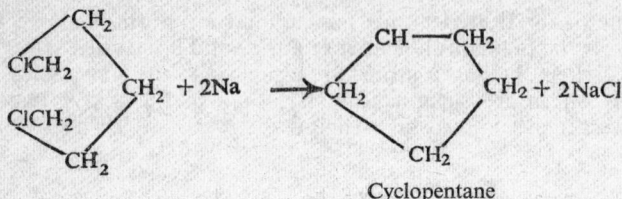

Cyclopentane

BAEYER'S STRAIN THEORY

On the tetrahedral model it has been calculated that the angle between the valence bonds is 109° 28'. Baeyer suggested that any pattern of a molecule which requires for its formation a distortion of this angle imposes strain on the molecule and consequent instability. For example, in the case of cyclopropane containing three carbon atoms in the ring, if the three carbon atoms are arranged in the form of an equilateral triangle the angle between adjacent carbon-to-carbon valences must be narrowed from 109° 28' to 60°. In other words, the angle of deviation for each C–C bond from its tetrahedral disposition is (109° 28' — 60°) divided by 2, or 24° 44'. In the case of cyclobutane this angle of deviation is 9° 44' and in the case of cyclopentane it is 0·44'.

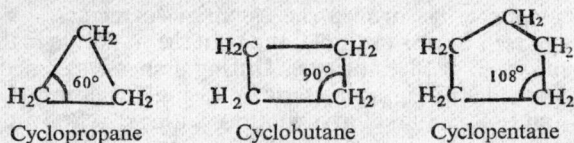

Cyclopropane   Cyclobutane   Cyclopentane

In the case of cyclopentane there is practically no departure from the normal tetrahedral angle. In the case of cyclohexane the angle of deviation is —5° 16', the negative sign indicating that the C–C bonds have to be widened to satisfy the geometry of the ring.

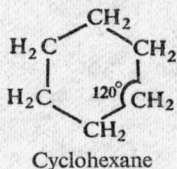

Cyclohexane

In the case of seven-membered rings, the angle of deviation is —9° 33', and this angle continuously increases with an increasing number of carbon atoms in the ring. A double bond may be considered a two-membered ring formed by the coalescence, as shown below, of two single bonds, one each from two carbon atoms linked to each other by a single bond.

In other words, the angle between the bonds constituting the double bond is 0°, and the bond angle deviation is (109° 28′ − 0) divided by 2 or 54° 44′.

Baeyer proposed the theory that the stability of a ring compound is dependent on the angle through which the valence bond must be bent away from its normal tetrahedral disposition. The greater this angle, the greater the strain and the less the stability of the compound. Judging from the angles of deviation given above, ethylene should be the most unstable. The great strain present in the ethylene molecule is reflected in its high reactivity in addition reactions, whereby the strain is released by the opening of the double bond with the formation of a strainless single bond. In the case of cyclopropane, which has a high degree of strain (24° 44′), the molecule may be expected to open up on the slightest provocation, thus releasing the strain within it. This expectation, based on Baeyer's theory, is actually substantiated by experiment. When cyclopropane is treated with chlorine the ring opens up readily to give 1,3-dichloropropane.

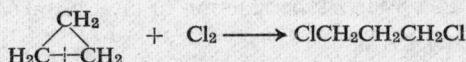

$$\begin{array}{c} CH_2 \\ H_2C{-}CH_2 \end{array} + Cl_2 \longrightarrow ClCH_2CH_2CH_2Cl$$

Compounds containing cyclobutane rings, though they do not open up so readily, are more reactive than five- and six-membered ring compounds. In cyclopentane and cyclohexane the rings, as is to be expected, according to Baeyer's theory, are remarkably stable, and no fission takes place.

Baeyer's theory infers that rings of more than six carbon atoms are unstable because of increasing deviation from the tetrahedral angle. However, it has been found that the perfume base muscone (from musk) and civetone (from civet) contain 15 and 17 carbons respectively in their rings. These and other large ring compounds not only occur in nature as quite stable compounds but have also been prepared in the laboratory by synthesis. Rings containing as many as 32 carbon atoms have been prepared by RUZICKA, and they were found to be quite stable, contrary to the expectations from Baeyer's strain theory. Again, Baeyer's theory requires cyclohexane to be less stable than cyclopentane. No such difference is observed, and there is nothing to distinguish the five-membered ring compounds from the six-membered ring compounds from the point of view of stability.

In order to explain the previous observations, which seem to militate against Baeyer's strain theory, SACHSE advanced a theory of non-planar strainless rings. Baeyer considered the rings to be planar, that is, he thought that the carbon atoms constituting the ring were all in one plane.

A planar ring of six carbon atoms with an internal angle of 120° requires that each C–C bond be bent outwards by 5° 16′ from its tetrahedral position, and hence is strained. Such a six-membered ring, Sachse pointed out, becomes absolutely free from strain if all the carbon atoms are not forced into one plane. If the ring assumes a warped or puckered condition, as shown in the diagrams below, the carbon atoms regain the normal valence angles. For a six-membered ring, two nonplanar strainless forms are possible, namely, the chair form and the boat form.

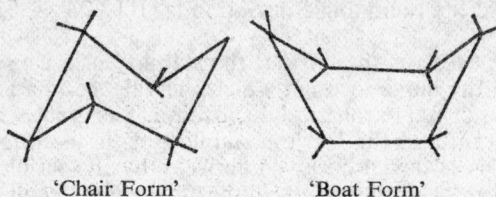

'Chair Form'   'Boat Form'
Fig. 14.

Such non-planar strainless rings in which the ring carbon atoms can have the normal angles are also possible for larger ring compounds. Actually, only one form of cyclohexane is known and not two forms as pictured above. The failure to isolate the two forms has been ascribed to rapid interconversions between them.

PREPARATION: 1. An alkylene dihalide can be treated with a metal-like sodium or zinc in a Wurtz type reaction to give cyclic compounds.

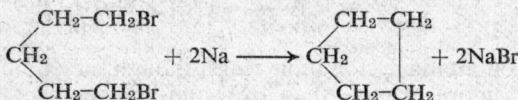

$$CH_2\begin{array}{c} CH_2-CH_2Br \\ \\ CH_2-CH_2Br \end{array} + 2Na \longrightarrow CH_2\begin{array}{c} CH_2-CH_2 \\ | \\ CH_2-CH_2 \end{array} + 2NaBr$$

Good yields are obtained by effecting debromination with zinc in formamide as a solvent.

2. Alicyclic compounds can be obtained from acetoacetic ester and malonic ester by condensation of the sodium salt of the ester with suitable alkylene dihalides. This method has been described in the sections on malonic and acetoacetic esters.

3. Pyrolysis of the barium salts of dibasic acids gives cyclic ketones in suitable cases. Ruzicka prepared several large ring ketones by this method. He found that the thorium salts gave better yields than those of calcium or barium. It was noted that five- and six-membered ketones were formed in high yields, whereas larger ring ketones were obtained in very low yields.

Adipic and pimelic acids can also be converted to cyclopentanone and cyclohexanone respectively by heating them with acetic anhydride.

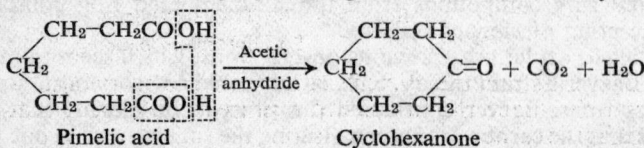

$$CH_2\begin{array}{c} CH_2-CH_2CO\lceil OH\rceil \\ \\ CH_2-CH_2\lfloor COO\rfloor H \end{array} \xrightarrow[\text{anhydride}]{\text{Acetic}} CH_2\begin{array}{c} CH_2-CH_2 \\ \\ CH_2-CH_2 \end{array}C=O + CO_2 + H_2O$$

Pimelic acid   Cyclohexanone

The lower dibasic acids give only the corresponding cyclic anhydrides when treated similarly with acetic anhydride.

4. **By Dieckmann Condensation.** This is an intramolecular Claisen-type ester condensation. When diethyl adipate is treated with sodium ethoxide 2-carbethoxycyclopentanone is obtained which when hydrolysed and decarboxylated gives the corresponding ketone.

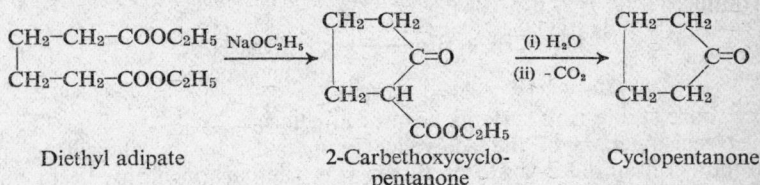

$$CH_2-CH_2-COOC_2H_5 \atop CH_2-CH_2-COOC_2H_5 \quad \xrightarrow{NaOC_2H_5} \quad \begin{array}{c} CH_2-CH_2 \\ \diagdown \\ C=O \\ \diagup \\ CH_2-CH \\ | \\ COOC_2H_5 \end{array} \quad \xrightarrow[\text{(ii) } -CO_2]{\text{(i) } H_2O} \quad \begin{array}{c} CH_2-CH_2 \\ \diagdown \\ C=O \\ \diagup \\ CH_2-CH_2 \end{array}$$

Diethyl adipate   2-Carbethoxycyclo-   Cyclopentanone
                         pentanone

Since the intermediate above is a β-keto ester, it can be alkylated with different halides, and the alkylated esters after hydrolysis and decarboxylation furnish 2-alkylcyclopentanones.

PROPERTIES: The cycloparaffins (like the alkanes) are inert towards oxidizing agents such as potassium permanganate and ozone. Cyclopropane alone undergoes rupture of the ring rather easily by the action of several reagents. Hydrogenation of the cycloparaffins often gives the open-chain paraffins. The temperatures at which the cycloparaffins react with hydrogen in the presence of nickel furnish a good illustration of Baeyer's strain theory. Ethylene with a high degree of strain reacts with hydrogen in the presence of nickel at room temperature to give ethane. The temperature at which cyclopropane undergoes a similar hydrogenation is higher (80° C.), and this temperature at which reduction takes place increases with the increasing size of the ring, as given below, thus reflecting the increase in stability as the rings approach a strainless condition.

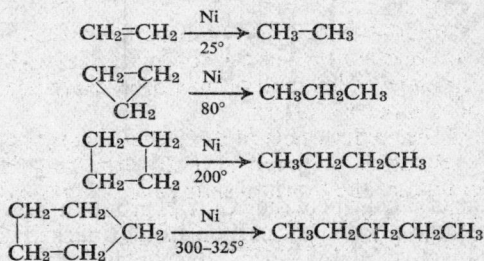

$$CH_2=CH_2 \xrightarrow[25°]{Ni} CH_3-CH_3$$

$$\begin{array}{c} CH_2-CH_2 \\ \diagdown \diagup \\ CH_2 \end{array} \xrightarrow[80°]{Ni} CH_3CH_2CH_3$$

$$\begin{array}{c} CH_2-CH_2 \\ | \quad\quad | \\ CH_2-CH_2 \end{array} \xrightarrow[200°]{Ni} CH_3CH_2CH_2CH_3$$

$$\begin{array}{c} CH_2-CH_2 \\ \diagup \quad\quad \diagdown \\ CH_2 \quad\quad CH_2 \\ \diagdown \quad\quad \diagup \\ CH_2-CH_2 \end{array} \xrightarrow[300–325°]{Ni} CH_3CH_2CH_2CH_2CH_3$$

Bromine easily converts cyclopropane into 1,3-dibromopropane.

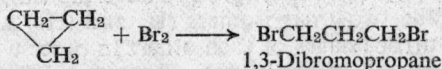

$$\begin{array}{c} CH_2-CH_2 \\ \diagdown \diagup \\ CH_2 \end{array} + Br_2 \longrightarrow BrCH_2CH_2CH_2Br$$
                                    1,3-Dibromopropane

This ring fission becomes difficult in the case of higher rings, and in the case of cyclohexane the ring is not ruptured at all and a substitution product is obtained.

$$CH_2-CH_2$$
$$CH_2 \qquad CH_2 + Br_2 \longrightarrow CH_2 \qquad CHBr + HBr$$
$$CH_2-CH_2 \qquad\qquad CH_2-CH_2$$

Bromocyclohexane

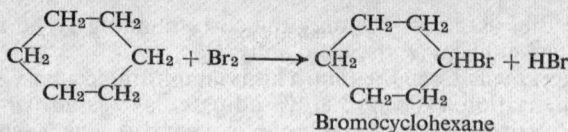

Hydrogen bromide ruptures the cyclopropane ring forming propyl bromide.

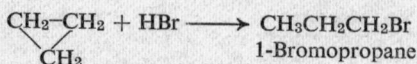

$$CH_2-CH_2 + HBr \longrightarrow CH_3CH_2CH_2Br$$
$$CH_2 \qquad\qquad\qquad 1\text{-Bromopropane}$$

The more stable ring systems are not affected.

Stereoisomerism of the Alicyclic Compounds. In the alicyclic compounds containing 3–5 atoms in the ring the carbon atoms lie in one plane and the hydrogen atoms project on opposite sides of this plane.

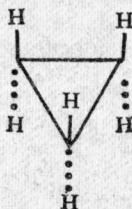

The ring acts in much the same fashion as a double bond. It hinders free rotation and fixes the position of the substituents in space. Hence geometrical isomerism should be possible. For instance, in the case of cyclopropane-1,2-dicarboxylic acid three forms (represented by (I), (II), and (III)) are possible.

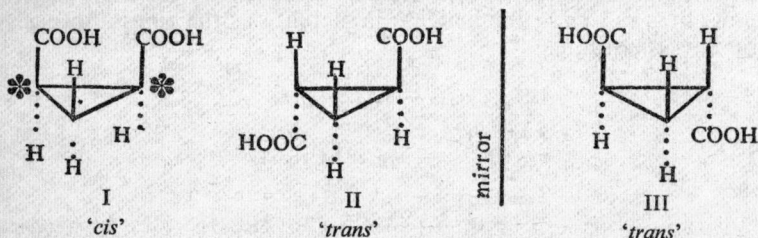

I	II	III
'cis'	'trans'	'trans'

mirror

(I) represents the cis acid where the –COOH groups are on the same side of the ring. It has two asymmetric carbon atoms (marked with asterisks), but is optically inactive (*meso*), since it has a plane of symmetry passing through the unsubstituted carbon atom and the two hydrogen atoms attached to it. With one carboxyl group above the ring and the other below it, we get the two *trans* isomers (II) and (III) which are non-superimposable mirror images of each other. In other words, the *trans* acid is asymmetric and exists in optically active forms. In alicyclic compounds *cis–trans* isomerism and optical isomerism become closely interwoven. As the ring size increases, the stereochemistry of such compounds becomes more and more complicated.

TERPENES

Many plants contain volatile oils in their leaves, blossoms, fruits, and barks, especially those belonging to the families *Coniferae* and *Myrtaceae* and the genus *Citrus*. The essential oils are obtained by steam distillation and have been used in perfumery since antiquity. These oils are called essential oils, not because they are absolutely necessary but because they are volatile essences. The essential oils consist of mixtures of unsaturated alicyclic compounds called terpenes, which have the empirical formula $(C_5H_8)_n$ and oxidation products of these. The terpenes are apparently built up of isoprene (C_5H_8) units.

The terpenes are classified into open-chain, monocyclic, bicyclic, sesqui-, and polyterpenes.

One familiar example of open chain terpenes is geraniol $(C_{10}H_{18}O)$ obtained from oils of rose and geranium. It has the formula

$$\underset{\underset{3}{\overset{}{}}}{CH_3}\overset{\overset{CH_3}{|}}{\underset{|}{C}}=CHCH_2CH_2\overset{\overset{CH_3}{|}}{\underset{2}{C}}=CH\underset{1}{CH_2OH}.$$

The 2,3 double bond gives rise to *cis–trans* isomerism. Geraniol is the *trans* isomer. The *cis* isomer is called Nerol. Nerol readily loses water to give the cyclic compound dipentene, demonstrating the *cis* arrangement of the −CH_2OH group with respect to the six-carbon substituent at carbon atom number 3.

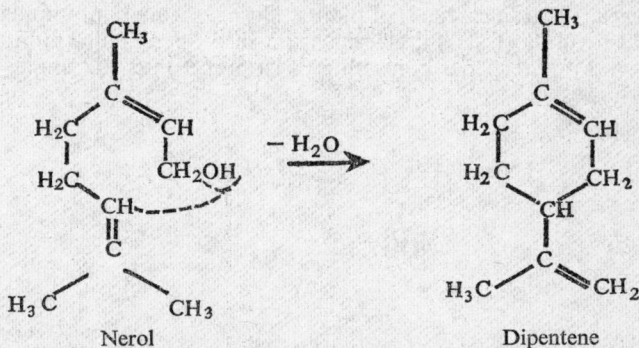

Nerol   Dipentene

Citral, the aldehyde, corresponding to geraniol, imparts to oil of lemon its characteristic odour. It is abundant in lemon grass oil. Citral is used as a starting material for preparing the ionones used in perfumery.

Menthol is a monocyclic terpene which occurs in peppermint oil. It is of value as a counter irritant for neuralgia and headaches. Its vapours, from nasal inhalers, give relief from cough and bronchitis. It is used in shaving lotions and cosmetic preparations.

SUMMARY

Alicyclic Compounds. These are derivatives of cycloparaffins (C_nH_{2n}) which are ring compounds. Though isomeric with alkenes, cycloparaffins resemble alkanes. Cycloparaffins can be pictured to be derived

by tying up the ends of carbon chains of suitable length which actually
have a zigzag pattern. Baeyer's strain theory, according to which the
strain in a ring compound is dependent on the deviation of the angle of
the ring from the tetrahedral angle. The theory is substantiated by the
increase in the order of stability from three- to six-member ring compounds. Modification of the theory in view of the occurrence of large-
ring compounds (muscone and civetone) in nature as well as their synthesis. Also the greater stability of cyclohexane as compared to cyclopentane is in conflict with Baeyer's theory. Sachse's hypothesis of non-
planar rings. Boat and chair forms of cyclohexane. Preparation of ring
compounds: (1) Alkylene dihalides of the type $Br-(CH_2)_n-Br$ reacted
with sodium. (2) Alkylation of ethyl malonate and ethyl acetoacetate
with dihalides: (3) Pyrolysis of barium or thorium salts of dibasic
acids, for example, barium adipate $\xrightarrow{\text{Heat}}$ cyclopentanone. Conversion
of dibasic acids to cyclic ketones by refluxing with acetic anhydride.
(4) Dieckmann condensation (an internal Claisen-type condensation):
Diethyl adipate $\xrightarrow{\text{NaOC}_2\text{H}_5}$ 2-carbethoxycyclopentanone $\xrightarrow[-\text{CO}_2]{\text{NaOH Heat}}$ cyclopentanone. Properties: ease of rupture of rings in accordance with strain
theory. Fixation of ring substituents in space due to the rigid nature of
the rings and consequent geometrical isomerism of ring compounds,
for example, cyclopropane—1,2-dicarboxylic acid. One *trans* (*dl*) and
one *cis* (*meso*) form.

   Terpenes. Volatile essences of plants. Derived from compounds of the
empirical formula $(C_5H_8)_n$; classification into open-chain, monocyclic,
sesqui-, and polyterpenes consisting of isoprene units. Examples of the
different types.

CHAPTER ELEVEN

REACTIVITY AND STRUCTURE—SOME ASPECTS

Organic chemistry is replete with instances where all kinds of reactions have been tried in order to find out if they would take place. It is not enough for us to know that certain reactions take place. The additional knowledge of what makes them take place and how they occur would be very valuable, especially in planning out the synthesis of new compounds. The subject of reaction mechanism is of recent origin and still far from completely developed. Nevertheless, progress has been made and we are in a position to draw some generalizations on the mechanisms of the simpler reactions involving electrovalent and covalent compounds. With respect to any particular reaction we would naturally desire both qualitative and quantitative information. We would like to know, for example, how a particular reaction takes place and also know how to assess the magnitude of the driving force of the reaction. The treatment of the quantitative aspects of reaction mechanism is outside the scope of the present book. Only a qualitative picture of some aspects of the subject will be presented. At this stage it will be profitable for the reader to review Chapter Two and refresh his ideas on the electronic structures of the different types of molecules.

Organic reactions result from electronic displacements occurring within the molecules of reactants. In order to understand these displacements we should know more about the nature and behaviour of electrons in atoms and molecules.

THE ELECTRONIC STRUCTURE OF CARBON AND ITS COVALENCE

One way of starting to explain how the electrons in the outer shell of carbon take part in covalent-bond formation is to consider what the electronic structure of an isolated carbon atom is like.

In Chapter Two we acknowledged that electrons possess wavelike properties as well as those of particles; that they can only have certain fixed amounts of energy; and that we cannot know the full details of any electron's motion. The best we can do is assess the probability of an electron in a particular energy level turning up at different places round the atomic nucleus.

The lowest-energy electrons occupy the first, or K 'shell'. This shell contains one orbital, spherically symmetrical about the nucleus: the probability of finding an electron occupying the orbital decreases as we move farther out from the nucleus in any direction.

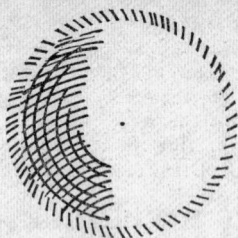

Fig. 15

Spherical atomic orbitals such as this are called **s-orbitals**.

Electrons show an electromagnetic effect which is like that which we observe on a larger scale when a charged sphere spins about an axis through its centre. Because we can do no better, we picture the electrons as if they are particles spinning about their own axes, too, always with the same amount of angular momentum. Not more than two electrons can occupy any one orbital and, when they do, they spin in opposite directions:

Fig. 16

Thus the first, or K, shell can contain two electrons with opposed spins, in an *s*-orbital.

The L shell contains electrons with more energy than those in the K shell. Two electrons occupy another *s*-orbital, like the one just described but larger, because the electrons have more energy. Two more

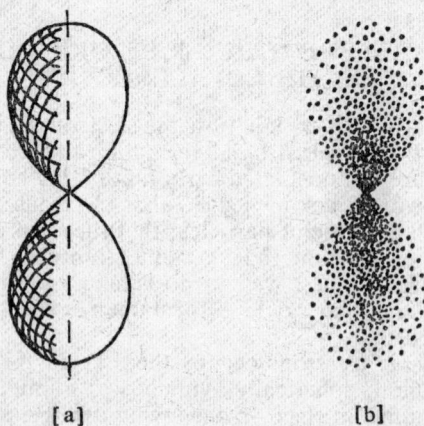

[a]   [b]

Fig. 17

singly occupy two out of three possible *p*-orbitals. A **p-orbital** is hour-glass-shaped with the nucleus between the two lobes, Fig. 17 (*a*). Fig. 17 (*b*) represents a cross-section of a *p*-orbital in a plane containing its axis of symmetry: the denser the dots in any little area, the greater the chance of finding there an electron occupying the orbital. There are three possible *p*-orbitals in any shell, and their axes of symmetry are mutually perpendicular, Fig. 18. If only we could see it, the valence

Fig. 18

shell of an isolated carbon atom in its state of lowest energy would look like Fig. 19, with two electrons in the *s*-orbital and one in each of the two *p*-orbitals.

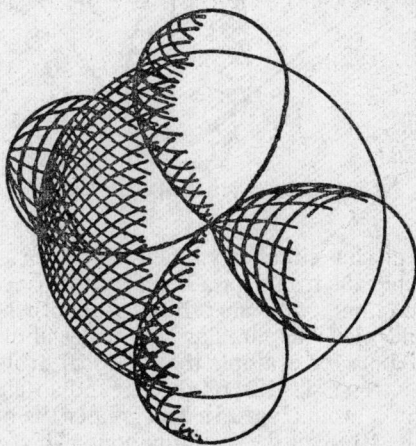

Fig. 19

In the next step of our theoretical treatment one of the L shell's *s*-electrons is excited into the vacant *p*-orbital. To form four covalent bonds with four other atoms these orbitals are then combined, or hybridized, to form four equivalent orbitals, called sp^3 **hybrid orbitals** because they were made from one *s*- and three *p*-orbitals. Although there is a slight chance of the electrons in these orbitals turning up anywhere in the universe, by far the largest portion of their time is spent within the boundaries of four fat, Indian-club-shaped orbitals pointing tetrahedrally outwards from the nucleus.

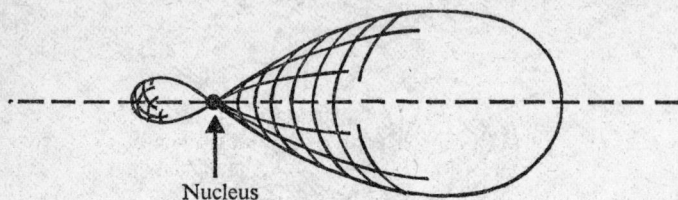

Nucleus

One sp^3 hybrid orbital
Fig. 20

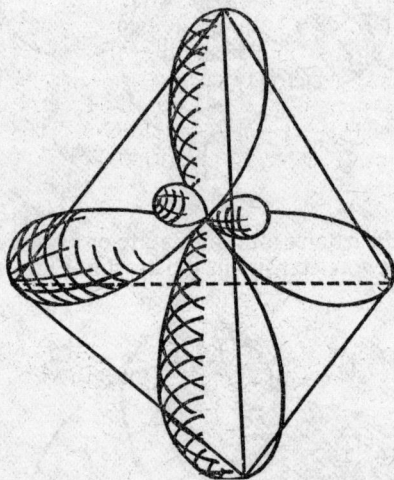

A **tetrahedrally prepared** carbon atom
Fig. 21

Each of these orbitals has circular symmetry about an axis going towards one corner of the tetrahedron. For a bond to be formed by one of these orbitals it must combine with an orbital on another atom which has similar energy and also has circular symmetry about what has now become the bond axis, for example the spherical orbital of a hydrogen atom's valence electron (Fig. 22); or another Indian-club-shaped orbital from another carbon atom (Fig. 23). The greater the overlap of the two atomic orbitals, the stronger the resulting bond. The resulting **molecular orbital**, containing two atomic nuclei, will hold two electrons and is then

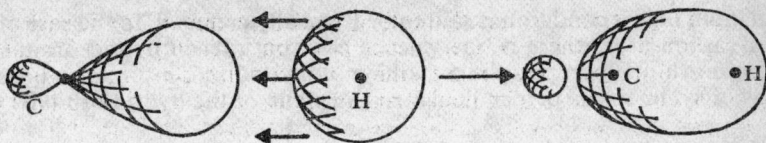

Fig. 22

Fig. 23

one covalent bond. Molecular orbitals such as these, with circular symmetry about the bond axis, are called **sigma-bonds** (σ-bonds).

The important point for the time being is that because σ-molecular orbitals have circular symmetry about the bond axis they permit free rotation of the two atoms so joined, about the bond axis, which reduces the possibilities of isomerism (cf. page 34).

In the homologous series of olefines and of those compounds containing a carbonyl group we encountered a carbon atom joined to another carbon or to an oxygen atom by a double bond. To describe the bonding in these compounds we first picture the *s*-orbital and two of the *p*-orbitals of the carbon atom's valence shell combined to form three *sp²*-hybrid orbitals. These are also Indian-club-shaped and their axes of symmetry are all in the same plane and at 120° to each other (Fig. 24).

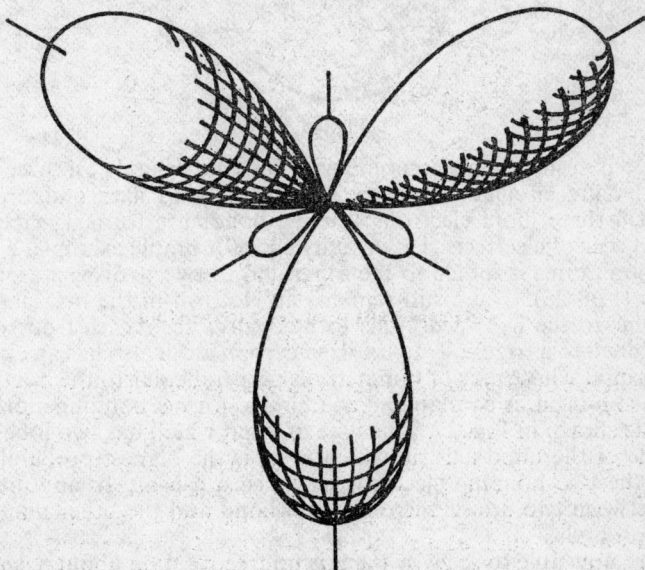

Fig. 24

An atom in this condition is said to be **trigonally prepared**. In the case of
our carbon atom there is one valence electron in each hybrid orbital.
The fourth one occupies the remaining, unhybridized, *p*-orbital, whose
axis of symmetry is perpendicular to the plane of the hybrid orbitals:

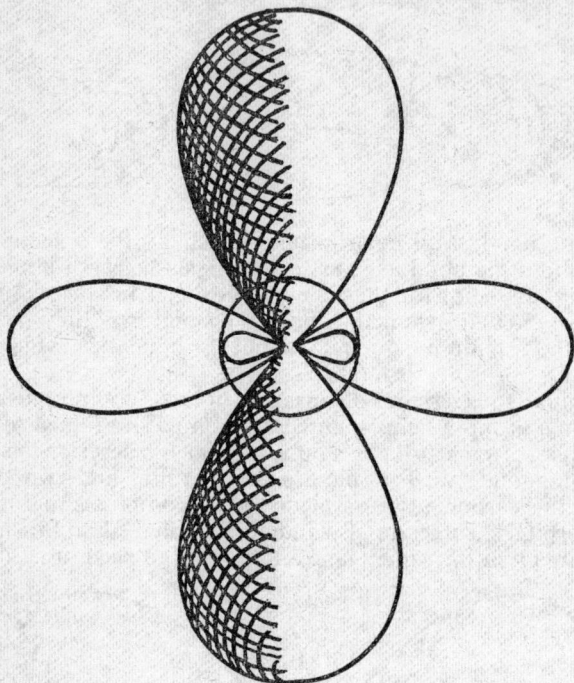

Fig. 25

The three hybrid orbitals combine with suitable orbitals on three other
atoms, having circular symmetry about the bond axes and bringing
with them three more electrons: three σ-bonds are formed, each containing a pair of electrons. Taking ethylene as a simple example, each of
the carbon atoms is joined to the other and to two hydrogen atoms by
σ-bonds (Fig. 26). There still remains an electron in the *p*-orbital perpendicular to the hybrid orbitals to be dealt with. At least one of the
atoms joined to a trigonally hybridized carbon atom also has an electron
in a *p*-orbital whose axis of symmetry is perpendicular to the bond axis.
These two *p*-orbitals overlap and combine to form a bonding molecular
orbital as shown in Fig. 27. This type of bond which has two lobes, one
each side of the bond axis, in which there is the highest probability of
finding the two bonding electrons, is called a **π-bond**. In any multiple
bond between two atoms there is one σ-bond and the other bonds are
always π-bonds.

   We are now able to see why there is no free rotation about a carbon–
carbon double bond (cf. page 121). The strongest bonds are formed

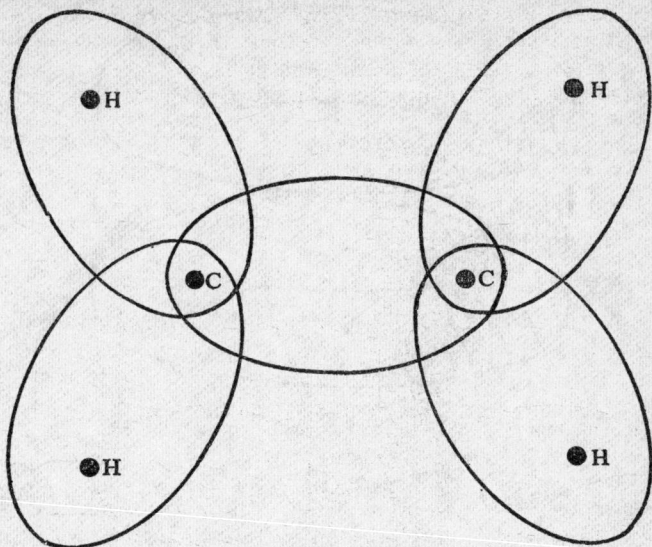

Fig. 26

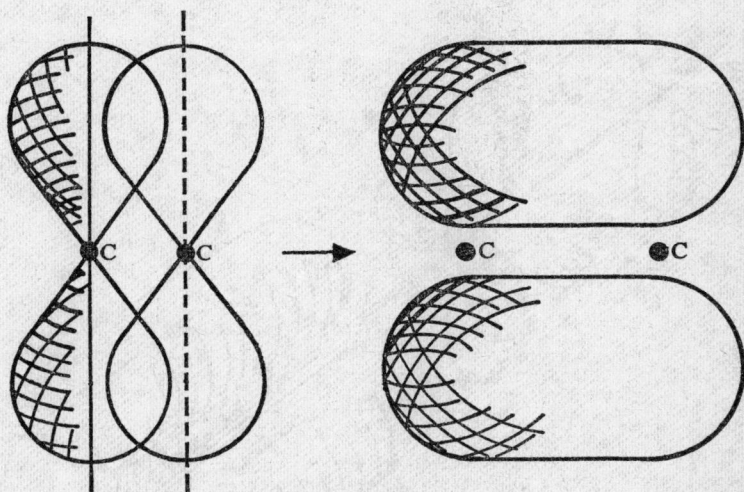

Fig. 27

when the component atomic orbitals overlap as much as possible. If we turn one of the carbon atoms in Fig. 25 about the bond axis the overlap of the two *p*-orbitals decreases. This represents a weakening of the bond, which means we had to do work in overcoming the strength of the bond. Thus the two atoms joined by the double bond will resist relative rotation, and the atoms or groups to which they are linked will stay rigidly

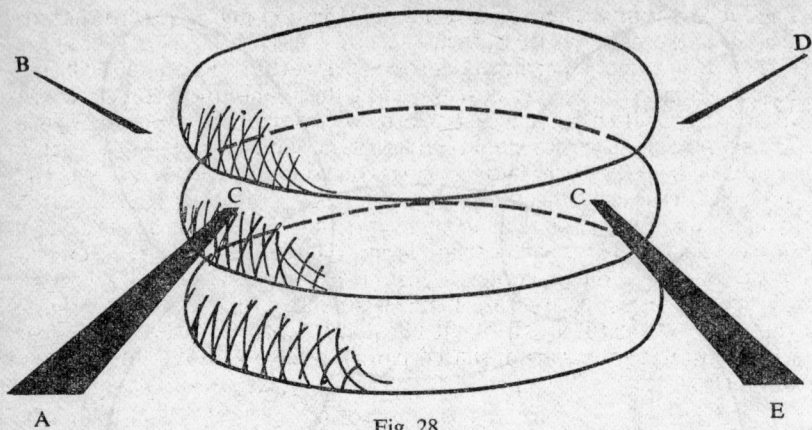

Fig. 28

Fig. 29

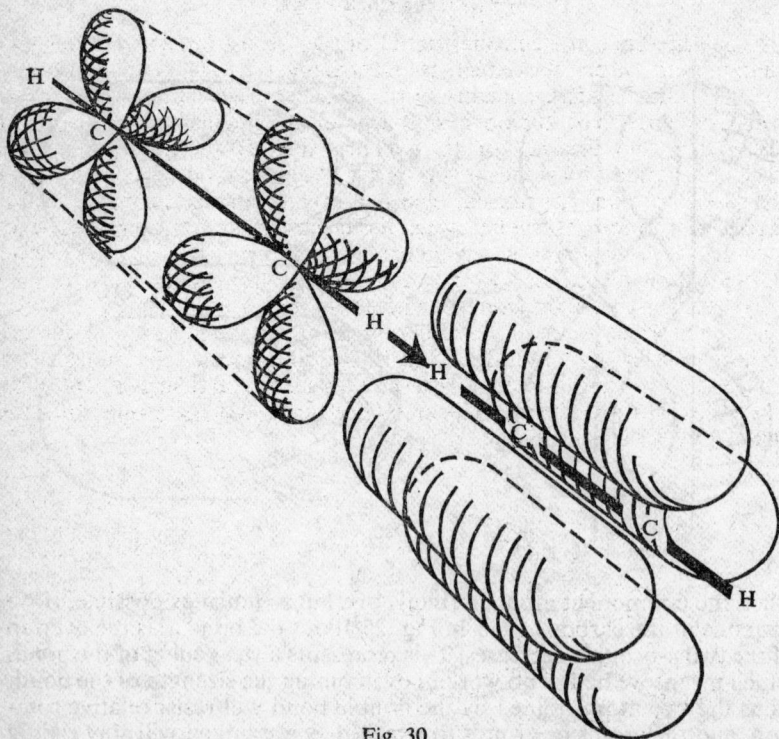

Fig. 30

fixed at the four corners of a rectangle (Fig. 28) giving rise to the possibility of geometrical isomerism.

The triple bond between two carbon atoms which occurs in the homologous series of acetylenes is achieved by the *s*- and one of the *p*-orbitals of the valence shell of each carbon atom hybridizing to give two **digonal, *sp*-hybrid** orbitals which extend outwards on opposite sides of the carbon nucleus. These can form two σ-bonds with suitable orbitals on two other atoms, e.g. in acetylene itself (Fig. 29). Two *p*-orbitals, whose axes of symmetry are perpendicular to each other and to the bond axis, remain in the valence shell of each carbon atom. Each of these orbitals contains one electron and they combine to form two π-bonds (Fig. 30). The nitrogen atom in a cyanide group behaves similarly, but, having one more valence electron than carbon, it is satisfied with a triple bond to the carbon atom and has an unshared pair of electrons in its other digonal orbital:

INDUCTIVE EFFECT

The difference in electronegativity of two atoms linked together by a single bond causes this effect. By electronegativity is meant the ability of the atom to attract electrons to itself. The electronegativity of an element depends on the positive charge on the nucleus of the atom and also upon the distance of the valence shell from the nucleus. The electronegativity increases in the order Carbon < Nitrogen < Oxygen < Fluorine, since the nuclear charge on the atom also increases in that order. In the case of the halogens, the electronegativity increases in the order Iodine < Bromine < Chlorine < Fluorine. This is so even though the nuclear charges increase, because the valance electrons are more and more distant from the nucleus in the series F, Cl, Br, I. When a carbon atom is linked to a different atom by a single bond, as, for example, in methyl chloride (CH_3Cl), the difference in electronegativity of the chlorine and carbon atoms causes an electron displacement in the direction indicated below. The arrowhead indicates the atom which has gained in electronic charge.

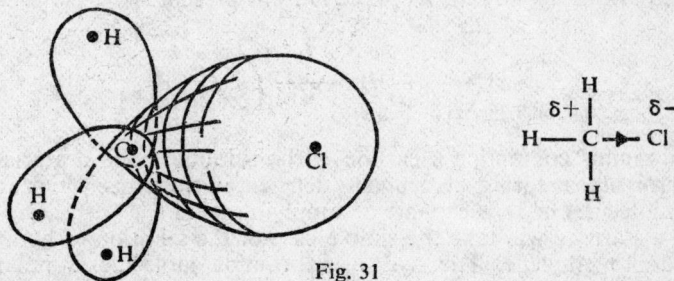

Fig. 31

This displacement manifests itself in the increased reactivity of methyl chloride as compared with methane. It is necessary to note that neither of the atoms, carbon or chlorine, has completely lost or gained an electron pair; the loss or gain is only partial. The initial point of attack is the carbon atom, which carries a small charge. This attracts a **nucleophilic reagent**, e.g. a hydroxyl ion, which is a reagent capable of donating a share of a pair of electrons to another atom short of electrons:

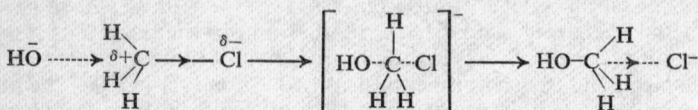

$$HO^{-} \cdots \rightarrow {}^{\delta+}C \longrightarrow Cl \longrightarrow \left[HO \cdots C \cdots Cl \right]^{-} \longrightarrow HO-C \cdots \rightarrow \cdots Cl^{-}$$

As the hydroxyl ion joins on one side of the methyl group so the chlorine atom leaves on the other side as a chloride ion. An electron displacement caused by the inductive effect makes itself felt along a chain of carbon atoms up to a certain distance, as follows:

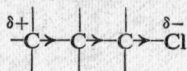

$$\overset{\delta+}{-}C \rightarrow C \rightarrow C \rightarrow \overset{\delta-}{Cl}$$

The primary and secondary alkyl halides are attacked in this way by a wide range of nucleophilic reagents (cf. Chapter Four). However, the methyl compounds are more reactive than the others. This is because all the others have one or two alkyl groups attached to the halogen-bearing carbon atom, which can release electrons to it and lower its positive charge. This makes it less attractive to a nucleophilic reagent.

The ability of carbon atoms to supply electrons through the inductive effect of a halogen atom does make another nucleophilic substitution mechanism possible for secondary alkyl halides and makes it very important for tertiary alkyl halides. For example, the chlorine atom in tertiary butyl chloride can be removed by a polar solvent as a chloride ion:

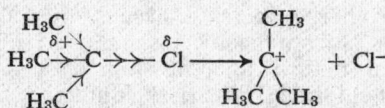

$$\begin{array}{c} H_3C \\ H_3C \rightarrow \overset{\delta+}{C} \rightarrow \overset{\delta-}{Cl} \longrightarrow \overset{CH_3}{C^+} + Cl^- \\ H_3C \qquad\qquad H_3C \quad CH_3 \end{array}$$

and a carbonium ion is formed in which the central atom is trigonally hybridized and carries a large share of the positive charge. This is then attacked swiftly by any nucleophilic reagent present:

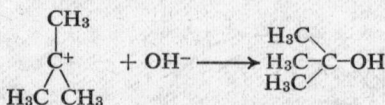

$$\begin{array}{c} CH_3 \\ C^+ \\ H_3C \quad CH_3 \end{array} + OH^- \longrightarrow \begin{array}{c} H_3C \\ H_3C-C-OH \\ H_3C \end{array}$$

Compounds containing a carbon–carbon double bond are attacked by **electrophilic reagents**, i.e. reagents deficient in electrons, which attack other molecules or ions capable of supplying them.

For a start we will take the simple case of the addition of hydrogen bromide to ethylene. The hydrogen bromide molecule is polarized because bromine is more electronegative than hydrogen, the positive hydrogen atom is attracted by the electrons in the π-bond:

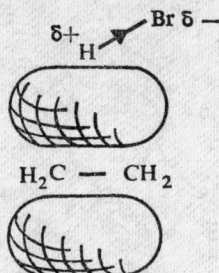

Under their influence the hydrogen bromide ionizes completely and the proton becomes attached to one of the carbon atoms by a σ-bond:

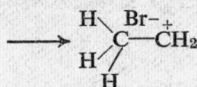

The positive charge thereby produced on the other carbon atom can then be neutralized by a bromide ion:

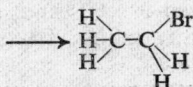

In an unsymmetrical olefine like propylene there is an inductive effect because the methyl group can supply electrons to the unsaturated carbon atom to which it is joined:

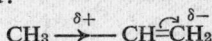

$$CH_3 \xrightarrow{\delta+} CH \overset{\delta-}{=} CH_2$$

The negative carbon atom will attract the electrophilic reagent, and the approaching reagent very likely polarizes the propylene molecule still further to something like

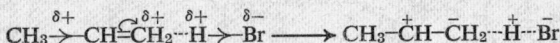

$$CH_3 \xrightarrow{\delta+} CH \overset{\delta+}{=} CH_2 \cdots H \xrightarrow{\delta+} Br \longrightarrow CH_3 - \overset{+}{C}H - \overset{-}{C}H_2 \cdots \overset{+}{H} \cdots \overset{-}{Br}$$

then the rest of the reaction proceeds as before

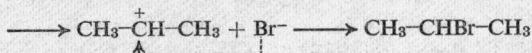

$$\longrightarrow CH_3 - \overset{+}{C}H - CH_3 + Br^- \longrightarrow CH_3 - CHBr - CH_3$$

and the bromine atom joins on to the carbon atom with the fewer hydrogen atoms, in accord with Markownikoff's law (see page 43), the underlying reasons for which we can now appreciate.

The inductive effect of the chlorine atom in α-chloro fatty acids was discussed in Chapter Seven.

DELOCALIZATION

If several trigonally hybridized carbon atoms are connected together with the axes of symmetry of the extra *p*-orbitals which contain one valence electron each, all parallel, then the *p*-orbital of any carbon atom overlaps equally the two on either side of it:

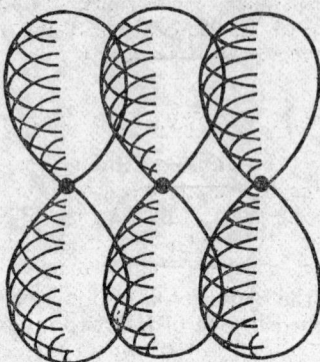

Fig. 32

All of these *p*-orbitals combine to form a series of π-molecular orbitals which are not restricted to two nuclei each, like the ones we have already met, but extend over all of the trigonally hybridized carbon atoms. These orbitals are called delocalized orbitals and, just like the localized ones we have already met, they can contain a maximum of two electrons with their spins opposed.

The simplest example to picture is butadiene whose formula is written classically as $H_2C=CH-CH=CH_2$. The four *p*-orbitals concerned (Fig. 33 (*a*)) combine to form two delocalized orbitals (Fig. 33 (*b*) and (*c*)) each of which contains two valence electrons

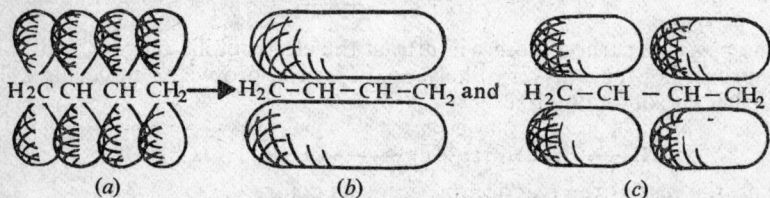

Fig. 33

The electrons in the delocalized bonding orbital of Fig. 33 (*b*) spend most of the time somewhere in the two lobes and can pass freely from one lobe to the other. The electrons in the orbital Fig. 33 (*c*) spend most of the time in the four lobes shown and use all four of them freely. Thus the electrons in either of the delocalized orbitals can turn up near any of the four carbon atoms, so the classical formula written above, using localized bonds only, gives a restricted, misleading impression of the true state of the molecule.

Compounds such as this, for which one could write a classical structure consisting of a system of alternate single and double bonds, are called conjugated compounds.

One consequence of delocalization is that it confers additional stability on the molecule, another is that it modifies the chemical properties of the substance so they are not what one would expect from a consideration of the molecule in terms of localized bonds only. The next section serves as an illustration.

ADDITION TO CONJUGATED SYSTEMS

Compounds which contain an alternate system of double bonds are said to have a conjugated system of double bonds. A substance having a conjugated system of double bonds behaves differently from one having the same number of double bonds but isolated from one another. Thus 1,4-pentadiene, having the structure $CH_2=CHCH_2CH=CH_2$ behaves normally and bromine adds on at both the ethylenic linkages. In 1,3-butadiene ($CH_2=CHCH=CH_2$) the double bonds are conjugated. When it is treated with bromine one equivalent of bromine is taken up rapidly and a dibromide, mainly 1,4-dibromo-2-butene ($BrH_2CCH=CHCH_2Br$), is easily isolated. Further addition of bromine is found to be very slow. The formation of the above dibromo compound essentially involves the ends of the delocalized π-orbitals being attacked, addition to carbon atoms 1 and 4 of the conjugated system and the π-orbital being shortened and localized to the 2,3-position.

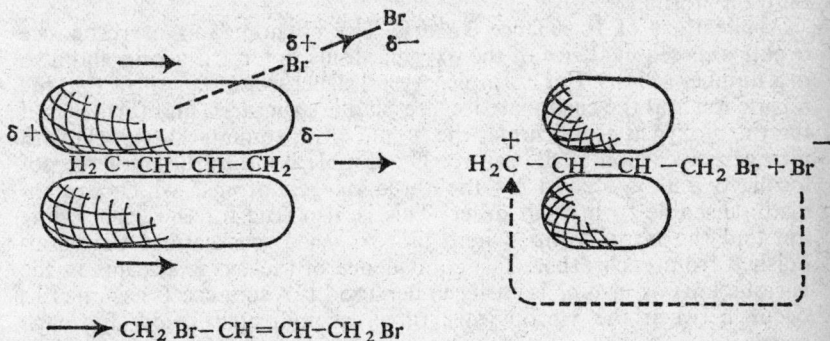

$$\longrightarrow CH_2 Br-CH=CH-CH_2 Br$$

The applications of the theory of resonance are very numerous and important. They are of considerable value in interpreting aromatic character and very useful in explaining several observations in the chemistry of the free radicals and triphenyl methane dyes. It is remarkable that the result of a theory originating from mathematics should have had such a dynamic impact upon the development of theoretical organic chemistry.

RESONANCE

An alternative approach to the explanation of structures and re-activities in terms of delocalized orbitals and the inductive effects of atoms of differing electronegativities is provided by the concept of resonance.

The idea of resonance is an important concept in chemistry. Its recognition was one of the significant milestones in the development of the subject, next only to the extension of the carbon atom into three-dimensional space by Van't Hoff. The idea was actually a result of the application of wave mechanics to chemical systems. In spite of the fact that wave mechanics is inseparably associated with abstruse mathematics, the conclusions obtained from it regarding resonance are simple. If more than one classical structure can be written for a molecule (or any other species like an ion or radical) without greatly altering the relative positions of the atoms, then the real structure is none of these but an intermediate one to which they all make contributions. This actual state, which cannot be simply represented on paper, is said to be a resonance hybrid of the classical structures or canonical forms.

The conditions for resonance to be possible between two structures are that: (1) the relative positions of the atoms in space must be nearly the same in both; (2) the number of shared electrons in both must be the same; and (3) the two must not differ greatly in stability.

As a result of resonance, the hybrid has, to some extent, the properties of each contributory structure, with the stable form contributing the most. It has a smaller energy content (i.e. more stable) than any of the individual structures would have. The distance between the atoms is less than the normal.

Applications of Resonance Theory. The phenomenon of resonance explains the equivalence of the oxygen atoms in a nitro group and also in a number of ions. For example, crystal structure studies show that the nitrate ion and the carbonate ion are planar structures, and further that the three oxygen atoms are at the points of an equilateral triangle with nitrogen or carbon at the centre. The double bond in these ions is not localized and is shared by the three oxygen atoms, which are indistinguishable from each other. This is also true for the carboxylate ion and the nitro group, where the oxygen atoms cannot be distinguished from each other. The equivalence of the oxygen atoms in the various ions or groups is easily understood if resonance is assumed to occur between the various structures, as indicated below for each species:

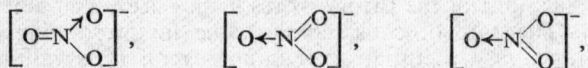

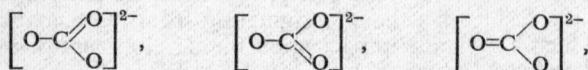

The addition reaction of an unsaturated compound can also be explained in terms of resonance. For example, the structure of acetaldehyde can be written

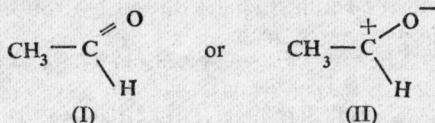

$$CH_3-C{\overset{O}{\diagup}}_{H} \qquad \text{or} \qquad CH_3-C{\overset{+}{\underset{H}{\diagup}}}{\overset{O^-}{}}$$

(I) (II)

Structure (II) takes into account the electronegativity of the oxygen atom. Resonance between the two forms produces a small positive charge on the carbon and a small negative charge on the oxygen atom:

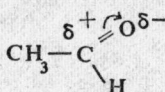

$$CH_3-C{\overset{\delta+}{\underset{H}{\diagup}}}{\overset{O^{\delta-}}{}}$$

The positive carbon atom is thus open to attack by nucleophilic reagents, e.g. CN^- or HSO_3^- or $:NH_3$.

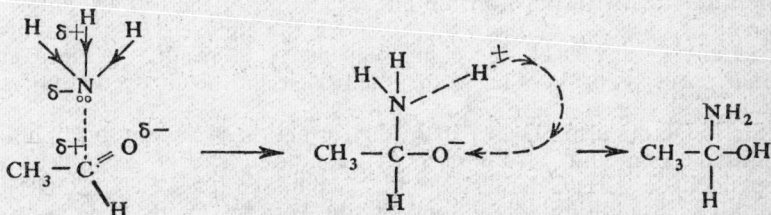

Another example of a molecule which is stabilized by resonance is vinyl chloride (CH_2CHCl). It has long been known that the chlorine atom in vinyl chloride is much harder to replace than the chlorine in ethyl chloride.

The following structures can be written for vinyl chloride:

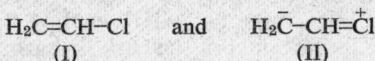

$$H_2C{=}CH{-}Cl \qquad \text{and} \qquad H_2\bar{C}{-}CH{=}\overset{+}{C}l$$

(I) (II)

In (II) one of the three lone pairs on the chlorine atom has taken part in dative covalence formation. According to the resonance concept, the real structure will be something between, like

$$H_2\overset{\delta-}{C}{=}CH{-}\overset{\delta+}{C}l$$

The carbon–chlorine bond is fractionally more than a single bond but less than a double bond, making the chlorine less labile in reactions that require its easy removal as Cl^-. As contrasted with vinyl halides, allyl halides are very reactive. For example, allyl chloride ($CH_2{=}CHCH_2Cl$) hydrolyses about eighty times as fast as propyl chloride ($CH_3CH_2CH_2Cl$), its corresponding saturated chloride. Other unsaturated halides, with the double bond in any other position relative to the halogen, are not as reactive. An explanation is possible on the basis of resonance theory. The hydrolysis of allyl chloride has been shown to depend on the concentration of the allyl carbonium ion ($CH_2=CHCH_2^+$). This carbonium ion is stabilized by resonance, as follows, and hence is easily formed, with the result that the rate of hydrolysis is increased.

$$CH_2=CH-\overset{+}{CH_2} \longleftrightarrow \overset{+}{CH_2}-CH=CH_2$$

STERIC HINDRANCE

Organic reactions, in addition to being governed by the polar factors discussed above, are often influenced by the space geometry of the molecules of the reactants or what are called **steric factors**. For example, when we compare the rates of esterification of primary and secondary alcohols with a particular acid we find the rates for the secondary alcohols to be less than those of the primary alcohols. In the case of secondary alcohols, because of an additional bulky substituent on the hydroxyl-bearing carbon atom, the hydroxyl group is more difficult to approach by the reactive species. The low basicity of trimethylamine as contrasted with mono and dimethylamine is also due to a steric factor. The three methyl groups on the nitrogen atom of the tertiary amine, essentially because of their bulk, render the lone electron pair inaccessible to any electron-seeking reagent. As a result, the electron donor property, or, in other words, the basicity of the nitrogen atom, is diminished.

Steric factors also play a part in the formation of sodium bisulphite addition compounds from carbonyl compounds. For example, *n*-heptaldehyde [$CH_3(CH_2)_5CHO$] and 2-heptanone [$CH_3(CH_2)_4COCH_3$] furnish such addition compounds readily. However, with 3-heptanone [$CH_3(CH_2)_3COCH_2CH_3$] the bulky groups attached to the carbonyl carbon atom block the approach of the bisulphite ion to the carbon atom and thereby prevent the formation of the addition product. Another case of steric hindrance in esterification reactions may also be mentioned. Of the aromatic acids represented by the following three structures, it is found that (I) and (II) are esterified easily with alcoholic hydrogen chloride, but compound (III) cannot be esterified under similar conditions. The carbonyl of the carboxyl group in (III) is sterically hindered by the two neighbouring methyl groups and thereby prevented from taking part in esterification.

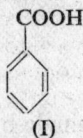

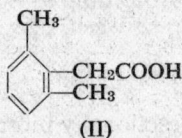

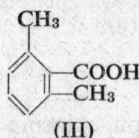

The presence of steric hindrance in substituted diphenyls of the type indicated has been firmly established by an extensive series of studies.

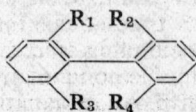

If the groups R_1, R_2, R_3, and R_4 are fairly bulky they get in each other's way and, as a result, both the phenyl nuclei and their substituents cannot lie in the same plane. The phenyl rings are forced to lie in planes which are at an angle to each other. In such a non-planar position the molecule lacks elements of symmetry and therefore is not superimposable on its mirror image. In other words, a diphenyl of the above type with sufficiently large substituents, though lacking an asymmetric carbon atom, is found to exist in *d* and *l* forms—an effect, undoubtedly, of the steric characteristics of the substituents.

In this very limited discussion of the effect of electronic and steric factors on reactivity and properties of organic compounds the examples given illustrate one or another of these factors. It should not be understood, however, that these factors operate to the mutual exclusion of each other or that they are the only factors influencing chemical reactivity. The latter in fact is a complex function of several variables associated with the structure of a compound.

SUMMARY

To understand the valence of carbon better we must first know more about the electronic structure of an isolated carbon atom. Two of the valence electrons are in an *s*-orbital, with their spins opposed (paired), and two are in *p*-orbitals in the valence shell. Before forming four single bonds these orbitals hybridize to form four, equivalent sp^3-hybrid orbitals which have circular symmetry about axes which are directed towards the corners of a tetrahedron. They combine with orbitals of other atoms, also having circular symmetry about the same axes and similar energies: the bigger the overlap, the stronger the bonds. The resulting σ-bonds are molecular orbitals with circular symmetry about the bond axis and containing two electrons with paired spins. In unsaturated compounds containing doubly bonded carbon atoms the carbon atom is trigonally prepared by combination of an *s*- and two *p*-orbitals to form three equivalent sp^2-hybrid orbitals whose axes are in the same plane and at 120° to each other. These orbitals form three σ-bonds using three suitable orbitals on three other atoms. The fourth valence electron of the carbon atom, in a *p*-orbital whose axis of symmetry is perpendicular to the plane of the three hybrid orbitals, takes part in the formation of a π-bond with an electron in a *p*-orbital on an adjacent atom. The absence of free rotation about the π-bond is the cause of geometrical isomerism. Carbon can also be digonally prepared: two equivalent orbitals pointing in opposite directions are formed by combination of an *s*- and a *p*-orbital (*sp*-hybridization). These can form two σ-bonds to two other atoms. The two remaining valence electrons of the carbon are in two *p*-orbitals whose axes are perpendicular to each other and to the σ-bond axes. They can combine with two *p*-orbitals on an adjacent atom to form two π-bonds containing two electrons each, e.g. in acetylene and the cyanide group.

Chemical reactivity depends on electronic displacements in a molecule as well as on steric factors. One type of electronic displacement is the inductive effect. This arises from difference in electronegativities of the atoms linked. For example, the reactivity of the halogen atoms in the alkyl halides is due to the attraction of a positive carbon atom for nucleophilic reagents: electrophilic reagents attack points of high electron-availability in olefines.

Delocalized molecular orbitals are formed when incompletely filled *p*-orbitals on a chain of adjacent atoms overlap significantly. The valence electrons occupying these π-orbitals can move freely along the whole chain of the atoms concerned. Compounds containing them are more stable than similar compounds containing localized orbitals only. Their chemical properties are modified too.

The concept of resonance is an alternative approach to the explanation of structure and reactivities. If more than one classical structure (canonical form) can be written for a molecule, ion, or radical, without greatly altering the relative positions of the atoms, then the real structure is none of these but an intermediate one to which they all make contributions: a resonance hybrid. Conditions for the occurrence of resonance: (i) should only involve redistribution of electrons without change in positions of atoms; (ii) the number of shared electrons in the different electronic structures must be the same; and (iii) the different structures should have nearly the same order of stability. Resonance results in greater stability of the molecule and shorter bond distances. Applications of resonance concept to: structures of nitrate and carbonate ions; nitro and carboxyl groups; and to addition reactions of the carbonyl group.

Steric Factors. These are also of importance in determining the reactivity of molecules. For example: (i) decreased rate of esterification of secondary versus primary alcohols; (ii) lower basicity of trimethylamine versus methyl and dimethyl amines; (iii) Bisulphite addition product with heptanal and 2-heptanone but not with 3-heptanone; (iv) difficult esterification of 2,6-disubstituted benzoic acids. Resolution of suitably substituted diphenyls provides firm evidence for steric hindrance in molecules.

PART III

AROMATIC COMPOUNDS

CHAPTER TWELVE

BENZENE AND ITS DERIVATIVES

Usually the study of organic chemistry is divided into four parts—aliphatic, alicyclic, aromatic, and heterocyclic. The aliphatic and alicyclic compounds have already been dealt with in Part II. Aromatic compounds will be studied in this part of the book.

The aromatic compounds are derived from a hydrocarbon benzene (C_6H_6) containing a closed ring of six carbon atoms. The word aromatic, to designate benzene and its derivatives, owes its origin to the fact that some of these compounds have pleasant odours. KEKULÉ, the great German chemist, who gave us the correct idea of the valence of the carbon atom, recognized the fact that certain substances from resins, balsams, and essential oils did not fit into the general classification of the carbon compounds known at that time (i.e. the open-chain compounds). He found that such substances as oil of bitter almonds, benzoic acid, and benzyl alcohol from gum benzoin and toluene from tolu balsam contained at least six carbon atoms and retained a six-carbon unit through ordinary chemical changes and degradation. These compounds, though they had a low hydrogen content, did not show the expected properties associated with unsaturation. The simplest member of the group was recognized as benzene.

COAL—SOURCE OF AROMATIC COMPOUNDS

The major source of aromatic chemicals for a long time has been coal. In recent years petroleum has become a significant alternative source of some of these compounds. We shall first outline the processes whereby aromatic compounds are obtained from coal. Coal when heated to about 1,000° C., in the absence of air in a retort, is converted into coke. Coke is used in metallurgical operations for reductions of ores and as a source of heat. On heating coal, volatile substances come out and are passed through a condenser into a sulphuric acid scrubbing solution. This retains ammonia and hydrogen sulphide gases and the coke-oven gas passes on. During this process, the organic substances form a separate layer and float on the sulphuric acid. Since this oil is light and floats on sulphuric acid, it is called light oil (b.p. 80–200° C.). The other volatile product which settles down to the bottom is called coal tar. The coal tar, which was once a nuisance in the coking industry, later turned out to be quite a veritable storehouse of organic compounds from which several types of drugs, medicinals, explosives, etc., could be prepared. Distillation of this coal tar furnishes three fractions referred to as middle oil (b.p. 200–250° C.), heavy oil (or creosote oil, b.p. 250–300° C.), and anthracene oil (b.p. 300–350° C.). The residue from coal-tar distillation is called pitch and is used in road making. The middle-oil fraction contains phenol and naphthalene, the heavy-oil fraction contains cresols and their homologues, and anthracene oil contains anthracene and related compounds. In addition, compounds containing nitrogen-like pyridine bases, carbazole, etc., and some containing sulphur-like thiophene are obtained from these various fractions.

Roughly speaking, a ton of coal yields about 11,200 cubic feet of illuminating gas. This illuminating gas is mainly a mixture of hydrogen and methane with smaller amounts of carbon monoxide, carbon dioxide, and some olefines. The ammonia which is absorbed in the sulphuric acid scrubbers gives about 20 lb. of ammonium sulphate. About 3 gallons of light oil are obtained, which on fractionation gives 2 gallons of benzene and about ½ gallon of toluene and smaller amounts of xylenes. The yield of coal tar is about 3% of the weight of the coal; that is to say, roughly about 8 gallons per ton. The yield of coke is about 1,500 lb. per ton of coal.

Coal $\longrightarrow$ Coke + light oil + coal tar
(1 ton)   (1,500 lb.)   (3 gallons)   (8 gallons)

+ Ammonium sulphate + iluminating gas
            (20 lb.)   (11,200 cu. ft.)

The ammonium sulphate obtained is a valuable fertilizer. Since enormous amounts of coal are coked every year, the value of the chemical products from coal tar alone runs into millions of pounds.

Benzene and related compounds are also obtained commercially by subjecting open-chain alkanes obtained from petroleum to cyclization followed by dehydrogenation. This conversion is brought about at high temperatures and pressures in the presence of suitable catalysts, for example:

n-Heptane   Methylcyclohexane   Toluene

Benzene is similarly obtained from *n*-hexane. The output of aromatic hydrocarbons from petroleum products has become increasingly significant in recent years.

BENZENE—ITS STRUCTURE

The parent hydrocarbon from which all aromatic compounds are derived is benzene. In 1825 FARADAY isolated it from condensates in pipes used for conducting coke-oven gas, which was then used as an illuminating gas. Benzene has some unusual properties. The molecular formula of benzene is C_6H_6, which seems to indicate a high degree of unsaturation comparable to that of acetylene, C_2H_2 (a saturated hydro-carbon having six carbon atoms will have the formula C_6H_{14}). But peculiarly enough, benzene does not display the reactivity associated with unsaturated compounds like alkenes or alkynes. We know that alkenes are affected by cold permanganate and that they add on bromine. Benzene is quite indifferent to these reagents. Under vigorous forcing conditions benzene reacts with bromine, sulphuric acid, and nitric acid, but the typical reaction products do not result from addition, as in the case of alkenes, but from substitution. Thus, for example, with bromine benzene gives bromobenzene; with sulphuric acid benzene gives sulphonic acid; and with nitric acid benzene gives nitrobenzene.

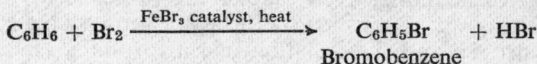

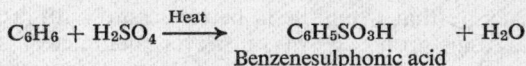

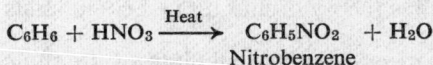

$$C_6H_6 + Br_2 \xrightarrow{\text{FeBr}_3 \text{ catalyst, heat}} C_6H_5Br + HBr$$
$$\text{Bromobenzene}$$

$$C_6H_6 + H_2SO_4 \xrightarrow{\text{Heat}} C_6H_5SO_3H + H_2O$$
$$\text{Benzenesulphonic acid}$$

$$C_6H_6 + HNO_3 \xrightarrow{\text{Heat}} C_6H_5NO_2 + H_2O$$
$$\text{Nitrobenzene}$$

Not only does benzene fail to give the usual results in tests for unsatura-tion (i.e. decolourization of permanganate or bromine water) but it also gives the substitution products previously mentioned, a property which we associate with saturated compounds. The several properties of ben-zene outlined above were hard to reconcile, and no satisfactory explana-tion of what we might call the contrary behaviour of benzene emerged until Kekulé turned his attention to the study of the nature and number of substitution products from benzene.

This line of attack seemed attractive to Kekulé, since the structure of a paraffin could be derived from a knowledge of the number of its substi-tution products. For example, ethylene gives only one monosubstituted product, thereby showing that all the hydrogen atoms in ethylene are equivalent. Propane gives two monosubstituted products and four disubstituted products, accounted for by the structure $CH_3CH_2CH_3$. Benzene was found to give only one monosubstitution product, and this was correctly interpreted by Kekulé as showing that all six hydrogen atoms in benzene are equivalent. He also found that only three disubsti-tuted products are obtainable from benzene. Kekulé conceived the brilliant idea that these facts could be accommodated by a cyclic struc-ture for benzene, represented as a regular hexagon built up of six carbon atoms each carrying a hydrogen atom. To conform to the quadrivalence of carbon he further suggested a system of alternate double bonds, as shown below.

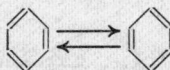

Kekulé's idea was that the double bonds flipped back and forth in the molecule at any instant, giving two types of molecules, indistinguishable in the case of benzene but capable of separate existence in the case of a derivative of benzene substituted on two adjacent carbon atoms like orthoxylene ((I) and (II)).

(I)   (II)

In compound (I) the methylated carbon atoms have a double bond between them, but in compound (II) the methylated carbon atoms are connected by a single bond. No evidence for the existence of these two isomers was forthcoming, even though the presence of three double bonds in the ring was confirmed by the absorption by benzene of three moles of hydrogen and the isolation of a triozonide. Hence Kekulé's structure was not completely accepted, and attempts were made to give benzene a structure that would be in better accord with facts. None of the structures thus proposed was, however, completely satisfactory. THIELE advanced a good explanation for the lack of olefinic properties in benzene by applying his theory of unsaturated compounds to Kekulé's formulae.

According to this theory, the two affinities of a double bond in an unsaturated compound do not completely saturate one another but leave a certain partial valence in excess on each carbon atom. This may be indicated by dotted lines. In addition reactions these partial valences are first used up in attachment to the addenda and then transformed into full valences with the simultaneous disappearance of the double bond, as follows:

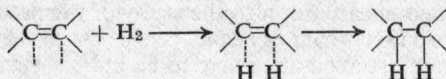

In a system of alternate double and single bonds, referred to as a system of conjugated double bonds, Thiele assumed that the two central partial valences saturate one another so that addition takes place only at the terminal 1,4-positions.

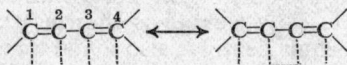

In a further extension of this theory, which provided a satisfactory explanation of the behaviour of many unsaturated compounds, Thiele suggested that the partial valences in the ring formulae of Kekulé mutually saturate one another, making benzene a very stable compound:

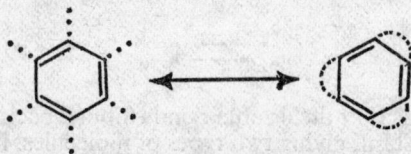

Thiele's explanation seemed quite convincing until a substance known as cyclo-octatetraene, containing a closed system of four conjugated double bonds, was synthesized.

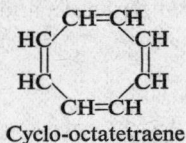

$$CH=CH$$
$$HC \quad CH$$
$$HC \quad CH$$
$$CH=CH$$

Cyclo-octatetraene

On the basis of Thiele's theory, this compound, too, must be stable and should display the aromatic character of benzene. On the contrary, cyclo-octatetraene was found to be quite reactive and gave the usual reactions of a substance containing double bonds. Therefore, Thiele's solution was not entirely accepted.

However, with the development of the principle of resonance, the difficulties encountered with Kekulé's formulae and Thiele's explanation disappeared. The electronic formulae of Kekulé's structures show that the carbon atoms are joined alternately by one and then by two pairs of shared electrons:

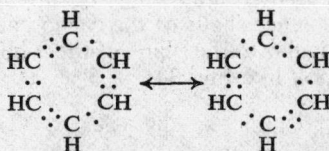

$$
\begin{array}{ccc}
& H & \\
& C & \\
HC & & CH \\
HC & & CH \\
& C & \\
& H &
\end{array}
\longleftrightarrow
\begin{array}{ccc}
& H & \\
& C & \\
HC & & CH \\
HC & & CH \\
& C & \\
& H &
\end{array}
$$

The two structures differ only in the distribution of electrons, and hence benzene is a resonance hybrid of Kekulé's two structures, with a structure intermediate between the two canonical forms. Also, when resonance occurs between equivalent structures the hybrid is characterized by great stability, which is the case with benzene. Further, a characteristic diagnostic test by which we may find whether or not resonance occurs is to see whether the C–C single-bond distance is decreased. Measurements by X-ray-diffraction methods show that the actual distance between two adjacent carbon atoms in the benzene ring is 1·40 Angstrom units. This value is intermediate between the carbon–carbon single bond (C–C) distance of 1·54 units found in the alkanes and the carbon–carbon double bond (C=C) distance of 1·34 units in alkenes. The best simple representation of such a structure is

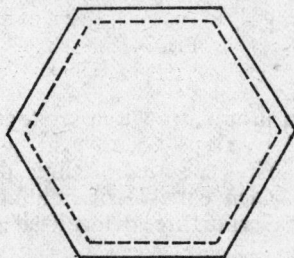

In terms of the molecular orbital approach: The carbon atoms in benzene are trigonally prepared, and each form two σ-bonds with two adjacent carbon atoms and one σ-bond to a hydrogen atom. This gives a strainless, planar, hexagonal structure and uses up all but six of the available valence electrons:

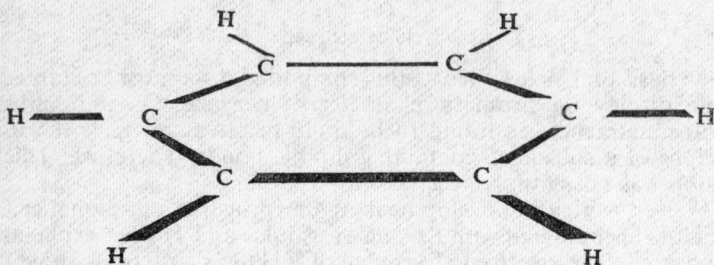

These can be pictured as originally singly occupying the six unhybridized *p*-orbitals in the valence shells of the carbon atoms, whose axes of symmetry are perpendicular to the plane of the σ-bonds. In an exploded diagram they would look like Fig. 34:

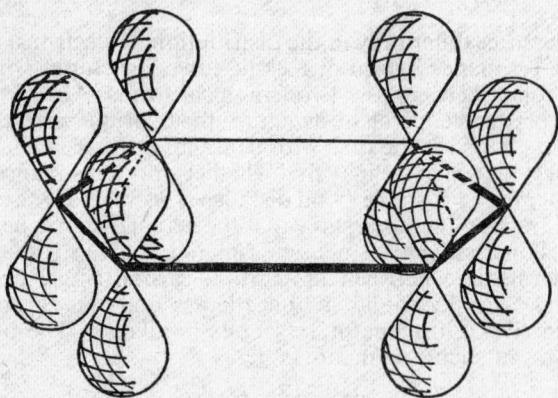

Fig. 34

If we picture the carbon atoms drawn closer together in their correct positions each *p*-orbital overlaps equally the *p*-orbitals on the two adjacent carbon atoms. Thus a series of three delocalized molecular orbitals can be formed, each capable of holding two electrons with opposed spins, giving the C_6 ring three delocalized π-bonds. Their shapes are something like:

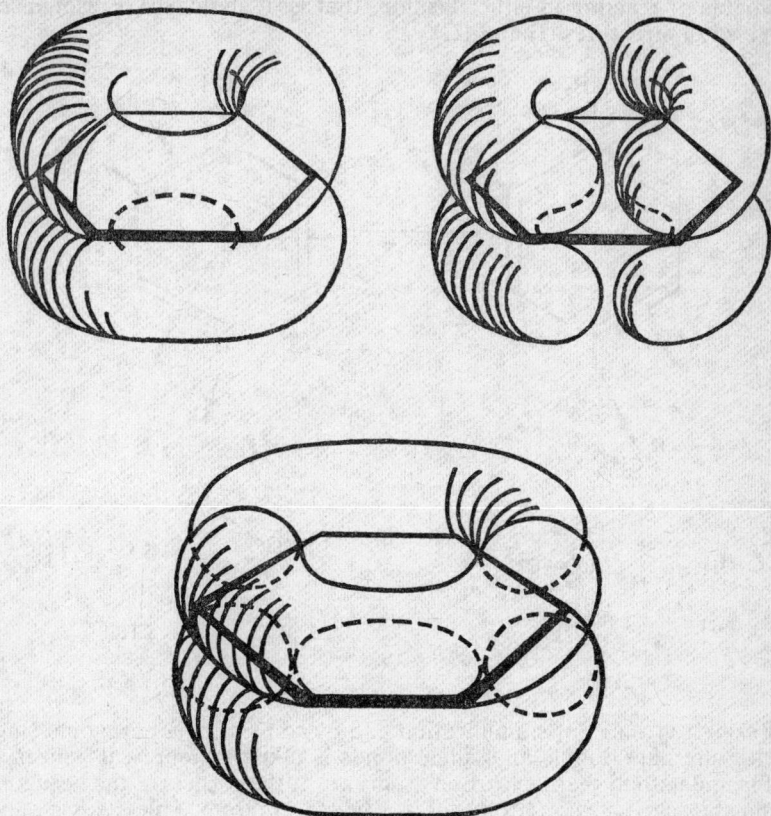

Fig. 35

With regard to the contrast in properties between benzene and cyclo-octatetraene, a satisfactory explanation may be advanced. Cyclo-octatetraene probably has a structure in which the eight carbon atoms do not lie in the same plane. Since the ring has eight carbon atoms, only a non-planar structure will be free from strain. However, even overlapping of the *p*-orbitals is impossible in such non-planar systems, and consequently, cyclo-octatetraene does not exhibit the aromatic characteristics of the benzene ring. This view is confirmed by the fact that the characteristic shortening of the carbon single bond is not observed in cyclo-octatetraene.

A purely chemical piece of evidence that directly supports the resonance structure of benzene may now be cited. On ozonization, ortho-xylene gives glyoxal (OHC–CHO), diacetyl (CH$_3$·CO·CO·CH$_3$), and methylglyoxal (CH$_3$CO·CHO). Neither structure (I) nor structure (II) gives rise to all three substances. Ortho-xylene must therefore be capable of reacting in either fashion, that is, it should be a resonance hybrid of structures (I) and (II).

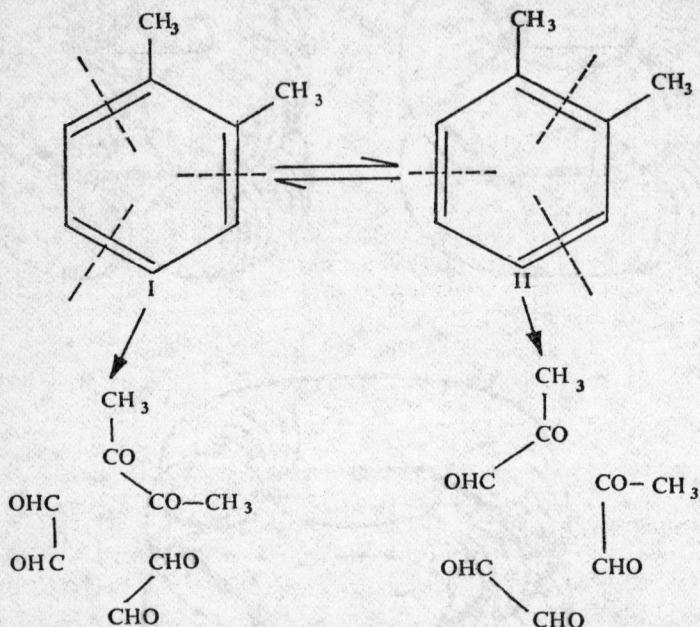

NOMENCLATURE: For simplification and by convention a hexagonal ring with alternate double and single bonds is taken to represent benzene. It is understood that the carbon atoms are at the corners of the hexagon and that each carbon atom holds a hydrogen atom, unless some substituent is shown. The mono-substituted derivatives of benzene offer no difficulty in formulation. The substituent is placed at any corner of the hexagon, since all six positions in the ring are equivalent. Thus, we have

chlorobenzene -Cl, nitrobenzene -NO₂, etc.

There are three isomeric di-substituted products of benzene. If the substituents are on adjacent carbon atoms the compound has a prefix ortho (*o*); if they are on alternate carbon atoms they have the prefix meta (*m*); and if they are on the diagonally opposite carbon atoms they carry the prefix para (*p*). Thus we have the following:

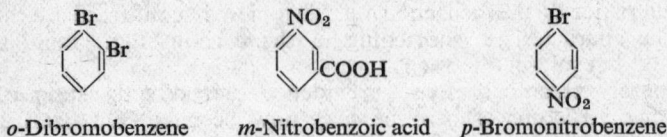

o-Dibromobenzene   *m*-Nitrobenzoic acid   *p*-Bromonitrobenzene

When more than two substituents are present, their positions are numbered as below:

COOH
NO$_2$
NO$_2$

2,3-Dinitrobenzoic acid

Br
Br
Br

1,2,4-Tribromobenzene

Br
NO$_2$
NO$_2$

2,4-Dinitrobromobenzene

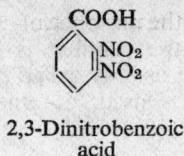

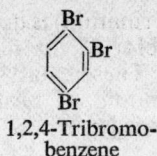

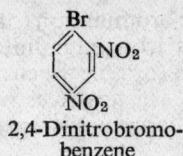

REACTIONS OF BENZENE

Benzene is a colourless liquid having a characteristic odour. It is a very stable substance and some of its reactions were mentioned in the discussion of its structure. We shall now give a general account of its reactions, most of which are typical of other aromatic compounds.

1. On catalytic hydrogenation in the presence of nickel at 200° C. benzene gives cyclohexane.

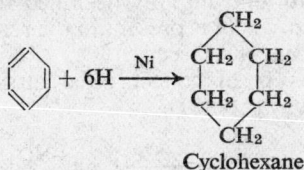

$$\bigcirc + 6H \xrightarrow{Ni} \text{CH}_2 (\text{ring})$$

Cyclohexane

2. **Oxidation.** Benzene is not affected by alkaline potassium permanganate or chromic acid at room temperature. Industrially benzene is oxidized by using vanadium catalyst at high temperatures to give maleic anhydride, as follows:

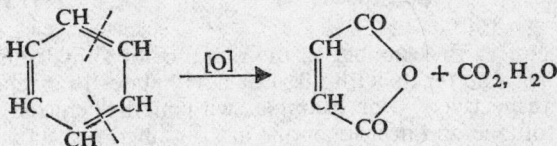

$$\text{benzene} \xrightarrow{[O]} \text{maleic anhydride} + CO_2, H_2O$$

Benzene gives a triozonide which on hydrolysis yields glyoxal.

3. **Halogenation.** It has been mentioned that the normal reaction of benzene is one of substitution. However, in direct sunlight benzene adds on three molecules of chlorine or bromine, giving an addition product $C_6H_6Cl_6$ or $C_6H_6Br_6$ respectively.

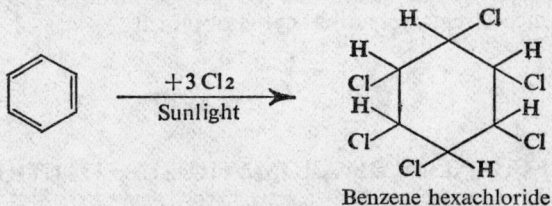

$$\bigcirc \xrightarrow[\text{Sunlight}]{+3Cl_2} \text{Benzene hexachloride}$$

Benzene hexachloride

One of the stereoisomeric forms of benzene hexachloride, called the γ isomer, is used as an insecticide known as Gammexane.

If the bromination (or chlorination) is done in the presence of catalysts such as iron, aluminium chloride, or iodine the product is bromo-benzene (or chlorobenzene). These catalysts are usually called halogen carriers. If an excessive amount of bromine is used, di- and poly-bromo compounds are obtained:

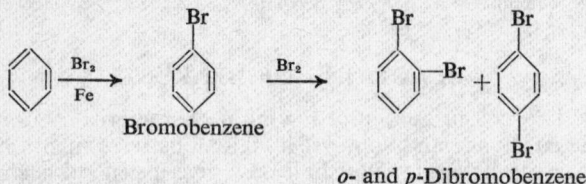

Bromobenzene

o- and *p*-Dibromobenzene

Iodine does not substitute directly owing to the fact that the hydrogen iodide formed during the reaction, being a strong reducing agent, reduces the iodo compound initially obtained to benzene. If a strong oxidizing agent like nitric acid is present in the reaction mixture it is pos-sible to prepare iodobenzene directly.

4. **Nitration.** A mixture of concentrated nitric and sulphuric acids converts benzene into nitrobenzene.

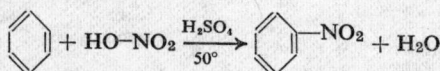

5. **Sulphonation.** Fuming sulphuric acid reacts with benzene to give benzenesulphonic acid.

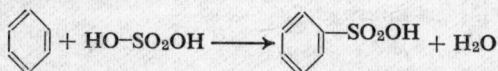

6. **Friedel–Crafts Reaction.** In the presence of anhydrous aluminium chloride, benzene reacts with alkyl or acyl halides to give alkyl or acyl benzenes respectively. For example, with methyl chloride and acetyl chloride, toluene and acetophenone are obtained respectively.

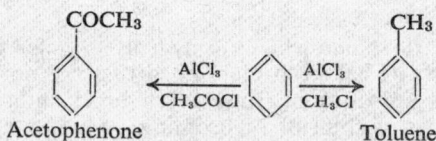

Acetophenone   Toluene

7. **Pyrolysis.** When benzene vapour is passed through a hot iron tube in the presence of catalysts diphenyl is obtained.

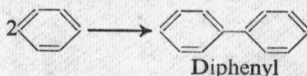

Diphenyl

MECHANISM OF AROMATIC SUBSTITUTION

As has already been mentioned, the normal reactions of benzene are those of substitution. The usual substituting agents are chlorine or bromine, nitric acid, and concentrated sulphuric acid. These reactions, namely halogenation, nitration, and sulphonation, are characteristic of all aromatic systems, and hence we shall deal with them in detail. An attempt will be made to give a mechanism of these reactions consistent with experimental observations. All these substances we have mentioned (chlorine, nitric acid, and sulphuric acid) have one common property, that is, they are all oxidizing agents: all these substitution reactions are in reality oxidation–reduction reactions in which the substituting group is the oxidizing agent (electron acceptor), and the benzene ring is the reducing agent (electron donor). The delocalized π-orbitals concentrate negative charges above and below the plane of six carbon atoms in the benzene ring. This shields the ring carbon atoms from attack by nucleophilic reagents and promotes attack by electrophilic reagents, for example cations.

BROMINATION

Benzene reacts with bromine in the absence of catalysts, but slowly, giving bromobenzene and hydrogen bromide. In the presence of iron the reaction takes place rapidly. The dipole moment of bromine, like that of benzene, is zero, and one wonders why there should be any affinity between these two molecules. Evidently, conditions are created by the presence of iron that enable the polar structures to come into operation. Iron reacts with bromine to form ferric bromide, which is a tricovalent compound having only six electrons in the outermost shell of the iron atom.

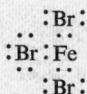

$$\overset{..}{\underset{..}{:}Br:} \\ :Br:Fe \\ \overset{..}{\underset{..}{:}Br:}$$

The molecule of ferric bromide can complete its octet by combining with a molecule of bromine. This polarises the bromine molecule.

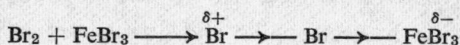

$$Br_2 + FeBr_3 \longrightarrow \overset{\delta+}{Br} \longrightarrow Br \longrightarrow \overset{\delta-}{FeBr_3}$$

The positive end is attracted by the π-electrons of a benzene molecule, readily forming a 'π-complex'.

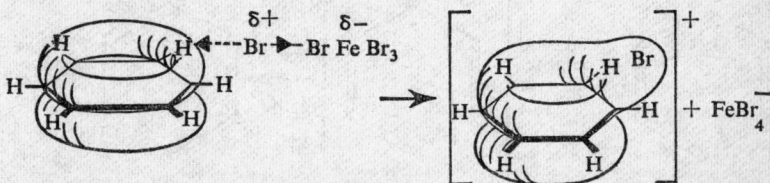

The next stage, conversion to a 'σ-complex', is difficult because the C_6 ring becomes less stable as the delocalized π-orbitals then only involve five carbon atoms and are less symmetrical: so there is an energy barrier to be surmounted in forming a σ-complex:

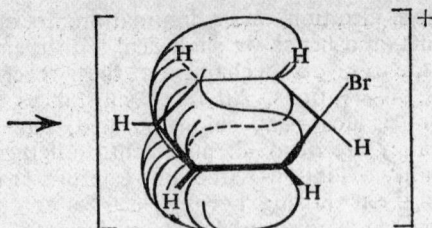

Symmetry and stability are largely regained when a proton is eliminated and an aromatic substitution product is formed.

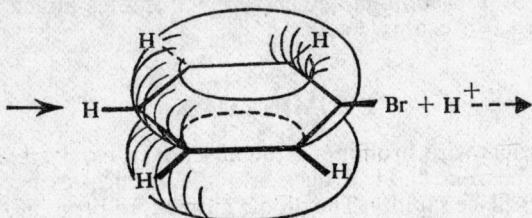

The formation of the intermediate π and σ complexes shown above is actually analogous to the first step in the addition of halogens to unsaturated compounds. The reason why the carbonium ion does not add a bromide ion (Br⁻) in the second step is that the resulting dibromide lacks the stabilization due to resonance of bromobenzene.

NITRATION AND SULPHONATION

These are also substitution reactions brought about by electrophilic attack similar to halogenation studied above. Nitric acid dissolved in sulphuric acid gives the positively charged nitronium ion (NO_2^+) as follows:

$$HNO_3 + 2H_2SO_4 \rightleftharpoons H_3O^+ + NO_2^+ + 2HSO_4^-$$

This nitronium ion is electrophilic and attacks any of the polarized benzene structures, as follows:

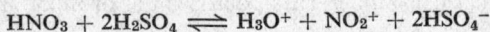

Sulphonation proceeds in much the same way, involving sulphur trioxide, which seems to react as

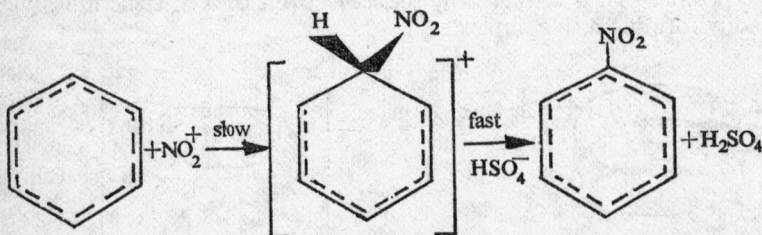

THE FRIEDEL–CRAFTS REACTION

The Friedel–Crafts reaction, whereby an alkyl halide or acyl halide reacts with benzene or other aromatic hydrocarbons in presence of aluminium chloride to give alkyl substituted hydrocarbons or ketones respectively, follows the same mechanism as bromination and nitration. The catalyst combines with R–X or R–COX to give, respectively, the complexes $(AlCl_3X)^- R^+$ and $(AlCl_3X)^- C^+OR$. Substitution takes place in the usual fashion after reaction is initiated by attack of R^+ or $R \cdot C^+ = O$, as follows:

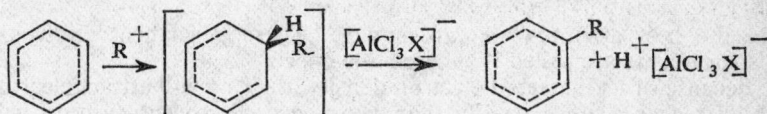

ORIENTATION OF AROMATIC SUBSTITUTION

On the introduction of a second substituent Y into a benzene derivative C_6H_5X, the position taken up by Y is directed by the substituent X, already present in the ring. Whether the disubstituted product obtained is mainly *ortho*, *para*, or *meta* depends mainly upon the speed of the substitution reaction. All rather fast substitution reactions give rise to mainly *ortho* and *para* derivatives, while reactions which are slow yield mostly the *meta* derivative. In both these cases it will be found that a satisfactory picture can be obtained by considering the electronic nature of the group already attached to the benzene ring.

While considering the mechanism of substitution in benzene it was noted that the benzene ring donates electrons to the substituting reagent. The facility with which a compound C_6H_5X will undergo a reaction to give C_6H_4XY may be expected to depend on the electronic structure of C_6H_5X, and in particular on whether group X increases or decreases the electron density in the ring. If electrons are more readily available in C_6H_5X than in benzene the rate of substitution in C_6H_5X will be proportionately faster. For example, phenol with a high electron density on the ring undergoes substitution reactions much faster than benzene. This is most simply interpreted in resonance terms. In addition to the benzenoid structure (I), we can write an *ortho*-quinonoid structure (II), in which one of the unshared pairs on the oxygen atom has formed a dative covalent bond to carbon. Similarly, we can also write a *para*-quinonoid structure (III).

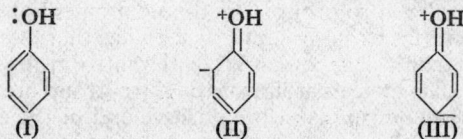

According to the concept of resonance, the real structure is none of these, but is an intermediate one to which they make contributions.

While it is difficult to represent this hybrid structure accurately on paper, two features are obvious: (1) the electron density in the ring will be greater than that of benzene; and (2) there will be small negative charges in the positions *ortho* and *para* to the hydroxyl group.

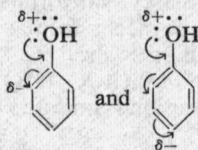

are the best attempts we can manage. These permanent electronic displacements are called the **mesomeric effect.**

Because of its increased electron density, the ring will attract electrophilic reagents more readily than benzene does, so substitution will be easier in phenol than in benzene. Because the negative charges are concentrated at the *ortho* and *para* positions, the reagent will attack there preferentially, so substitution will be mainly *ortho* and *para*.

This is actually the case, since bromination of phenol takes place readily, even in dilute aqueous solutions and in the absence of catalysts. When bromine water is added to phenol an immediate decolourization of the bromine takes place, with the precipitation of 2,4,6-tribromophenol. When the electrophilic reagent approaches close to the C$_6$ ring the permanent mesomeric effect is increased through the attraction of electrons into the ring by the reagent. This time-variable effect is called the **electromeric effect.** The sum of the two effects is called the **tautomeric effect.**

Let us now study another type of substituent, the nitro group, with reference to its directing influence. The nitro group (unlike the −OH group) is an electron acceptor and decreases the electron density in the ring. The following four resonance structures may be written for nitrobenzene:

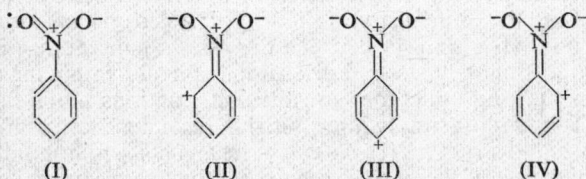

These structures show that the ring is deficient in electrons and that the induced positive charge is located in position 2, 4, or 6, all of which are *ortho* or *para* to the nitro group. The dipole moment of nitrobenzene is 3·95 units, a fact that strongly supports the view that the ring is partially depleted of electronic charges, and that the nitro group is correspondingly enriched. The oxygen atoms of the nitro group are at the negative end, and the benzene ring is at the positive end of the dipole. The low electron density in the ring slows down substitution reactions of nitrobenzene and, further, since the *meta* positions are relatively more electron-rich than the other positions, substitution will take place preferably in the *meta* positions. Similarly, other electron-attracting groups, such as the sulphonic acid group, the cyano group, the aldehyde group, etc., are *meta* orienting.

Theoretical considerations like the above, taken in conjunction with the experimental behaviour of benzene derivatives on substitution, have led to some empirical rules regarding orientation. One such generalization, which is very useful in predicting the orienting effect of a particular group, may be mentioned here. Like all other good rules, this one also has its exceptions. According to this rule, groups attached to the aromatic ring by atoms forming only single bonds are *ortho–para* orienting, and groups attached by atoms forming multiple bonds are *meta* orienting. For example, the groups $-OH$, $-NH_2$, $-NR_2$, $-CH_3$, and

$-X$ are all *ortho–para* orienting, whereas $-\overset{\underset{|}{OH}}{C}{=}O$, $-\overset{\underset{|}{OR}}{C}{=}O$, $-C{\equiv}N$, $-SO_2OH$,

and $-N\overset{\diagup O}{\diagdown_{O}}$ groups are *meta* orienting. It must be noted that it is the atom attached to the ring which is the key atom to be considered in applying the above rule. Thus $-CH_2-COOH$, $-CH_2-NO_2$ are *ortho–para* orienting, although there are multiple bonds in the group, because the atoms attached to the ring in each case form only single bonds.

HOMOLOGUES OF BENZENE AND POLYNUCLEAR HYDROCARBONS

The homologues of benzene can be prepared by the Friedel–Crafts method of alkylation using anhydrous aluminium chloride as a catalyst. We have already considered the mechanism of such alkylations. The general reaction is:

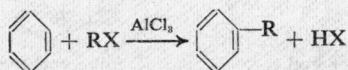

$$\text{benzene} + RX \xrightarrow{AlCl_3} \text{benzene}-R + HX$$

The reaction succeeds also with *ortho–para*-directing substituents in the benzene ring. However, in the presence of *meta*-orienting groups in the ring there is no reaction. For example, nitrobenzene cannot be alkylated by the Friedel–Crafts reaction. Though aluminium chloride is the favoured catalyst, other catalysts, like hydrogen fluoride, zinc chloride, ferric chloride, stannic chloride, sulphuric acid, and boron trifluoride, are often used.

The physical properties of monoalkyl benzenes are essentially like those of benzene. The alkyl groups are *ortho–para* orienting. Hence alkyl benzenes undergo the usual reactions, such as nitration, halogenation, sulphonation, and alkylation, more readily than benzene. In the case of alkyl benzenes reaction may also occur in the alkyl groups (side chains). The side chain has aliphatic properties, and hence may be halogenated under the same conditions as alkanes. For example, when toluene is chlorinated at its boiling point in the presence of light the halogen enters the side chain, and progressive substitution takes place, as represented below:

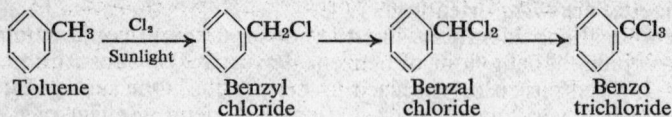

| Toluene | Benzyl chloride | Benzal chloride | Benzo trichloride |

On the other hand, if we use iron or aluminium chloride as a catalyst chlorination gives a mixture of *ortho*- and *para*-chlorotoluenes.

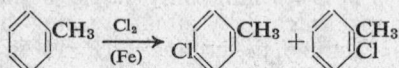

Monoalkyl benzenes when oxidized with alkaline permanganate or other suitable oxidizing agents give benzoic acid. In other words, the side chain, whatever its length or structure, is preferentially oxidized to $-COOH$. This oxidation indicates the stability of the aromatic ring.

In addition to the above homologues of benzene, there are also some polynuclear hydrocarbons which may be mentioned here. These may be divided into two classes: (1) those in which two or more benzene rings are linked together directly or through one or more carbon atoms, and (2) those in which the rings are condensed together in such a way that each pair has two carbon atoms in common. Diphenyl, in which two benzene rings are linked directly together, and triphenylmethane, in which three benzene rings are attached to a common carbon atom, belong to the former class.

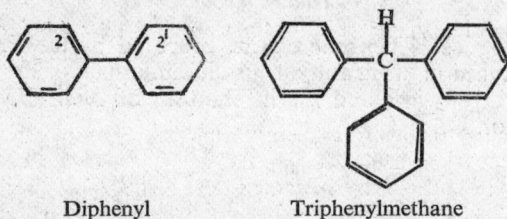

| Diphenyl | Triphenylmethane |

Naphthalene, anthracene, and phenanthrene belong to the latter category of hydrocarbons with condensed ring systems:

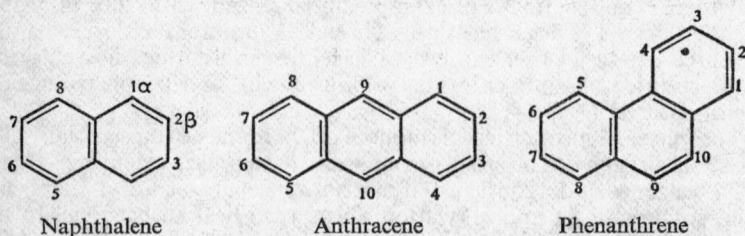

| Naphthalene | Anthracene | Phenanthrene |

The structures and the method of numbering the carbon atoms in them are shown above.

Diphenyl derivatives can be prepared by several methods. Diphenyl is a white crystalline solid which because of its high boiling point (225° C.) and stability is used industrially as a heat-transfer medium. Evaporators, reaction kettles, and other pieces of equipment can be heated to a high temperature using diphenyl, where the use of steam will be found to necessitate the employment of high pressure. Diphenyl for these purposes is manufactured by passing benzene vapour over a catalyst in iron tubes. Diphenyl can also be prepared by heating iodobenzene with copper powder (Ullmann reaction).

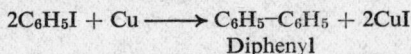

$$2C_6H_5I + Cu \longrightarrow C_6H_5-C_6H_5 + 2CuI$$
$$\text{Diphenyl}$$

The chemical properties of diphenyl are very similar to those of benzene and diphenyl undergoes nitration, sulphonation, and halogenation in a typical aromatic manner. The existence of optical isomerism in the case of diphenyl derivatives, with suitable substituents in 2, 2', 6, and 6' positions, has already been mentioned.

Of the other polyphenyls, mention may be made of the discovery of the triphenylmethyl radical by GOMBERG of the University of Michigan. He was attempting to prepare hexaphenylethane by an action of finely divided silver, zinc, or mercury on triphenylchloromethane.

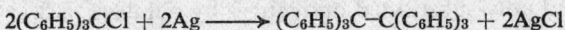

$$2(C_6H_5)_3CCl + 2Ag \longrightarrow (C_6H_5)_3C-C(C_6H_5)_3 + 2AgCl$$

The product he obtained was, however, an oxygenated compound. It was found that the same oxygenated compound could be obtained by the action of sodium peroxide on triphenylchloromethane.

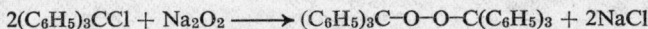

$$2(C_6H_5)_3CCl + Na_2O_2 \longrightarrow (C_6H_5)_3C-O-O-C(C_6H_5)_3 + 2NaCl$$

Repetition of the first experiment in the absence of air gave the expected hexaphenylethane as a colourless crystalline solid. Exposure of the benzene solution of this substance gave the peroxide originally obtained. Also, a solution of the hydrocarbon, initially colourless, soon turned yellow. The yellow solution, after removal of the solvent, yielded the colourless hydrocarbon indicating reversible change. Gomberg further noted that hexaphenylethane reacted with bromine and iodine almost instantaneously—a fact quite out of keeping with the properties normally associated with a saturated ethane molecule. He concluded, therefore, that the hydrocarbon dissociated into two triphenylmethyl free radicals in which the carbon atom to which the rings are linked is trivalent.

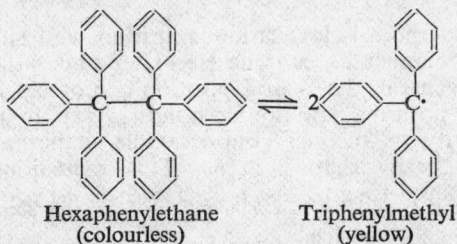

Hexaphenylethane   Triphenylmethyl
(colourless)   (yellow)

The existence of triphenylmethyl radical, though contrary to the concept of the tetrahedral carbon atom, has received an explanation in terms of the resonance theory developed subsequently. The triphenylmethyl radical has an unshared electron which can be moved around to the various *o*- and *p*-positions of one ring as shown below.

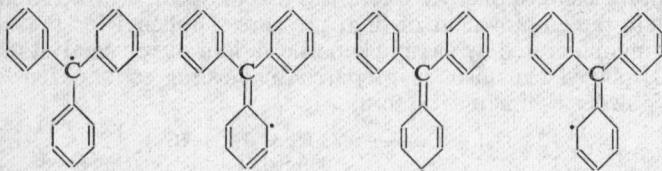

Resonance between these structures results in the peculiar stability of the above free radical.

Naphthalene ($C_{10}H_8$). Of the aromatic hydrocarbons containing condensed ring systems and present in coal tar, naphthalene is the most abundant. It is important as the starting material for the manufacture of phthalic anhydride and other intermediates for the dye industry. As a moth repellent, it has been supplanted by other more effective substitutes.

Mono-substituted naphthalene derivatives are referred to by the prefix α or β and poly-substituted derivatives by numbers.

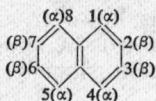

Naphthalene gives two mono-substituted products of the type $C_{10}H_7X$ and ten di-substituted products of the type $C_{10}H_6X_2$.

REACTIONS: Air oxidation of naphthalene in the vapour phase with vanadium pentoxide as catalyst is the basis for the manufacture of phthalic anhydride:

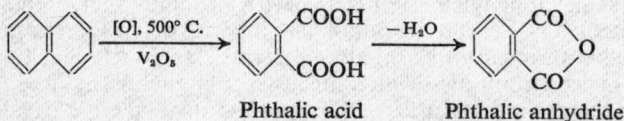

Naphthalene undergoes halogenation, nitration, and sulphonation in much the same manner as benzene (see later under the appropriate derivatives). In general, the α- or 1-position is more easily substituted than the β- or 2-position. The double bonds in naphthalene exhibit in part the reactivity of olefines. For example, naphthalene adds on hydrogen more easily than benzene. This reduction can be accomplished in stages to give *di-*, *tetra-*, and *deca*-hydronaphthalenes respectively.

Anthracene ($C_{14}H_{10}$). This hydrocarbon is obtained from the anthracene oil fraction of coal-tar distillation. The synthesis of anthracene from *o*-bromobenzyl bromide lends support to the structure of the compound as a condensed linear system of three benzene rings, as shown below:

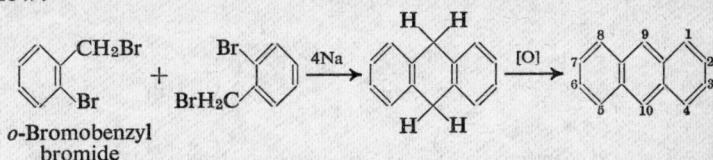

o-Bromobenzyl bromide

Anthracene is considered to be a resonance hybrid of the following structures. Of these, (III) and (IV) contain two truly aromatic benzene rings each, in contrast to (I) and (II), which contain only one such ring, and hence contribute predominantly to the resonance hybrid.

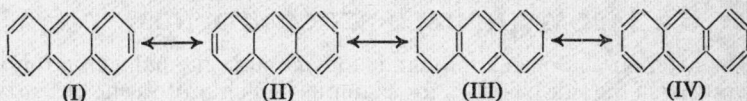

(I) (II) (III) (IV)

Since all three rings do not simultaneously have a conjugated system of single and double bonds, anthracene exhibits a certain degree of aliphatic character. The 9 and 10 positions in the anthracene molecule, referred to as *meso* positions, are the usual points of attack during such reactions as halogenation, oxidation, and reduction. Nitration and sulphonation, however, give mixtures of subsituted anthracenes. Oxidation of anthracene gives anthraquinone and reduction gives 9,10-dihydroanthracene.

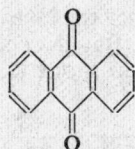

Anthraquinone   9,10-Dihydroanthracene

Anthraquinone is important because of its relationship to alizarin and other naturally occurring dyes, on the one hand, and to the purgative principles of cascara, rhubarb, aloes, etc., which are glycosides of hydroxy derivatives of anthraquinone.

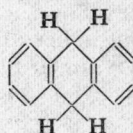

Phenanthrene ($C_{14}H_{10}$). This hydrocarbon is isomeric with anthracene and occurs along with it in coal tar. It has the following structure:

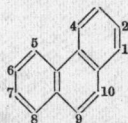

The three benzene rings in phenanthrene are truly aromatic as compared with anthracene, and hence phenanthrene is more stable and less reactive. For instance, phenanthrene from coal tar is easily freed from contaminated anthracene by selective oxidation of the latter to the quinone. The usual aromatic substitution reactions proceed poorly with phenanthrene and give mixtures of several products. Only bromination gives a good yield of 9-bromophenanthrene.

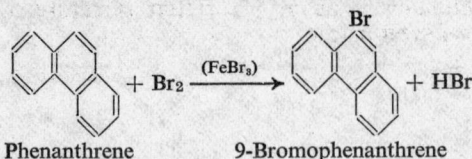

$$\text{Phenanthrene} + Br_2 \xrightarrow{\text{(FeBr}_3\text{)}} \text{9-Bromophenanthrene} + HBr$$

Phenanthrene   9-Bromophenanthrene

The 9,10 double bond is the point of attack in reduction and oxidation reactions. The phenanthrene ring is found in such natural products as sterols, bile acids, sex hormones, and morphine alkaloids.

AROMATIC HALOGEN COMPOUNDS

The aromatic halogen compounds may contain the halogen atom in the ring or in the side chain as, for example, in *o*-chlorotoluene or benzyl chloride.

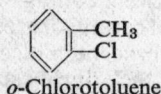

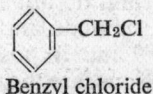

o-Chlorotoluene   Benzyl chloride

The position of the halogen atom in the compound is of great importance in deciding its reactivity. Therefore, the chlorine atom in chlorobenzene is strongly attached to the ring, and in contrast with aliphatic chlorides shows no particular tendency to react with potassium hydroxide (except under drastic conditions). But the chlorine atom in benzyl chloride is much more labile. Even in moist air, it is rapidly hydrolysed to benzyl alcohol.

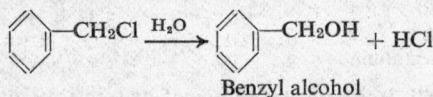

$$\text{—}CH_2Cl \xrightarrow{H_2O} \text{—}CH_2OH + HCl$$

Benzyl alcohol

This difference in reactivity may be explained satisfactorily on the basis of the electronic theory. In the chlorobenzene molecule a situation occurs which recalls the case of vinyl chloride, discussed in Chapter Eleven. A lone pair of electrons in a *p*-orbital of the chlorine atom, whose axis of symmetry is perpendicular to the plane of the C_6 ring, extends the system of delocalized-molecular orbitals (Fig. 36) or, from

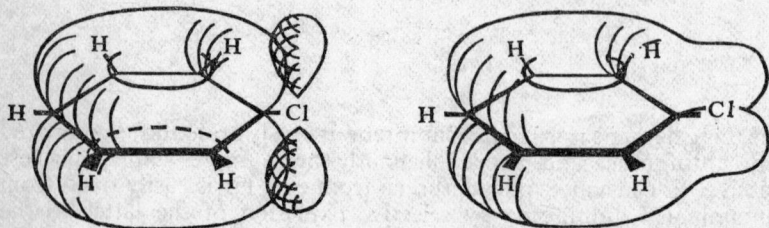

Fig. 36

the resonance point of view, participates in resonance with the electrons
of the ring, with the consequent shortening of C–Cl distance and firmer
attachment of the chlorine atom. The actual structure is intermediate
between the four shown below.

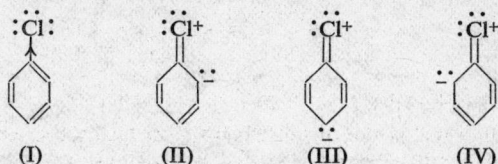

(I)   (II)   (III)   (IV)

Benzyl chloride, however, is analogous to allyl chloride. Its reactivity
may be accounted for by its tendency to form an ion analogous to allyl
carbonium ion, namely benzyl carbonium ion, which is stabilized by
resonance as follows:

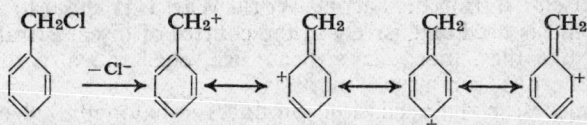

Chlorobenzene is prepared in the laboratory by direct chlorination in
the presence of a carrier (iron, aluminium chloride, or iodine). The
mechanism of this reaction has already been discussed. Referring back
to the resonance structures for chlorobenzene, in formula (I), the
electronegative chlorine atom is shown to exert an attraction for the
electrons of the ring. The dipole moment of chlorobenzene is 1·55 units,
which is less than that of methyl chloride (1·85 units). This decrease
in the dipole moment of chlorobenzene is evidence of the contribution
to the resonance hybrid of structures (II), (III), and (IV). These states
not only reduce the reactivity of the chlorine atom but also lead to an
ortho–para orienting effect. But in structure (I) chlorine withdraws
electrons inductively from the ring, and hence should be *meta* orienting.
Indeed, such *meta* orientation takes place when chlorobenzene is
chlorinated; but the *ortho* or *para* substitution due to structures (II),
(III), and (IV) is a faster reaction. These two forces oppose each other,
and the net result is that the chlorine is chiefly *ortho–para* orienting, but
the rate of substitution is less than in the case of benzene.

Industrially, chlorobenzene is used for the manufacture of phenol by
heating it with sodium hydroxide solution at about 300° C., under a
high pressure of about 200 atmospheres:

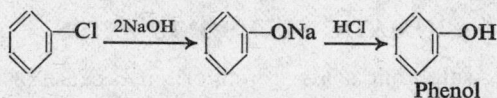

Phenol

Similarly, chlorobenzene is treated with ammonia at about 200° C., at a
pressure of about 60 atmospheres in the presence of a catalyst to furnish
aniline.

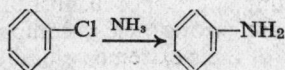

Chlorobenzene is also used for the preparation of the well-known insecticide DDT (Dichlorodiphenyltrichloroethane). DDT is obtained by the reaction of one molecule of chloral with two molecules of chlorobenzene in the presence of sulphuric acid, as follows:

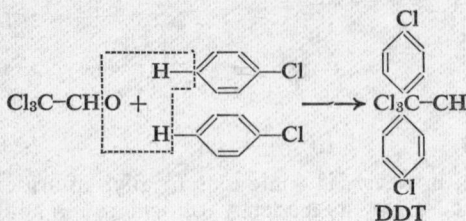

$$Cl_3C-CH O + \begin{array}{c} H-\!\!\!\!-\!\!\!\!\langle \rangle\!\!-Cl \\ H-\!\!\!\!\langle \rangle\!\!-Cl \end{array} \longrightarrow Cl_3C-CH$$

DDT

Though first synthesized in 1874, DDT first attained great prominence as an insecticide during the Second World War. It is lethal to a variety of insects and is used extensively in the control of diseases transmitted by insects like flies, mosquitoes, fleas, lice, etc., and also to protect agricultural crops from attack by pests.

Naphthalene on halogenation produces α-halonaphthalenes. For example, α-bromonaphthalene is obtained in excellent yield by bromination of naphthalene in boiling carbon tetrachloride solution. On bromination anthracene gives the 9,10-dibromide, which readily loses hydrogen bromide to give 9-bromoanthracene:

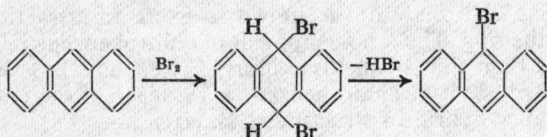

Similarly, phenanthrene may be brominated to give either the 9,10-dibromide or 9-bromophenanthrene.

Arylhalides are important intermediates in the syntheses of many complicated organic compounds. Their use in the synthesis of diphenyls by utilization of the ULLMANN reaction has already been mentioned. Grignard reagents (RMgX) are fairly easily prepared, particularly from aryl bromides and iodides, and display the versatility of their aliphatic analogues.

AROMATIC SULPHONIC ACIDS

The aromatic sulphonic acids are important because of their usefulness as starting materials for the preparation of other aromatic compounds. They are prepared by direct sulphonation of aromatic hydrocarbons with concentrated or fuming sulphuric acid. The sulphonic acid group is *meta* orienting, and hence disubstitution is more difficult.

The sulphonic acids are strong acids, comparable with sulphuric acid in strength. In view of their extreme solubility in water, sulphonic acids are usually isolated in the form of their sodium or barium salts. The sulphonic acids form salts, esters, acid chlorides, amides, etc., in the same fashion as carboxylic acids. When benzenesulphonic acid is treated with phosphorus pentachloride it forms benzenesulphonyl chloride:

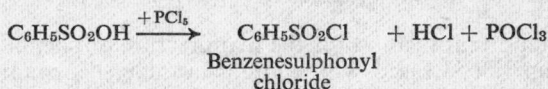

$$C_6H_5SO_2OH \xrightarrow{+PCl_5} C_6H_5SO_2Cl + HCl + POCl_3$$

Benzenesulphonyl
chloride

When the acid chloride thus obtained is treated with ammonia, benzene-sulphonamide is obtained. Sulphanilamide, the sulpha drug so widely used for the treatment of infectious diseases, is *p*-aminobenzene-sulphonamide (see Chapter Eighteen).

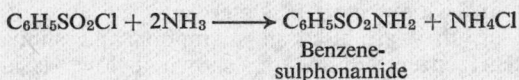

$$C_6H_5SO_2Cl + 2NH_3 \longrightarrow C_6H_5SO_2NH_2 + NH_4Cl$$

Benzene-
sulphonamide

When *p*-toluenesulphonamide is treated with sodium hypochlorite, the sodium salt of *p*-toluenesulphonchloramide, known as '**chloramine-T**', is obtained:

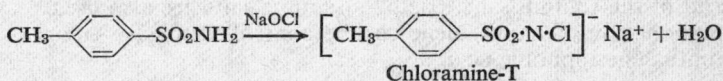

$$CH_3-\underset{}{\bigcirc}-SO_2NH_2 \xrightarrow{NaOCl} \left[CH_3-\underset{}{\bigcirc}-SO_2 \cdot N \cdot Cl\right]^- Na^+ + H_2O$$

Chloramine-T

Chloramine-T is extensively used as an antiseptic and a disinfectant. Halazone, another disinfectant, is related to chloramine-T and is more stable.

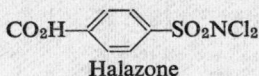

$$CO_2H-\underset{}{\bigcirc}-SO_2NCl_2$$

Halazone

The sulphonation of toluene gives a mixture of *o*- and *p*-toluenesulphonic acids. The *p*-isomer is used in the manufacture of chloramine-T or halazone, while the *o*-compound serves as the starting material for the preparation of the sweetening agent saccharin. The methyl group of *o*-toluenesulphonamide, on oxidation, is converted into the carboxyl group to give *o*-sulphonamidobenzoic acid, which readily loses water and gives saccharin.

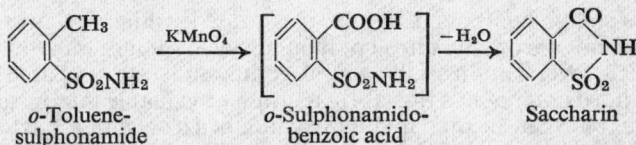

o-Toluene-   *o*-Sulphonamido-   Saccharin
sulphonamide   benzoic acid

Saccharin and its sodium salt are about 500 times sweeter than cane sugar. They are used by diabetic patients from whose diet sugar is excluded. Saccharin has no nutritional value. It is used extensively in the sweetening of toothpastes, aerated waters, and so on.

When the sodium salt of benzenesulphonic acid is fused with sodium hydroxide phenol is obtained.

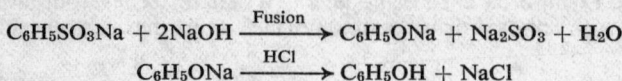

$$C_6H_5SO_3Na + 2NaOH \xrightarrow{\text{Fusion}} C_6H_5ONa + Na_2SO_3 + H_2O$$
$$C_6H_5ONa \xrightarrow{\text{HCl}} C_6H_5OH + NaCl$$

This is an important method for the manufacture of phenol. When the same sodium salt is fused with sodium cyanide, benzonitrile, shown below, is obtained.

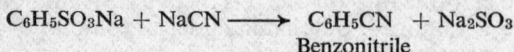

$$C_6H_5SO_3Na + NaCN \longrightarrow C_6H_5CN + Na_2SO_3$$
Benzonitrile

The sodium salts of some substituted benzene and naphthalene sulphonic acids are widely used as detergents. Because of their good lathering power, they are used in shampoos and similar products. Sodium dodecylbenzenesulphonate, shown below, is an excellent detergent.

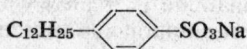

$$C_{12}H_{25}-\underset{}{\bigcirc}-SO_3Na$$

Instead of the sodium salts, triethanol amine salts are also used.

Sulphonation of naphthalene can be controlled to give either α- or β-naphthalenesulphonic acid.

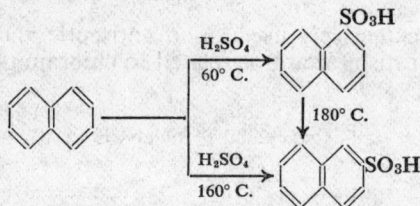

Low-temperature sulphonation gives the α-sulphonic acid, which at higher temperatures in the presence of sulphuric acid goes over into the β-isomer. The latter also results from direct sulphonation at 160° C.

AROMATIC NITRO COMPOUNDS

Most aromatic compounds are easily nitrated, while aliphatic compounds are not. Aromatic nitro compounds yield amines on reduction, from which several industrially useful compounds are prepared. The aromatic nitro compounds thus form a group of valuable intermediates.

Nitrobenzene can be prepared by reacting benzene with a mixture of concentrated nitric acid and sulphuric acid, known as mixed acid, at about 50° C. Sulphuric acid serves to ionize nitric acid to produce the nitronium cation (NO_2^+), which is the species involved in nitration (see mechanism previously discussed). If the temperature of nitration is raised and fuming nitric acid is used, *m*-dinitrobenzene results.

$$\text{C}_6\text{H}_6 + \text{HNO}_3 \xrightarrow{\text{H}_2\text{SO}_4} \text{C}_6\text{H}_5\text{NO}_2 + \text{H}_2\text{O}$$

$$\text{C}_6\text{H}_5\text{NO}_2 \xrightarrow[\text{100° C.}]{\text{Fuming HNO}_3 + \text{H}_2\text{SO}_4} \text{C}_6\text{H}_4(\text{NO}_2)_2$$

m-Dinitrobenzene

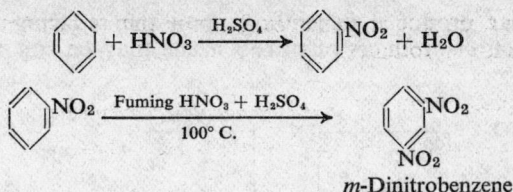

The nitric–sulphuric acid mixture is the usual agent employed for nitration of aromatic compounds. Some modifications may have to be effected, depending on the nature of the material to be nitrated. For nitrating a compound of the formula C_6H_5X, where X is an electron-withdrawing group such as $-NO_2$, $-SO_3H$, $-COOH$, owing to the low electron density on the ring, strong nitrating agents and vigorous conditions are needed. But if X is an electron-releasing group like $-OH$, $-CH_3$, etc., the electron density of the ring is relatively high and comparatively mild agents may suffice. For example, the conditions applicable for nitration of nitrobenzene may be contrasted with the use of dilute nitric acid at room temperature to nitrate phenol. Other nitrating agents sometimes used include a mixture of alkali nitrate and sulphuric acid and acetyl nitrate, obtained by dissolving fuming nitric acid in acetic anhydride. Acetyl nitrate has the advantage, because with alkyl-benzenes it almost exclusively yields mononitro derivatives which are largely *ortho* isomers.

The effect of the nitro group, whereby the ring is deactivated as a whole and the *o* and *p* positions are made relatively more electron-deficient than the *m* position, has already been referred to. This deactivation renders the ring susceptible to a type of substitution reaction different from the electrophilic substitution reactions encountered so far. For example, when nitrobenzene is heated with solid potassium hydroxide, potassium salts of *o*- and *p*-nitrophenols are obtained in limited yields in addition to oxidation products derived from them.

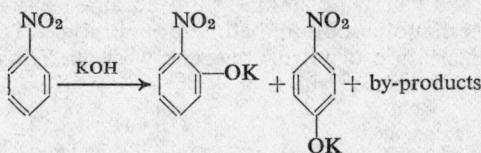

This reaction involves attack of the positive centres at the *o* and *p* positions by an anion (OH^-) and is referred to as a nucleophilic substitution.

Reduction of Nitro Compounds. The most important reaction of nitro compounds is their reduction to amines. There are several intermediate stages involved in this reduction, and by proper choice of the conditions of the reduction these intermediate products could be obtained. For example, with nitrobenzene, reduction in acid medium with tin and hydrochloric acid, or zinc and hydrochloric acid gives the final product, aniline. Reduction in neutral medium with zinc and ammonium chloride gives phenylhydroxylamine, probably through the intermediate nitrosobenzene. In alkaline medium the reduction of nitrobenzene gives rise to bimolecular products, depending upon the reducing agents. The various reduction products of nitrobenzene may be summarized in a chart, as follows:

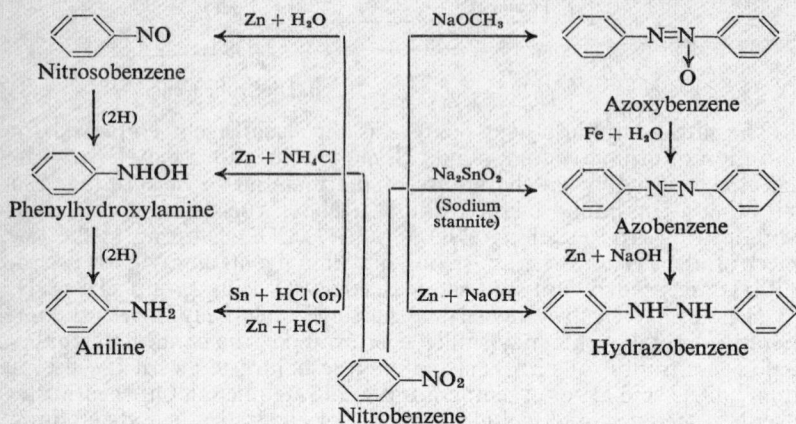

Nitrosobenzene can be obtained by reducing nitrobenzene with ion powder or zinc dust and water. But it is best obtained by oxidation of phenylhydroxylamine with sodium dichromate. In the presence of hydrochloric acid, phenylhydroxylamine rearranges to give *p*-amino-phenol.

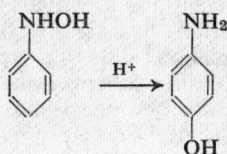

Azoxybenzene is a yellow crystalline solid obtained by refluxing nitro-benzene with methanolic sodium methoxide solution. It is formed by an aldol-type condensation of nitrosobenzene with phenylhydroxylamine, followed by elimination of water.

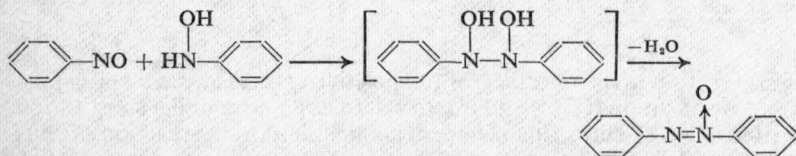

On heating azoxybenzene with iron powder and water **azobenzene**, a red crystalline substance, is obtained. The latter is also prepared by reducing nitrobenzene with zinc dust and alcoholic sodium hydroxide solution or sodium stannite. A point of theoretical interest attaches to the azo-benzene molecule. The presence of a double bond between the two nitro-gen atoms of azobenzene gives rise to the existence of the following geometrical isomers:

C₆H₅—N
‖
C₆H₅—N

syn-Azobenzene
(Dipole moment 3·0 units)

C₆H₅—N
‖
N—C₆H₅

anti-Azobenzene
(Dipole moment 0 units)

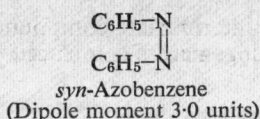

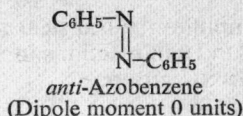

The *anti* compound has been known for a long time, and the *syn* compound was prepared recently by irradiation of the *anti* compound. The dipole moments support the structures given. The terms *syn* and *anti* correspond to *cis* and *trans*.

Reduction of nitrobenzene with zinc dust and sodium hydroxide in more concentrated solutions than required for the production of azobenzene gives hydrazobenzene. The latter undergoes a remarkable rearrangement when treated with acid to give benzidine.

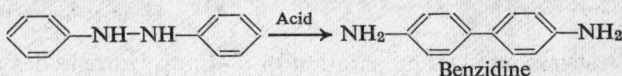

⟨ ⟩—NH—NH—⟨ ⟩ —Acid→ NH₂—⟨ ⟩—⟨ ⟩—NH₂

Benzidine

Benzidine is an important intermediate for the manufacture of dyes like Congo Red.

Activating Effect of Nitro Group. The nitro group exerts a loosening effect upon a halogen atom in the *ortho* or *para* position. A halogen atom attached to a benzene ring, as noted earlier, is inert and cannot be readily displaced. However, in *o*- and *p*-nitrochlorobenzene the chlorine atom is exchanged readily for hydroxyl, amino, or other groups.

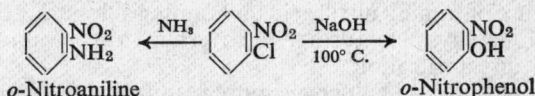

⟨ ⟩NO₂/NH₂ ←NH₃— ⟨ ⟩NO₂/Cl —NaOH 100° C.→ ⟨ ⟩NO₂/OH

o-Nitroaniline   *o*-Nitrophenol

The chlorine atom in 2,4-dinitrochlorobenzene is even more reactive in such reactions. If the halogen is *meta* with respect to the nitro group it does not show any comparable reactivity. The carbon atom to which the halogen is attached tends to become a positive centre in the presence of an *o*- or *p*-nitro group. This tendency is at variance with the electron-withdrawing nature of the halogen atom, which is consequently displaced by a nucleophilic species like OH⁻ or NH₃. We can write the following canonical forms

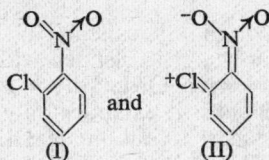

(I)   and   (II)

for *o*-nitrochlorobenzene, and

(III)   (IV)

for *p*-nitrochlorobenzene. The electron-attracting nature of the nitro group will make the contribution of structures (II) and (IV) important.

This simplifies the transient formation of intermediate compounds, like (V) and (VI), in reactions in which the halogen atom is replaced, e.g. by the hydroxyl group:

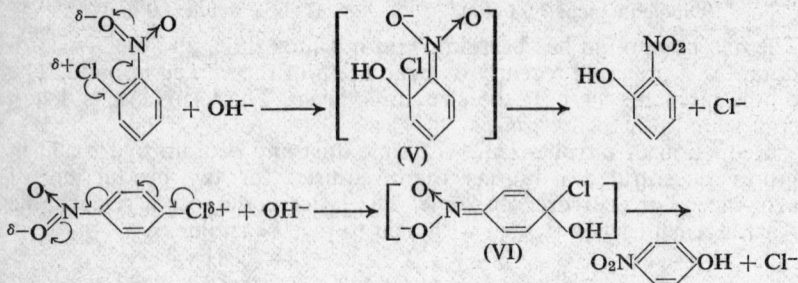

(V)

(VI)

Polynitrobenzenes. The preparation of *m*-dinitrobenzene has already been mentioned. By reduction with ammonium hydrosulphide, one of the nitro groups is selectively reduced to yield *m*-nitroaniline.

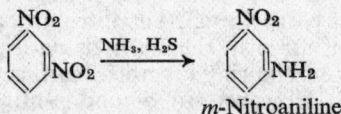

m-Nitroaniline

1,3,5-Trinitrobenzene (T.N.B.) is obtained by heating *m*-dinitrobenzene with a mixture of fuming nitric acid and fuming sulphuric acid at 100° C. for about four days. No further nitro groups can be introduced into benzene by direct nitration.

Nitration of Toluene. Toluene is more readily nitrated in *o* and *p* positions than benzene because of the electron-releasing nature of the methyl group. The progressive nitration of toluene gives, successively, mono (*o* and *p*), di (2,4), and 2,4,6-trinitrotoluenes.

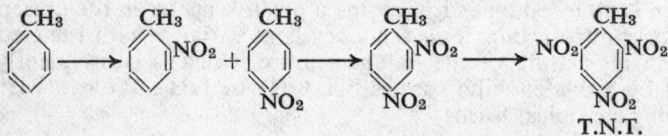

T.N.T.

2,4,6-Trinitrotoluene, commonly called T.N.T., is one of the most important explosives known. It is blended with dynamite to lower the freezing point of nitroglycerine. Ordinarily, it is a stable compound and unaffected by normal shocks. It explodes only when detonated in presence of a detonator like mercury fulminate. Other nitro compounds are also used as explosives. These include T.N.B., picric acid, and tetryl.

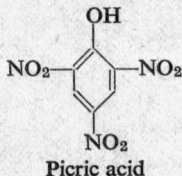

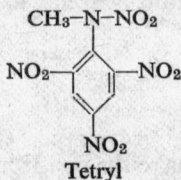

Picric acid

Tetryl

Nitration of Naphthalene. Of the nitro compounds of polynuclear hydrocarbons, those of naphthalene should be mentioned. Nitration with mixed acid gives chiefly α-nitronaphthalene. Since naphthalene is more readily substituted in general than benzene, the conditions of nitration are relatively milder than in the preparation of nitrobenzene. There is no satisfactory procedure available for making β-nitronaphthalene in good yield, though it can be obtained indirectly from β-naphthylamine.

PROPERTIES AND USES OF NITRO COMPOUNDS: While some nitro compounds are pale yellow liquids, a large number are light yellow solids. Nitro compounds generally have characteristic odours. Nitrobenzene, also referred to as oil of mirbane, has an odour similar to that of almond oil and has often been used in soaps and polishes. Benzene derivatives containing nitro and *t*-butyl groups have musklike odours and are used in perfumery. The vapours of many nitro compounds are toxic and should not be inhaled.

SUMMARY

Aromatic compounds are derived from benzene (C_6H_6). Coal tar and petroleum as sources of aromatic compounds. Benzene characterized by great stability in spite of apparent unsaturation and also by tendency to undergo substitution reaction. Kekule structures involving conjugated double bonds. Thiele's explanation of inertness of benzene. Electronic interpretation of benzene structure as a resonance hybrid and physical and chemical evidence for it. Molecular orbital interpretation, delocalized π-bonds. Nomenclature of substitution products.

Reactions of Benzene. (i) Hydrogenation to cyclohexane. (ii) Oxidation to maleic anhydride. (iii) Addition of halogens in presence of sunlight to give benzene hexachloride. Substitution of benzene by halogen in presence of carriers. (iv) Nitration to nitrobenzene. (v) Sulphonation to benzenesulphonic acid. (vi) Reaction with RX or R−COX in presence of $AlCl_3$ to give alkyl or acyl benzenes (Friedel–Crafts reaction). (vii) Pyrolysis to give diphenyl. Aromatic substitutions involve attack of the nucleus by electrophilic reagents, e.g. bromination of benzene in-involves the following steps:

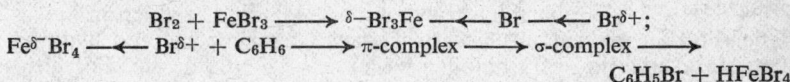

$$Br_2 + FeBr_3 \longrightarrow {}^{\delta-}Br_3Fe \longleftarrow Br \longleftarrow Br^{\delta+};$$
$$Fe^{\delta-}Br_4 \longleftarrow Br^{\delta+} + C_6H_6 \longrightarrow \pi\text{-complex} \longrightarrow \sigma\text{-complex} \longrightarrow$$
$$C_6H_5Br + HFeBr_4$$

Nitration involves NO_2^+ and sulphonation SO_3. Friedel–Crafts reaction involves R^+ or $R\overset{+}{C}O$. Orientation of aromatic substitution; mesomeric and electromeric effects; electron-releasing groups render *o* and *p* position electron-rich, and hence *o, p*-orienting. Also rate of substitution as compared to benzene increased. Electron-attracting groups like NO_2 deactivate the ring and the rate of substitution is slowed down. Since *m* position is less deactivated than *o* and *p* positions, substitution is largely *meta*. Orientation rule states that groups like −OH, −CH_3, etc., attached to the ring by atoms which form only single bonds, are *o*, *p*-orienting, while groups like $-NO_2$, $-C=O$, etc., attached by atoms forming multivalent bonds are *m*-orienting.

Other Aromatic Hydrocarbons. Benzene homologues prepared by the Friedel–Crafts reaction. Distinctive chemical properties include: (i) side-chain substitution in addition to *o*- and *p*-ring substitution, e.g. toluene $\longrightarrow C_6H_5CH_2Cl$, $C_6H_5CHCl_2$, etc.; (ii) oxidation of side chains to $-COOH$, e.g. toluene to benzoic acid. Polynuclear hydrocarbons are of two types: (i) those in which the benzene rings are connected through one or more carbons, for example, diphenyl; and (ii) those containing pairs of benzene rings fused through two common carbons, for example, naphthalene, anthracene, etc. Diphenyls by Ullmann reaction. Existence of triphenylmethyl free radical and explanation for it. **Naphthalene** occurs in coal tar. 2-monosubstituted derivatives ($C_{10}H_7X$) and 10-substituted derivatives of the type $C_{10}H_6X_2$. Reactions: air oxidation with vanadium pentoxide to phthalic anhydride. α position more reactive than β. Reduction to di-, tetra-, and deca-hydro derivatives. **Anthracene.** Occurs in coal tar. From *o*-bromobenzyl bromide. Resonance forms and reactive positions. Oxidation to anthraquinone and reduction to 9,10-dihydroanthracene. **Phenanthrene.** Less reactive than anthracene. Substitutions proceed poorly. Bromination to 9-bromo- and reduction to 9,10-dihydrophenanthrenes.

Aromatic Halogen Compounds. Nuclear and side chain derivatives. Difference in reactivities and explanation. *o*, *p*-orientation of chlorine in chlorobenzene with decrease in rate of substitution. Chlorobenzene as intermediate in industry for manufacture of phenol, aniline, and DDT. Bromination of naphthalene, anthracene, and phenanthrene.

Aromatic Sulphonic Acids. From hydrocarbons by sulphonation. Isolation through Ba salts. Conversion to sulphonyl chlorides and then to sulphonamides. Chloramine-T and Halazone, the disinfectants from *p*-toluenesulphonamide. Preparation of saccharin from *o*-toluenesulphonamide. Phenols and nitriles from sulphonic acids. Salts of sulphonic acids as detergents. α- and β-sulphonic acids from naphthalene.

Aromatic Nitro Compounds. Mono- and di-nitration of benzene using mixed acid. Other nitrating agents. Nucleophilic substitutions of nitrobenzene by OH^-. Reduction of nitrobenzene: $Sn + HCl \longrightarrow$ aniline; $Zn + NH_4Cl \longrightarrow$ phenylhydroxylamine. Alkaline reduction gives, depending on reagents used, azoxybenzene, azobenzene, and hydrazobenzene. Activation of a halogen by an *o*- or *p*-NO_2 group. **Polynitrobenzenes.** *m*-dinitrobenzene from nitrobenzene (fuming $HNO_3 + H_2SO_4$ at 100° C.). Reduction to *m*-nitroaniline (ammonium sulphide). 1,3,5-Trinitrobenzene by drastic nitration. Progressive nitration of toluene to T.N.T., the important explosive and an ingredient in dynamite. Other explosives include picric acid and tetryl. Nitration of naphthalene with mixed acids gives mainly α-nitronaphthalene. Properties and uses of nitro compounds.

Problem Set No. 9

1. Name the following hydrocarbons:

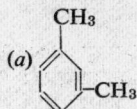

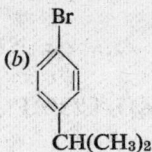

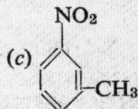

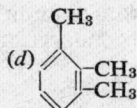

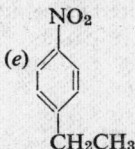

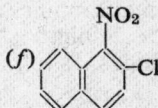

(a) [structure: benzene with CH₃ and –CH₃]

(b) [structure: benzene with Br and CH(CH₃)₂]

(c) [structure: benzene with NO₂ and –CH₃]

(d) [structure: benzene with CH₃, –CH₃, –CH₃]

(e) [structure: benzene with NO₂ and CH₂CH₃]

(f) [structure: naphthalene with NO₂ and Cl]

2. What are the organic products obtained in the following reactions?

(a) Toluene and one equivalent of bromine in sunlight without catalyst.
(b) Nitrobenzene and sulphuric acid.
(c) Mixed acid and bromobenzene.
(d) Fusion of sodium salt of *p*-toluenesulphonic acid with sodium hydroxide followed by acidification.
(e) Permanganate oxidation of ethylbenzene.
(f) Toluene and one equivalent of acetylchloride in the presence of aluminium chloride.
(g) Heating 2,4-dinitrochlorobenzene with a solution of sodium cyanide.
(h) Heating *o*-nitrobromobenzene with copper powder.

3. How may the following compounds be prepared from benzene?

(a) *m*-nitrobromobenzene
(b) *p*-nitrobenzoic acid
(c) *m*-nitroaniline
(d) *p*-chlorobenzenesulphonic acid
(e) ethylbenzene
(f) phenol

4. How would you differentiate between the following?

(a) *p*-Bromotoluene and benzyl bromide.
(b) Benzene and hexene-1.
(c) Toluene and *p*-dimethylbenzene.

CHAPTER THIRTEEN

AROMATIC AMINES, DIAZO COMPOUNDS, AND DYES

The aromatic amines can be pictured as being derived from ammonia, just like the aliphatic amines. There can be primary, secondary, and tertiary amines, as follows:

Aniline (Primary)   Diphenyl amine (Secondary)   Triphenyl amine (Tertiary)

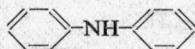

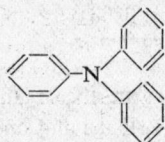

All the groups attached to the nitrogen atom need not be aromatic, and there can be mixed amines like the following:

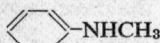

Methylaniline   Dimethylaniline   Methyldiphenyl amine

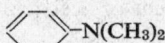

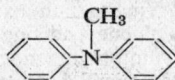

Amines containing the $-NH_2$ group in the side chain, such as benzylamine, $C_6H_5CH_2NH_2$, and 2-phenylethyl amine, $C_6H_5CH_2CH_2NH_2$, behave like aliphatic amines and are generally not included under aromatic amines.

ANILINE

Aniline is the simplest of the aromatic amines. A study of its chemistry will show the general behaviour of primary aromatic amines as a class.

The word aniline is derived from the Spanish word *Anil*, meaning indigo. Perhaps it is also related to the word *Nila*, used in India to describe indigo, the brilliant blue dye for which India was famous and from which aniline was first isolated by distillation with lime. Coal tar also contains some aniline. Hofmann established the identity of samples of aniline from these sources in 1843.

PREPARATION: Aniline and other primary aromatic amines are made generally by reduction of nitro compounds, as shown below:

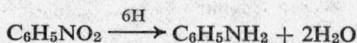

$$C_6H_5NO_2 \xrightarrow{6H} C_6H_5NH_2 + 2H_2O$$

The common reducing agents are: (1) tin and hydrochloric acid; (2) stannous chloride and hydrochloric acid; (3) iron and dilute hydrochloric acid; (4) zinc and acetic acid; and (5) hydrogen and a catalyst. In the laboratory tin and hydrochloric acid form the favoured reducing agent. When the reaction is complete the free base (aniline) is liberated by the addition of sodium hydroxide solution and then isolated by steam distillation.

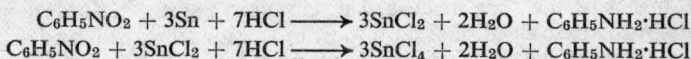

$$C_6H_5NO_2 + 3Sn + 7HCl \longrightarrow 3SnCl_2 + 2H_2O + C_6H_5NH_2 \cdot HCl$$
$$C_6H_5NO_2 + 3SnCl_2 + 7HCl \longrightarrow 3SnCl_4 + 2H_2O + C_6H_5NH_2 \cdot HCl$$

The reduction requires three g. atoms of tin for two moles of nitrobenzene. In industry this reduction is carried out with iron and dilute hydrochloric acid. The process for the manufacture of aniline was referred to in Chapter Twelve.

PROPERTIES AND REACTIONS: (1) **Basicity and Salt Formation**. The aromatic amines differ from the aliphatic amines in the marked decrease in basicity of the former. Ammonia and the aliphatic amines in general are basic to litmus with a dissociation constant, K_b, of the order of 10^{-5}, while for aniline K_b is of the order of 10^{-10}. Aniline is a resonance hybrid of the following electronic structures:

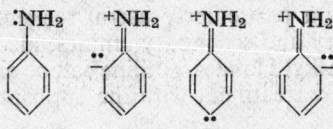

Some basicity is to be expected in aniline because there is a pair of unshared electrons at the nitrogen atom. This basicity is, however, less than in the case of ammonia due to the displacement of the electrons towards the ring. Aniline is basic enough to precipitate slightly soluble metallic hydroxides from solutions of their salts and to form salts with acids. Thus it precipitates iron from solution as ferric hydroxide and combines with hydrogen chloride to form aniline hydrochloride. The hydrochlorides of amines form double salts with stannic chloride ($SnCl_4$) and platinic chloride ($PtCl_4$) of the type $(RNH_2)_2 \cdot MCl_4 \cdot 2HCl$(M=Sn or Pt). The double salts with platinic chloride are particularly useful, since on ignition they leave a residue of platinum. Knowing the weight of the double salt taken, and the weight of the platinum obtained therefrom, one can calculate the molecular weight of the amine.

(2) **Acyl Derivatives**. Aromatic primary amines (and also secondary amines like their aliphatic counterparts) can be acylated with acid halides or acid anhydrides. The acetylation of aniline with acetyl chloride or acetic anhydride yields, for example, acetanilide.

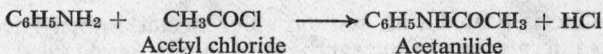

$$C_6H_5NH_2 + \underset{\text{Acetyl chloride}}{CH_3COCl} \longrightarrow \underset{\text{Acetanilide}}{C_6H_5NHCOCH_3} + HCl$$

Benzoylation, similarly, furnishes benzanilide.

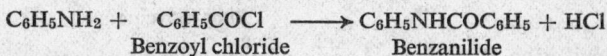

$$C_6H_5NH_2 + \underset{\text{Benzoyl chloride}}{C_6H_5COCl} \longrightarrow \underset{\text{Benzanilide}}{C_6H_5NHCOC_6H_5} + HCl$$

These acylations are carried out in the presence of base (either sodium hydroxide or pyridine) to take up the hydrogen chloride formed. The acyl derivatives are crystalline high-melting solids which are easily purified and are, therefore, useful in the identification of amines.

Both primary and secondary aromatic amines furnish benzenesulphonamides of the type discussed in Chapter Eight. These derivatives aid, as in the case of aliphatic amines, in the identification and separation of different classes of amines. Aniline reacts with carbonyl chloride (phosgene) to give phenyl isocyanate or diphenyl urea, depending upon whether one or two moles of aniline react with one mole of phosgene.

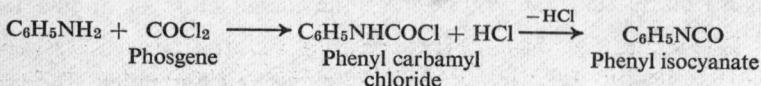

$$C_6H_5NH_2 + COCl_2 \longrightarrow C_6H_5NHCOCl + HCl \xrightarrow{-HCl} C_6H_5NCO$$

Phosgene   Phenyl carbamyl   Phenyl isocyanate
                        chloride

The phenyl isocyanate initially reacts with an additional molecule of aniline to give diphenyl urea.

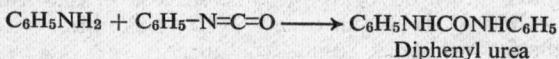

$$C_6H_5NH_2 + C_6H_5-N=C=O \longrightarrow C_6H_5NHCONHC_6H_5$$

Diphenyl urea

Phenyl isocyanate is a particularly useful reagent for preparing solid derivatives of primary and secondary amines, alcohols, and phenols.

(3) **Reaction with Alkyl Halides.** Aniline can be alkylated to give mono and dialkylanilines by treatment with the appropriate alkyl halide.

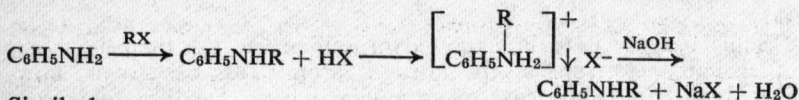

$$C_6H_5NH_2 \xrightarrow{RX} C_6H_5NHR + HX \longrightarrow \left[C_6H_5\overset{\overset{R}{|}}{N}H_2 \right]^+ X^- \xrightarrow{NaOH}$$

$$C_6H_5NHR + NaX + H_2O$$

Similarly

$$C_6H_5NHR \xrightarrow{RX} C_6H_5NRR$$

By controlling the proportions of the reactants and the temperature of the reaction, either the mono- or the di-alkylaniline may be made the major product. In industry, the same alkylations are effected in an autoclave by heating aniline with an alcohol R–OH in the presence of sulphuric acid at high temperatures and pressures. For example, methyl and dimethylanilines are made under the conditions stated below:

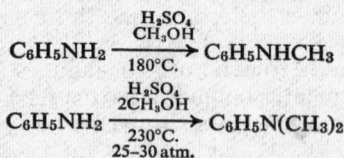

$$C_6H_5NH_2 \xrightarrow[180°C.]{\overset{H_2SO_4}{CH_3OH}} C_6H_5NHCH_3$$

$$C_6H_5NH_2 \xrightarrow[\substack{230°C. \\ 25-30\,atm.}]{\overset{H_2SO_4}{2CH_3OH}} C_6H_5N(CH_3)_2$$

When aniline is heated with sodium chloroacetate, phenylglycine, an important intermediate in the synthesis of indigo, is obtained.

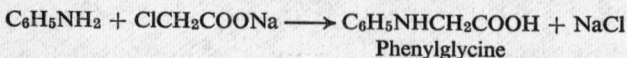

$$C_6H_5NH_2 + ClCH_2COONa \longrightarrow C_6H_5NHCH_2COOH + NaCl$$

Phenylglycine

(4) **Substitution Reactions.** As is to be expected from the resonance structures of aniline, the increase in electron density of the ring favours ease of substitutions in *o*- and *p*-positions. In fact, substitution occurs

very rapidly. When aniline is shaken with bromine water an immediate precipitation of 2,4,6-tribromoaniline is obtained—the yield being almost quantitative.

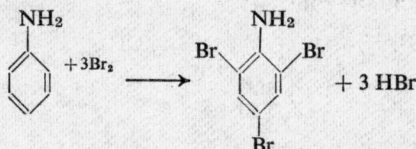

$$
\underset{}{\text{(aniline)}} + 3Br_2 \longrightarrow \underset{}{\text{(tribromoaniline)}} + 3\ HBr
$$

Direct nitration of aniline is not possible, since such reactions give rise to coloured and tarry products and even explosive oxidation. Nitration can be accomplished by protecting the amino group by acetylation and then performing the nitration.

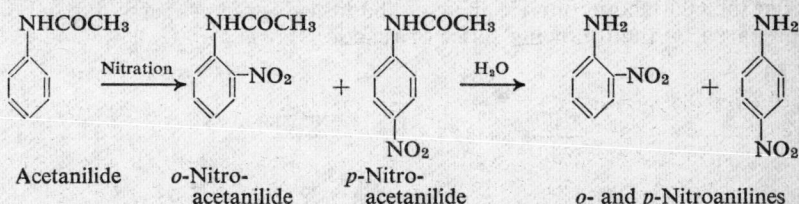

$$
\text{Acetanilide} \xrightarrow{\text{Nitration}} \underset{\text{o-Nitro-}\atop\text{acetanilide}}{\text{NHCOCH}_3\text{-NO}_2} + \underset{\text{p-Nitro-}\atop\text{acetanilide}}{\text{NHCOCH}_3} \xrightarrow{\text{H}_2\text{O}} \underset{}{\text{NH}_2\text{-NO}_2} + \underset{\text{o- and p-Nitroanilines}}{\text{NH}_2}
$$

The acetyl group can be removed by hydrolysis subsequent to nitration.

Two derivatives of aniline, sulphanilic acid and arsanilic acid, require special mention because of their use in medicine. Sulphanilic acid is made by heating aniline with concentrated sulphuric acid at 180° C. for 5 hours. The aniline sulphate, which is first formed, loses water during heating, with the probable formation of phenylsulphamic acid, which rearranges to sulphanilic acid, as follows:

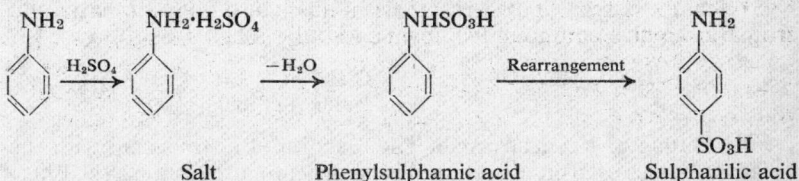

$$
\underset{}{\text{NH}_2} \xrightarrow{\text{H}_2\text{SO}_4} \underset{\text{Salt}}{\text{NH}_2\cdot\text{H}_2\text{SO}_4} \xrightarrow{-\text{H}_2\text{O}} \underset{\text{Phenylsulphamic acid}}{\text{NHSO}_3\text{H}} \xrightarrow{\text{Rearrangement}} \underset{\text{Sulphanilic acid}}{\overset{\text{NH}_2}{\underset{\text{SO}_3\text{H}}{}}}
$$

Sulphanilic acid is a solid with a high melting point. It is more acidic than basic and forms salts with bases. It exists largely in the form of an inner salt or zwitterion like the amino acids.

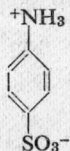

$$
\overset{+\text{NH}_3}{\underset{\text{SO}_3{}^-}{\bigcirc}}
$$

Amides derived from sulphanilic acid are the well-known sulpha drugs, which are so effective in the treatment of bacterial infections. Sulphanilic acid is also a valuable dye intermediate.

A reaction parallel to that between aniline and sulphuric acid takes place when aniline is treated with arsenic acid. First, aniline forms a salt with the arsenic acid, then this salt dehydrates to give arsanilide and finally arsanilic acid, as follows:

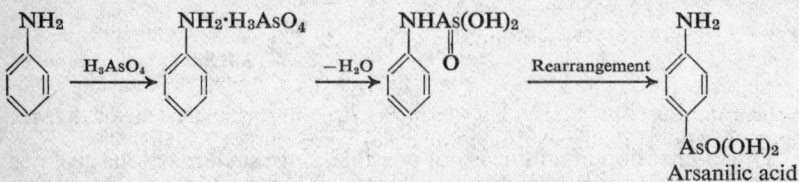

Arsanilic acid

In a series of investigations of arsenical organic compounds which would be effective in the treatment of syphilis, DR. PAUL EHRLICH found that the 606th compound, Salvarsan, had the desired property. It can be prepared by the following series of reactions:

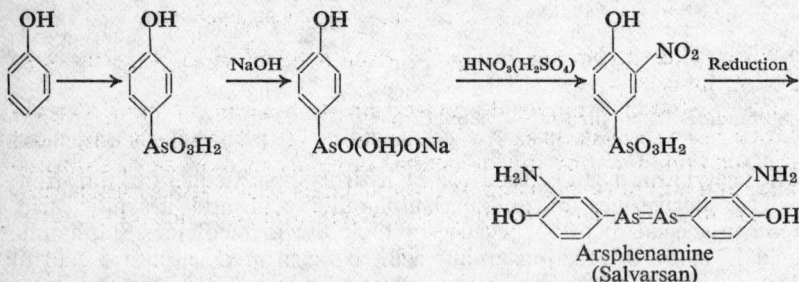

Arsphenamine
(Salvarsan)

Several other aromatic derivatives containing arsenic have been prepared for the treatment of syphilis.

(5) **Schiff's Bases.** Aniline reacts with aldehydes, both aliphatic and aromatic, to give compounds known as SCHIFF'S bases, as follows:

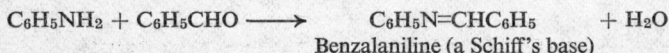

$$C_6H_5NH_2 + C_6H_5CHO \longrightarrow \quad C_6H_5N=CHC_6H_5 \quad + H_2O$$
Benzalaniline (a Schiff's base)

(6) **Reaction with Nitrous Acid.** The reaction of nitrous acid with the different classes of aromatic amines is partly similar to and partly different from that with aliphatic amines. An aromatic primary amine, when treated with nitrous acid in hydrochloric acid *in the cold*, gives a diazo compound, for example:

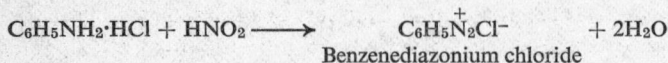

$$C_6H_5NH_2 \cdot HCl + HNO_2 \longrightarrow \quad C_6H_5\overset{+}{N}_2Cl^- \quad + 2H_2O$$
Benzenediazonium chloride

The above diazo compound is hydrolysed when heated, as follows:

$$C_6H_5\overset{+}{N}_2Cl^- + H_2O \longrightarrow HCl + N_2 + C_6H_5OH$$

The net effect is similar to that which occurs in the case of an aliphatic primary amine. The diazo compounds are versatile intermediates for the synthesis of a variety of compounds and are dealt with separately later on. With nitrous acid, secondary aromatic amines give *N*-nitroso compounds, for example:

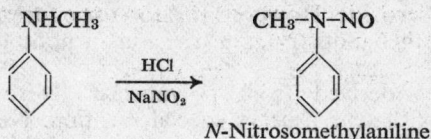

N-Nitrosomethylaniline

Whereas aliphatic tertiary amines are unreactive towards nitrous acid, those of the aromatic series undergo nitrosation in the ring.

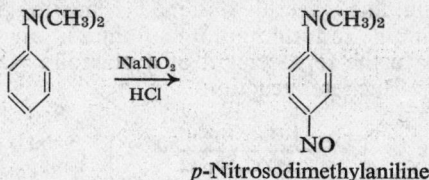

p-Nitrosodimethylaniline

p-Nitrosodimethylaniline is a grass-green solid of great importance in the synthesis of dyes.

(7) **Oxidation.** Aniline, depending on the conditions used, gives a number of products on oxidation. Aniline black, commonly employed as a stain and a cotton dye, is obtained by using a mixture of a dichromate and sulphuric acid for oxidation. Drastic oxidation furnishes *p*-benzoquinone (or simply *p*-quinone), a yellow crystalline solid.

p-Benzoquinone (or *p*-Quinone)

OTHER COMMON AROMATIC AMINES

Diphenylamine is the most common secondary amine. It is used in the preparation of dyes and as a stabilizer for explosives. It is obtained by heating a mixture of aniline and aniline hydrochloride, as follows:

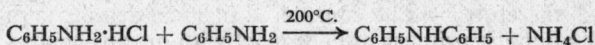

$$C_6H_5NH_2 \cdot HCl + C_6H_5NH_2 \xrightarrow{200°C.} C_6H_5NHC_6H_5 + NH_4Cl$$

The preparation of methyl and dimethylaniline has already been mentioned. Amines with two amino groups in the ring, called phenylenediamines, are also known. *o*- and *p*-Phenylenediamines are obtained by reduction of the corresponding nitroanilines, e.g.:

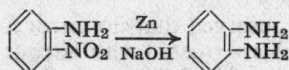

The *meta* isomer is obtained by reduction of the easily accessible *m*-dinitrobenzene. The phenylenediamines are easily oxidized and become discoloured on exposure to air. *p*-Phenylenediamine has been used as a hair dye by applying a water solution of it to the hair and then oxidizing it with hydrogen peroxide. However, it is too toxic for prolonged use. It is used in the leather industry as a dye and in photography as a developer.

Among amines derived from polynuclear hydrocarbons, those composed of naphthalene deserve special mention. Naphthylamine is prepared by the reduction of α-nitronaphthalene, the reaction being parallel to the reduction of nitrobenzene to aniline. This is a colourless solid, but rapidly discolours on exposure to air and has a disagreeable odour. β-Naphthylamine is readily obtained by heating β-naphthol with a solution of ammonia and ammonium sulphite in an autoclave. This reaction, resulting in the replacement of a phenolic −OH by the NH_2 group, is called the Bucherer reaction.

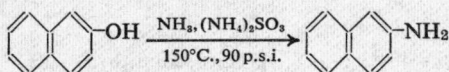

$$\text{(naphthol)}-OH \xrightarrow[\text{150°C., 90 p.s.i.}]{NH_3, (NH_4)_2SO_3} \text{(naphthalene)}-NH_2$$

The process can be reversed; that is, the amino group can be converted into the phenolic −OH by heating the amine with sodium bisulphite solution, adding sodium hydroxide and removing the ammonia by distillation of the solution. Bucherer reaction is not applicable for the preparation of benzene derivatives. On sulphonation, α-naphthylamine gives 1-aminonaphthalene-4-sulphonic acid, known as naphthionic acid.

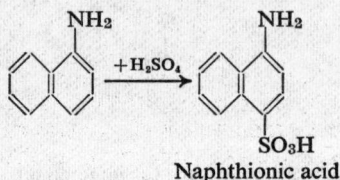

$$\underset{NH_2}{\text{(naphthalene)}} \xrightarrow{+H_2SO_4} \underset{SO_3H}{\overset{NH_2}{\text{(naphthalene)}}}$$

Naphthionic acid

Naphthionic acid is used for the preparation of Congo Red and other dyes.

DIAZOTIZATION

Diazotization is the most important reaction of primary aromatic amines and, therefore, will be considered in some detail. As previously mentioned, aliphatic primary amines react with nitrous acid to give the corresponding alcohol with the elimination of nitrogen.

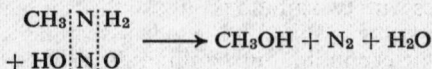

$$CH_3|N|H_2 \atop + HO|N|O \longrightarrow CH_3OH + N_2 + H_2O$$

PETER GRIESS (1860); observed that no nitrogen is evolved in the case of aromatic amines which are treated with nitrous acid if the solution is cold. Nitrous acid, when required, is prepared *in situ*, that is, in the reaction vessel itself by the addition of an acid like sulphuric acid or hydrochloric acid and sodium nitrite solution. The reaction between an acidified solution of aniline and the nitrous acid takes place as follows:

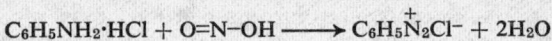

$$C_6H_5NH_2{\cdot}HCl + O{=}N{-}OH \longrightarrow C_6H_5\overset{+}{N}_2Cl^- + 2H_2O$$

The compound formed is called **benzenediazonium chloride**. The prefix *diazo* indicates that there are two nitrogen atoms per molecule of benzene, and the suffix *onium* suggests its saltlike character, analogous to ammonium chloride. This process is called diazotization, and is applicable to most aromatic primary amines. In practice, diazotization is carried on by dissolving the amine in dilute hydrochloric acid or sulphuric acid, cooling to 5° C., and then gradually adding a well-cooled solution of sodium nitrite in water. Temperature control is essential: below 5° C. the reaction is slow and above 10° C. a phenol is formed. The solution of the diazonium salt thus obtained may be used in all the reactions to be described hereafter, there being no need to isolate the free diazonium salt, since isolation is neither necessary nor desirable due to the fact that these salts are liable to explode. Though the equation depicting the diazotization indicates that two molecules of acid are required for diazotization per molecule of aniline (one mole of acid for forming the salt with aniline and another to liberate nitrous acid from the nitrite), in actual practice, roughly three moles of acid are employed.

As has already been indicated, the isolation of the diazonium salt is not necessary for most purposes. The crystalline diazonium salt can, however, be isolated. The amine is dissolved in ethyl alcohol containing hydrogen chloride, the solution is cooled in ice, and the diazotization is carried out by gradually adding an organic nitrite like amyl nitrite. After diazotization is complete the solution is diluted with ether and vigorously cooled in a freezing mixture of ice and salt to precipitate out the only ether-insoluble product, the diazonium chloride. The salt dissolves in water, giving a neutral conducting solution.

We shall now consider two of the older structures proposed for diazonium salt: one proposed by KEKULÉ (1866) and the other by BLOOMSTRAND (1869).

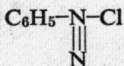

$$C_6H_5{-}N{=}N{-}Cl \qquad\qquad C_6H_5{-}N{-}Cl$$
$$\qquad\qquad\qquad\qquad\qquad\qquad \underset{N}{\|\|}$$

(Kekulé)   (Bloomstrand)

The above formulae fail to account for the ionic nature of the salt. The modern formula pictures the diazonium ion as a resonance hybrid of the two following contributory structures:

$$[C_6H_5{-}\ddot{N}{=}\ddot{N}]^+ \longleftrightarrow [C_6H_5{-}N{\equiv}\ddot{N}]^+$$

The diazonium salts are highly reactive and, due to their versatility, form one of the most useful classes of organic compounds. We may recognize two important types of reaction of the diazonium salt, one in which the nitrogen atoms are removed and the replacement of the diazo group effected, and the other in which the nitrogen atoms are retained in the new compound formed. We shall consider both these types of re-actions, taking benzenediazonium chloride as a typical diazo salt.

(I) **Conversion to Phenols.** The diazonium group can be replaced easily by the hydroxyl group by simply warming an aqueous solution of the diazonium salt with 40–50% sulphuric acid, as shown below.

$$C_6H_5\overset{+}{N_2}Cl^- + H_2O \longrightarrow C_6H_5OH + N_2 + HCl$$

The production of phenol recalls the formation of an alcohol from an aliphatic primary amine by the action of nitrous acid. This reaction is used occasionally for the preparation of some phenols when conventional methods are unsatisfactory.

(II) **The Sandmeyer Reaction.** The replacement of the diazonium group by the halogen or cyano group under the catalytic influence of cuprous halides or cyanide is known as the SANDMEYER reaction. When a solution of cuprous salt is added to a diazonium salt a double salt is usually obtained as a precipitate. This double salt decomposes when warmed with the liberation of nitrogen and the formation of the halide or the cyanide, as follows:

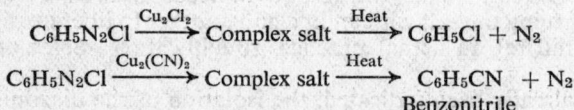

$$C_6H_5N_2Cl \xrightarrow{Cu_2Cl_2} \text{Complex salt} \xrightarrow{\text{Heat}} C_6H_5Cl + N_2$$

$$C_6H_5N_2Cl \xrightarrow{Cu_2(CN)_2} \text{Complex salt} \xrightarrow{\text{Heat}} \underset{\text{Benzonitrile}}{C_6H_5CN} + N_2$$

The following preparations illustrate the method:

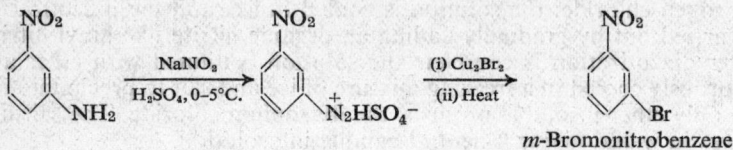

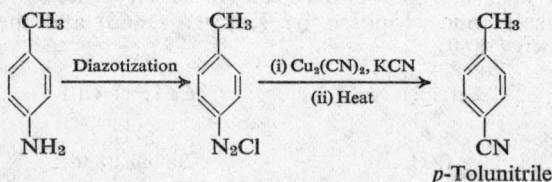

m-Bromonitrobenzene

p-Tolunitrile

The Sandmeyer reaction necessitates the use of cuprous salt. In GATTERMANN'S modification cuprous salts are not needed and the replacement is effected by allowing the diazonium salt to react with a halogen acid in the presence of finely divided copper. In contrast to replacement by chlorine or bromine, the replacement of the diazonium group by iodine is carried out with facility, no copper being necessary. When potassium iodide is added to a solution of the diazonium salt iodobenzene is readily obtained.

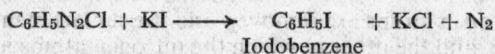

$$C_6H_5N_2Cl + KI \longrightarrow \underset{\text{Iodobenzene}}{C_6H_5I} + KCl + N_2$$

Aromatic compounds of fluorine can be obtained by treating the diazonium salt with hydrofluoboric acid (HBF_4).

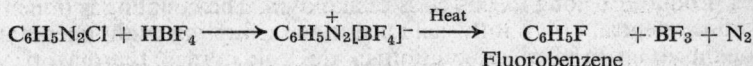

$$C_6H_5N_2Cl + HBF_4 \longrightarrow C_6H_5\overset{+}{N}_2[BF_4]^- \overset{Heat}{\longrightarrow} \underset{Fluorobenzene}{C_6H_5F} + BF_3 + N_2$$

(III) Deamination. The diazonium group can also be replaced by hydrogen. An early procedure consisted of treating the diazonium salt with ethanol.

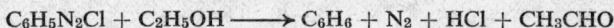

$$C_6H_5N_2Cl + C_2H_5OH \longrightarrow C_6H_6 + N_2 + HCl + CH_3CHO$$

This reaction makes possible the preparation of certain *meta*-substituted derivatives of benzene in which the substituents are *ortho–para*-orienting groups. For example, 1,3,5-tribromobenzene is made available by the following route, where the last step involves deamination through a diazonium salt:

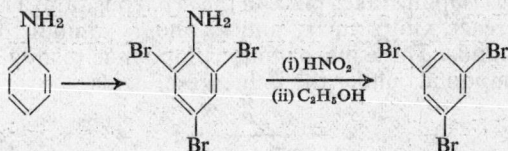

A side-reaction in deaminations with ethanol is the formation of an ether, shown below.

$$C_6H_5N_2Cl + C_2H_5OH \longrightarrow C_6H_5OC_2H_5 + HCl + N_2$$

Hypophosphorous acid (H_3PO_2) is also used in effecting deaminations of the above type.

(IV) Reduction. The reduction of benzenediazonium salt may give one of three types of products. If both the nitrogen atoms are retained a derivative of hydrazine is obtained. If the nitrogen–nitrogen bond is broken ammonia and a primary amine are obtained. If both the nitrogen atoms are eliminated a hydrocarbon results. When benzenediazonium chloride is reduced with a mild reducing agent, like sodium sulphite, phenylhydrazine hydrochloride is obtained.

$$C_6H_5N_2Cl + 2Na_2SO_3 + 2H_2O \longrightarrow C_6H_5NHNH_2 \cdot HCl + 2Na_2SO_4$$

Phenylhydrazine is obtained as a yellow oil from the hydrochloride by the addition of sodium hydroxide. It will be remembered that phenylhydrazine was used by Emil Fischer with consummate skill in determining the identity and structures of sugars in the course of his classical investigations of carbohydrates. He rightfully called phenylhydrazine his first chemical love. Phenylhydrazine affects the skin and is quite toxic. When the solution of the diazonium chloride is reduced catalytically with hydrogen in the presence of platinum the N–N bond is severed, resulting in the formation of aniline and ammonium chloride.

$$C_6H_5N_2Cl + 3H_2 \longrightarrow C_6H_5NH_2 + NH_4Cl$$

(V) Coupling Reaction—Formation of Azo Compounds. Diazonium salts combine with a variety of aromatic amines and phenols to give azo dyes. This reaction is called **coupling**, and during this reaction both the nitrogen atoms of the diazonium salt are retained. About 50% of all the dyes produced belong to this class of azo dyes. The coupling is done in alkaline, neutral, or faintly acid solution. Coupling with phenol is brought about in alkaline solution in the cold. When the diazotized solution is added to an alkaline solution of phenol coupling occurs with facility and an orange dye is obtained.

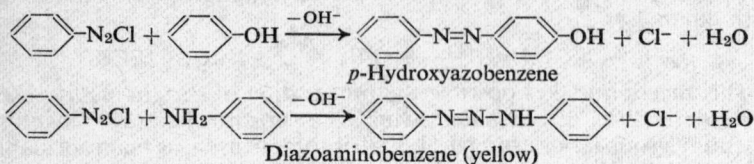

$$\langle\rangle\text{-N}_2\text{Cl} + \langle\rangle\text{-OH} \xrightarrow{-\text{OH}^-} \langle\rangle\text{-N=N-}\langle\rangle\text{-OH} + \text{Cl}^- + \text{H}_2\text{O}$$

p-Hydroxyazobenzene

$$\langle\rangle\text{-N}_2\text{Cl} + \text{NH}_2\text{-}\langle\rangle \xrightarrow{-\text{OH}^-} \langle\rangle\text{-N=N-NH-}\langle\rangle + \text{Cl}^- + \text{H}_2\text{O}$$

Diazoaminobenzene (yellow)

The coupling takes place in the *para* position unless it is occupied, in which case the coupling takes place in the *ortho* position. The diazonium salts usually react with primary and secondary amines to give diazo-amino compounds. These diazoamino compounds readily rearrange to aminoazo compounds under the influence of acids.

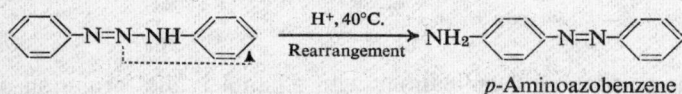

$$\langle\rangle\text{-N=N-NH-}\langle\rangle \xrightarrow[\text{Rearrangement}]{\text{H}^+, 40°\text{C.}} \text{NH}_2\text{-}\langle\rangle\text{-N=N-}\langle\rangle$$

p-Aminoazobenzene

Diazonium salts couple readily with tertiary amines to give azo dyes. Methyl orange is obtained by coupling diazotized sulphanilic acid with dimethylaniline.

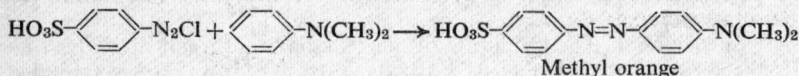

$$\text{HO}_3\text{S-}\langle\rangle\text{-N}_2\text{Cl} + \langle\rangle\text{-N(CH}_3)_2 \rightarrow \text{HO}_3\text{S-}\langle\rangle\text{-N=N-}\langle\rangle\text{-N(CH}_3)_2$$

Methyl orange

Some general conclusions regarding coupling reactions may be drawn from the above examples. The diazonium salts couple with aromatic compounds in which a strong electron donor (*o*, *p*-orienting group) is attached to the benzene nucleus. A generalized equation, as shown below, can be written for these coupling reactions. The reaction is pictured as being brought about by an attack of the electron-rich *para*-carbon atom in an amine or a phenol by the positive diazonium ion.

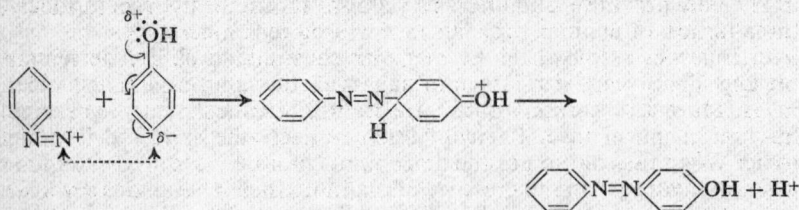

$$\langle\rangle + \langle\rangle\overset{\delta+}{\text{:OH}} \longrightarrow \langle\rangle\text{-N=N-}\langle\rangle\overset{+}{\text{=OH}} \longrightarrow$$
$$\text{N=N}^+ \qquad\qquad \overset{\delta-}{} \qquad\qquad \text{H}$$

$$\langle\rangle\text{N=N}\langle\rangle\text{OH} + \text{H}^+$$

Since coupling occurs with the loss of a proton, such reactions should be conducted in alkaline or faintly acidic solutions and not in strongly acidic solutions. The substituent in the aromatic ring should be a good electron donor like $-\text{OH}$ or $-\text{N(CH}_3)_2$, which would increase the electron density on the ring. Thus, for example, no coupling will take place with chlorobenzene or nitrobenzene, where the substituents reduce the electron density on the ring. Coupling reactions are also facilitated by substituents which increase the positive charge on the diazonium ion. Thus, for example, diazonium salts with nitro groups in *o*- or *p*-positions couple more readily. When solid benzenediazonium chloride is dissolved in distilled water it is found that the aqueous solution is neutral to litmus and does not show any appreciable degree of hydrolysis. Since hydrochloric acid is a strong acid, and the diazonium salt does not hydrolyse, the basic moiety of the salt, that is, benzenediazonium hydroxide, must be a strong base. But when an aqueous solution of benzenediazonium chloride is treated with excess potassium hydroxide, benzenediazonium hydroxide initially formed rearranges to benzenediazoic acid (potassium salt).

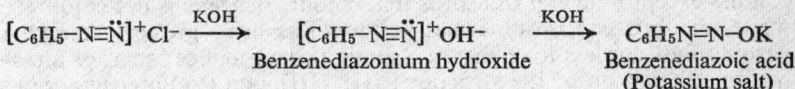

$$[C_6H_5-N\equiv\ddot{N}]^+Cl^- \xrightarrow{\text{KOH}} [C_6H_5-N\equiv\ddot{N}]^+OH^- \xrightarrow{\text{KOH}} C_6H_5N=N-OK$$

Benzenediazonium hydroxide   Benzenediazoic acid (Potassium salt)

Benzenediazonium hydroxide can be isolated by treating benzenediazonium chloride with moist silver oxide.

$$C_6H_5N_2Cl + \text{'AgOH'} \longrightarrow C_6H_5\overset{+}{N}_2OH^- + AgCl$$

In the case of benzenediazoic acid, there is a double bond between the two nitrogen atoms, and hence geometrical isomerism should be possible. This is actually the case, and the *syn*- and the *anti*-forms of the potassium salt of benzenediazoic acid have been prepared and are found to have different properties.

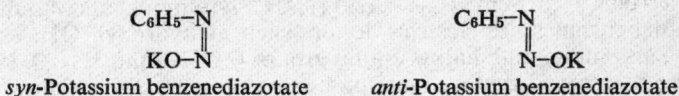

$$\begin{array}{cc} C_6H_5-N & C_6H_5-N \\ \| & \| \\ KO-N & N-OK \end{array}$$

syn-Potassium benzenediazotate   *anti*-Potassium benzenediazotate

DYES

As has been previously mentioned, about 50% of all the dyes manufactured come under the class of azo dyes prepared from diazonium salts. Hence this seems an appropriate place to discuss the dyes as a class, though not all of them may fall in the category of azo dyes.

The art of colouring cloth or other materials is as old as human vanity. There were very few materials available for dyeing in the olden days. Most of the ancient dyes were derived from vegetable and animal sources. Some of them, like the Tyrian purple, were so costly that they were only available to kings and the very wealthy, and therefore such names as Royal Purple were derived. In contrast with our ancestors, we seem to be quite comfortably off with respect to dyes. We have at our command a host of dyes of every conceivable shade and of varying degrees of fastness, to suit almost any purpose. A dye by definition is a coloured compound that adheres to cloth and retains its colour against the attack of light, moisture, and soap under normal conditions of wear. A fast dye remains permanently; a fugitive dye fades or washes off. Dyes are applied to cloth in several different ways. The so-called direct or substantive dyes dye fibres directly and in the absence of any other chemical. Some dyes form insoluble coloured compounds (called lakes) with metallic salts called **mordants**. The cloth to be dyed is first soaked in the mordant and the dye applied when the coloured dye-mordant complex is deposited in the fibres. This process is referred to as mordant dyeing. Vat dyeing refers to the process whereby cloth is soaked in an alkaline solution of the reduced form of a dye and then exposed to air. Oxidation serves to fix the dye within the fibres of the cloth.

Not all coloured compounds are dyes, since the dyes must have the additional property which makes possible their fixation on fibres. The colour of a compound depends upon its ability to absorb part of the light falling on it. A compound will appear blue if it absorbs all the colours except blue and transmits that colour. Benzene is not coloured, while nitrobenzene is somewhat yellow. The nitro group is called a chromophore or a colour bearer. Among the resonance forms of nitrobenzene shown on p. 194 structures (II), (III), and (IV) are quinonoid (see Chapter Fifteen) in nature. If an electron donor group were attached to the ring at a position *ortho* or *para* to the nitro group the quinone structures would contribute to a greater extent to the resonance of the molecule, since the electron donor group would take the positive charge more readily than the carbon atom. Hence, the colour may be expected to be intensified. Thus in the case of *o*-nitroaniline its colour (orange) is much deeper than that of nitrobenzene, which is pale yellow. The amino group aids the colour in this situation and is called an auxochrome. The chromophoric groupings are unsaturated and are usually electron acceptors. Examples of the chromophoric groupings are nitro (NO_2), nitroso ($N=O$), carbonyl ($C=O$), azo ($N=N$), *o*-quinone, and *p*-quinone. The auxochromes are electron donors such as hydroxyl, OH, amino, $-NH_2$, and mono and dialkyl amino groups ($-NHR$ and $-NR_2$). It may also be mentioned that a compound containing a long system of conjugated double bonds develops colour, as, for instance, in the case of the carotenoids.

Most of the dyes are salts of acids or bases. In either case the ion derived from such a dye has a conjugated system of alternate single and double bonds and is a resonance hybrid of several electronic structures. The auxochromes, in addition to influencing the colour of the dye, aid in fastening it to the fabric. When a dye is reduced the system of alternate double and single bonds is disturbed and the colour fades or disappears. Such reduced or faintly coloured forms of dyes are called **leuco dyes**, and on oxidation give the coloured dyes. The colour of a dye is influenced by the substitution of certain groups in the auxochrome. Those which deepen the colour are called **bathochromic** groups, and those which lighten the colour are called **hypsochromes**. Alkyl and aryl groups are bathochromes, while acetyl and other acyl groups are hypsochromes. The acyl groups hinder electronic shifts, which are necessary for the production of colour in the dye ion. Hypsochromes are useful in the case of dyes which have several auxochromes, since the acetylation of part of them produces a dye of lighter shade. The usefulness of a dye also depends on the nature of the fibres to be dyed. Fibres of animal origin (silk and wool) are essentially proteins, and direct salt formation holds the dye to the $-NH_2$ or $-COOH$ groups of the protein. In other words, a dye in order to be directly applied to wool or silk must contain either basic or acidic groups. The same dye will not be useful for cotton and rayon, which are composed of neutral cellulose fibres. The dyes which are directly applied to cotton are held to the fibres by either hydrogen bonding or absorption. Nylon and cellulose acetate are also neutral fibres, but they contain no $-OH$ groups, necessary for formation of hydrogen bonds. They are dyed in the presence of a dispersing re-agent in medium with water-insoluble dyes.

The dyes are usually classified by the organic chemist on the basis of their chemical structure as follows: (1) azo dyes; (2) triphenylmethane dyes; (3) phthalein dyes; (4) anthraquinone dyes; and (5) vat dyes.

SUMMARY

Aromatic amines contain nitrogen directly attached to the ring. Primary amines (RNH_2) from RNO_2 by reduction. Reduction effected with metal and acid or catalytically. **Reactions**: (I) Basicity: weaker than NH_3 due to displacements of electron pair towards ring. Formation of simple and double salts. (II) Acylation in the presence of base, for example, $C_6H_5NH_2 \xrightarrow{RCOCl} C_6H_5NHCOR$. Reaction of aniline with $COCl_2$ to give phenyl isocyanate and diphenyl urea. (III) Alkylation of amines; industrial method of preparation of methylaniline and dimethyl-aniline. (IV) Substitution reactions: (*a*) aniline to 2,4,6-tribromoaniline; (*b*) direct nitration gives tar, and hence carried out after protection of NH_2—aniline $\longrightarrow$ acetanilide; (*c*) sul-phonation to sulphanilic acid—importance of sulphanilamide and other derivatives; (*d*) preparation of arsanilic acid—Salvarsan and its use. (V) Schiff's bases from aldehydes. (VI) Reaction with nitrous acid: aniline $\longrightarrow$ benzenediazonium chloride. Secondary amines give *N*-nitroso compound. With tertiary amines, *p*-nitrosation occurs. (VII) Oxidation of aniline to quinone.

Other Common Aromatic Amines. (1) Diphenylamine from aniline hydrochloride by heating with aniline at 200° C. (2) Diamines: *o*- and *p*-nitroaniline to *o*- and *p*-phenylenediamines by reduction with alkali and zinc; *m*-isomer by mono-reduction of *m*-dinitrobenzene. Uses of phenylenediamines. (3) α-Naphthylamine from α-nitronaphthalene; β-isomer from β-naphthol by Bucherer reaction.

Diazotization. $C_6H_5NH_2 \xrightarrow{HNO_2} C_6H_5N_2Cl$: Structure of diazo com-pounds: $[C_6H_5-\overset{..}{N}=\overset{..}{N}]^+ \longleftrightarrow [C_6H_5-N\equiv\overset{..}{N}]^+$. Reactions: (I) Benzene-diazonium chloride to phenol using 40–50% sulphuric acid. (II) Sand-meyer Reaction. $C_6H_5N_2Cl \longrightarrow$ complex $\longrightarrow C_6H_5X(X=Br, Cl, CN)$. Use of copper powder and HX in Gattermann's modification; benzene-diazonium chloride to iodobenzene by treating with potassium iodide. (III) Deamination: benzenediazonium chloride to benzene by treatment with ethanol. Application to preparation of 1,3,5-tribromobenzene. (IV) Reduction of benzenediazonium chloride to phenylhydrazine.

(V) Coupling with amines and phenols to give azo dyes. Coupling occurs with compounds having electron-releasing substituents. Rearrangement of benzenediazonium hydroxide to benzenediazoic acid.

Dyes. A dye is a coloured compound which adheres to cloth and re-retains its colour under normal conditions of wear and tear. Direct (substantive), mordant, and vat dyeing. Chromophoric and auxo-chromic groups responsible for colour. Most dyes are salts in which the organic ion has a system of conjugated double bonds and is a resonance hybrid of several electronic structures. Bathochromic and hypsochromic effects of substituents. Usefulness of a dye dependent on the nature of the fibres to be dyed. Chemical classification of dyes.

Problem Set No. 10

1. Starting from benzene, how can the following compounds be prepared: (*a*) *m*-nitrochlorobenzene; (*b*) *o*-nitroaniline; (*c*) nitrosobenzene; (*d*) phenylhydrazine; (*e*) phenylisocyanide?
2. How may the following conversions be achieved using diazonium salts: (*a*) *o*-dichlorobenzene from *o*-chloroaniline; (*b*) 3,4-dichlorobenzoic acid from *o*-dichlorobenzene; (*c*) 1,3,5-tribromobenzene from aniline; (*d*) *m*-bromonitrobenzene from benzene; (*e*) *p*-chlorobenzoic acid from chlorobenzene; (*f*) toluene from *p*-toluidine?

CHAPTER FOURTEEN

PHENOLS AND AROMATIC ALCOHOLS

Phenol is the monohydroxy derivative of benzene. In general, any hydroxy compound in which the hydroxyl group is attached to an aromatic nucleus belongs to the class of phenols. In the case of aromatic alcohols, the hydroxyl group is present in the side chain. The word *phenol* shows that it is an *enol*, that is a compound containing an $^-$OH group attached to a doubly bound carbon atom. The phenols may be classified into monohydric, dihydric, and trihydric groups, depending upon the number of hydroxyl groups in the nucleus. The monohydric phenols will be considered first. The preparation and properties of phenol will illustrate the chemistry of phenols as a class.

PREPARATION: Phenol (also called carbolic acid) is obtained by extraction with alkali of the middle oil fraction of coal-tar distillation. But the supply from this source is not adequate to meet the great demand for phenol, and large quantities of it are prepared synthetically. When studying sulphonic acids we noted that sodium hydroxide fusion of benzenesulphonic acid gives sodium phenolate. This is one method of introducing the hydroxyl group, as used industrially. Another commercial method involves the reaction of chlorobenzene with alkali under vigorous conditions. Phenol may be prepared in the laboratory by decomposition of benzenediazonium chloride.

PROPERTIES AND REACTIONS OF PHENOL: Phenol is a colourless, low-melting, highly corrosive solid with a characteristic odour. It is slightly acidic, having an acid dissociation constant $K_A=10^{-10}$. This is due to contributions of structures like (II) and (III) to the resonance hybrid

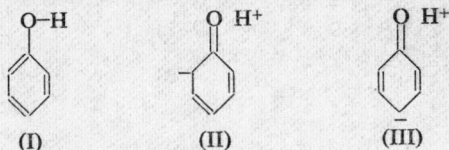

(I)   (II)   (III)

structure something like (IV) and (V).

(IV)   (V)

227

It is weaker than carbonic acid, whose first dissociation constant is
4.3×10^{-7}. Phenol can be liberated from its salts by passing carbon
dioxide, while the salts of carboxylic acid are not affected by carbon
dioxide. The dipole moment of phenol is 1·7 units, with the ring negative
and the ⁻OH group positive. In the case of phenol, the enolic group, that
is ⁻CH=C⁻OH, is present. One may, therefore, expect this enolic form to
be in tautomeric equilibrium with the keto forms, as follows:

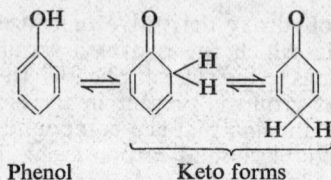

Phenol   Keto forms

Phenol, however, exists almost exclusively in the enol form. The
resonance structures for phenol (p. 227) indicate that the *ortho* and *para*
positions are relatively electron-rich, and hence substitution occurs at
these positions with great ease. The reactions of phenol involve either
the OH group or the ring. Reactions centring around the hydroxyl will
be dealt with first.

(I) **Salt Formation.** Phenol dissolves in dilute alkali to give phenolate
salts, but it does not dissolve in solutions of sodium carbonate. The
lack of reaction with sodium carbonate is the basis of a test to distinguish
phenols from carboxylic acids (which dissolve with effervescence).

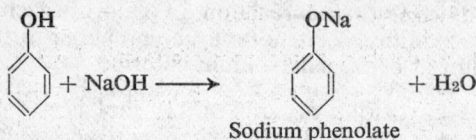

Sodium phenolate

(II) **Conversion to Ethers.** Sodium phenolate gives ethers on treatment
with alkyl halides or alkyl sulphates (Williamson's Synthesis).

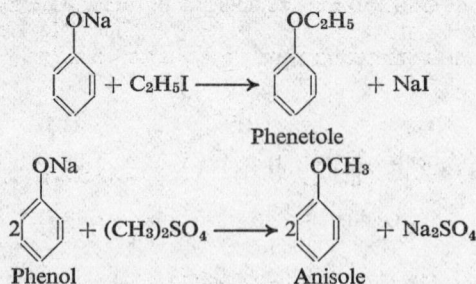

Phenetole

Phenol   Anisole

(III) **Esterification.** Phenol can be acetylated with acetic anhydride or
acetyl chloride in the presence of pyridine or other bases to give phenyl
acetate.

Phenyl acetate

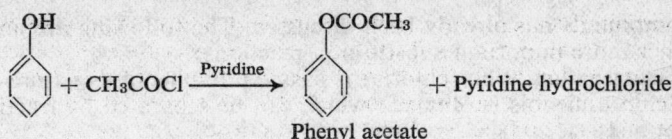

Phenol cannot be esterified by treatment directly with acids, though alcohols can be esterified in this manner. This difference has been attributed to the poor electron-donating ability of the phenolic oxygen atom as contrasted with that of the oxygen atom in an aliphatic ⁻OH group.

(IV) **Exchange of –OH for Other Groups.** Phenol, unlike the alcohols, reacts poorly with phosphorus trihalides to give halobenzenes. When phosphorus pentachloride or oxychloride reacts with phenol the major product obtained is triphenyl phosphate.

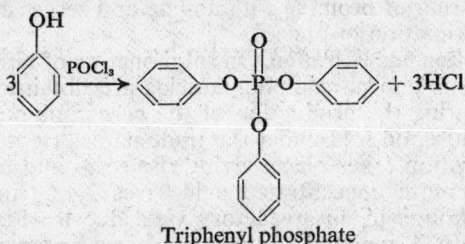

Triphenyl phosphate

Triphenyl phosphate and tricresyl phosphates are used as plasticizers. The latter (TCP) are also used as petrol additives. With phosphorus pentasulphide, phenol gives thiophenol:

$$5C_6H_5OH + P_2S_5 \longrightarrow 5C_6H_5SH + P_2O_5$$

(V) **Reaction with Ferric Chloride.** Phenols as a rule give intense colours, usually violet or blue, with aqueous ferric chloride solution. The test is often used to detect the presence of a phenolic hydroxyl group in an unknown compound.

(VI) **Reduction.** When phenol is heated with zinc dust reduction takes place and benzene is obtained. If phenol is reduced with hydrogen in the presence of nickel, cyclohexanol is obtained. The latter can be oxidized to adipic acid, an intermediate for production of nylon.

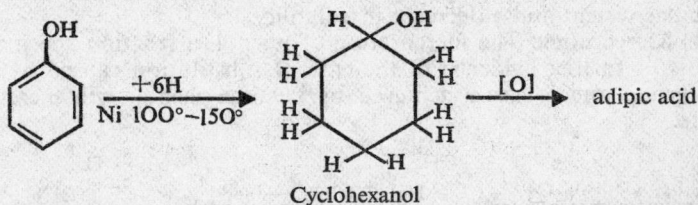

Cyclohexanol

SUBSTITUTION REACTIONS OF PHENOL: The *o*-, *p*-directing ⁻OH group causes facile substitutions of the nucleus. The coupling of diazonium compounds with phenols in alkaline solution to give *o*- or *p*-substituted azo compounds has already been discussed. The following are among the other more important substitution reactions:

(I) **Halogenation.** When chlorine is passed into phenol a mixture of *o*- and *p*-chlorophenols is obtained which can be separated by fractional distillation:

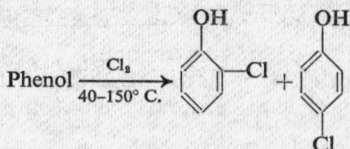

When phenol is brominated at low temperatures in carbon disulphide solution the main product is *p*-bromophenol. However, with bromine water, 2,4,6-tribromophenol is rapidly and quantitatively formed. This reaction recalls that of bromine with aniline and serves as the basis for a volumetric estimation of phenol.

(II) **Sulphonation and Nitration.** On sulphonation of phenol, a mixture of orthosulphonic and parasulphonic acids are obtained, higher temperatures favouring the production of the *para* compound. The *ortho* isomer is an antiseptic sold under the trade name Aseptol. With dilute nitric acid, nitration takes place, giving rise to *o*- and *p*-nitrophenols. The action of more concentrated acid gives 2,4,6-trinitrophenol or picric acid, although in unsatisfactory yield due to side reactions involving oxidation. The *o*- and *p*-nitrophenols can be separated by steam distillation, the *ortho* product being steam-volatile while the *para* isomer is not. In the case of *o*-nitrophenol, internal hydrogen bonding takes place and the molecule is in a monomolecular state and as a consequence more volatile.

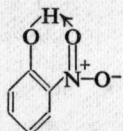

o-Nitrophenol
(Internal hydrogen bonding)

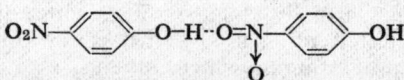

p-Nitrophenol
(Intermolecular hydrogen bonding)

In the case of *p*-nitrophenol, on the other hand, only intermolecular association is possible by hydrogen bonding, resulting in an increase in molecular weight and a decrease in volatility.

(III) **Mercuration.** The mercuration of phenol, a reaction specific to phenols, is further evidence of the ease of substitution caused by the ⁻OH group and is easily achieved by heating phenol with mercuric acetate.

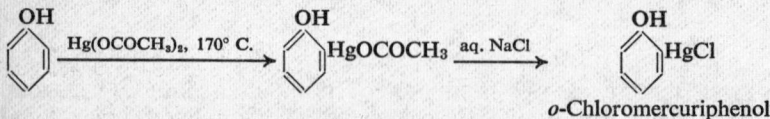

o-Chloromercuriphenol

The *o*-acetoxymercuriphenol is converted into chlormercuri derivative by pouring the reaction mixture into a sodium chloride solution. Such organic mercury compounds are used as antiseptics.

(IV) **Reaction with Formaldehyde.** Phenol and formaldehyde react together in a complex manner to give the well-known plastic, Bakelite. The initial step is the condensation of one mole of formaldehyde with one mole of phenol at the reactive *o*- or *p*-position.

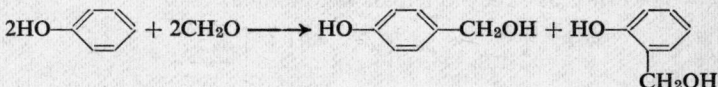

$$2HO-\underset{}{\bigcirc} + 2CH_2O \longrightarrow HO-\underset{}{\bigcirc}-CH_2OH + HO-\underset{CH_2OH}{\bigcirc}$$

Condensation of these products with a further molecule of phenol takes place as follows:

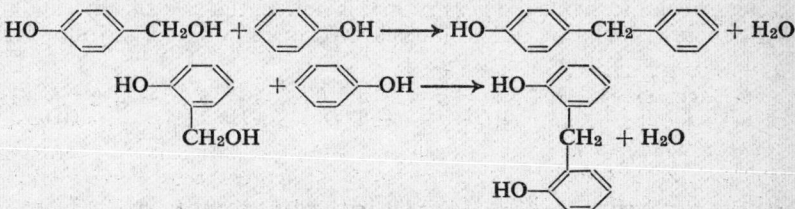

$$HO-\underset{}{\bigcirc}-CH_2OH + \underset{}{\bigcirc}-OH \longrightarrow HO-\underset{}{\bigcirc}-CH_2-\underset{}{\bigcirc} + H_2O$$

$$HO-\underset{CH_2OH}{\bigcirc} + \underset{}{\bigcirc}-OH \longrightarrow HO-\underset{\underset{HO-\bigcirc}{CH_2}}{\bigcirc} + H_2O$$

The process can be continued to form, under suitable conditions, linear polymers of the type shown below, containing alternate methylene units and benzene rings with CH_2OH groups in *ortho* or *para* positions. Such chains can also be cross-linked to give three-dimensional polymers.

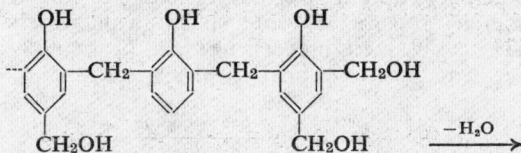

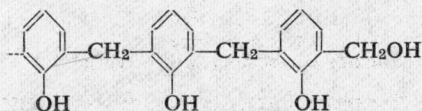

Linear polymers

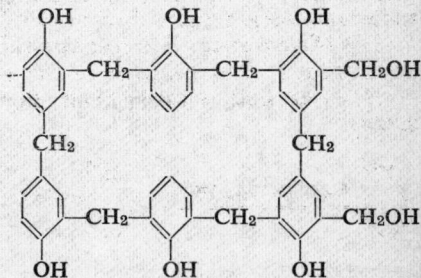

Cross-linked polymer

Phenol–formaldehyde and other plastics are separately considered in Chapter Nineteen.

(V) **Reimer–Tiemann Reaction.** This reaction was discovered during experiments to ascertain what would happen if one used phenol instead of aniline in the carbylamine test for a primary amine. When phenol, chloroform, and sodium hydroxide are allowed to react a phenolic aldehyde known as salicylaldehyde is obtained.

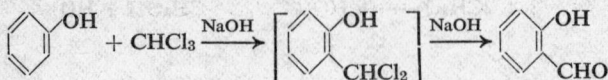

Some *para* isomer is also formed. The two isomers can be separated by taking advantage of the fact that salicylaldehyde (the *ortho* compound) is steam-volatile, while the *para* isomer is not. In the case of salicylalde-hyde there is chelation due to internal hydrogen bonding, as follows:

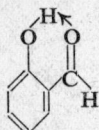
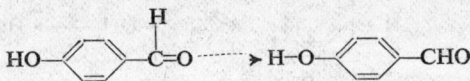

Chelated form of salicylaldehyde

Intermolecular association in *p*-hydroxybenzaldehyde

Hence salicylaldehyde maintains its monomolecular state and is more volatile than the *para* isomer, where the hydrogen bonding is not intra-molecular but intermolecular (cf. *o*- and *p*-nitrophenols).

(VI) **Kolbe Reaction.** When dry sodium phenolate is heated to 130° C. with carbon dioxide under a pressure of about six atmospheres the sodium salt of salicylic acid is obtained in excellent yield, probably by rearrangement of the intermediate sodium phenyl carbonate.

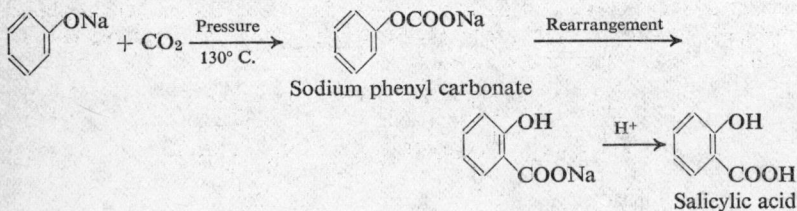

Sodium phenyl carbonate

Salicylic acid

OTHER SPECIFIC PHENOLS

The cresols are the next higher homologues of phenol.

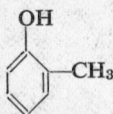
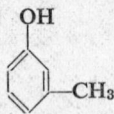
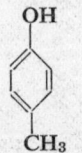

o-Cresol   *m*-Cresol   *p*-Cresol

They are obtained from coal tar. They can be prepared in the laboratory by diazotizing the corresponding toluidines and by heating the diazonium compounds in water.

Among the naturally occurring phenols are thymol, carvacrol, and eugenol. Thymol occurs in oil of thyme and mint and carvacrol in oil of caraway. Both of them have germicidal properties. Thymol is an ingredient of toothpaste and mouth lotions. Thymol and carvacrol are related to *p*-cymene (1-methyl-4-isopropylbenzene) and have the following structures:

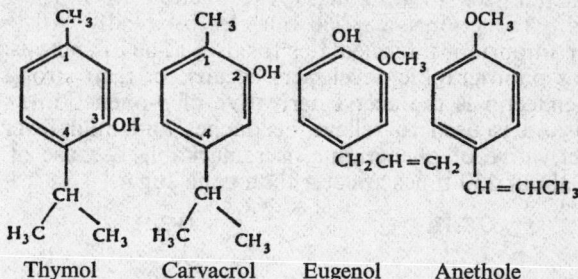

Thymol   Carvacrol   Eugenol   Anethole

Eugenol is a constituent of clove oil. It is used in perfumery and also in the manufacture of vanillin. It is also useful as a local anaesthetic. Anethole is a methyl ether of considerable use as a flavouring material and is obtained from the oil of aniseed.

Among nitrophenols *o*- and *p*-nitrophenols have been studied. Picric acid or 2,4,6-trinitrophenol may be prepared by direct nitration of phenol with the usual nitric acid–sulphuric acid mixture. But during this process considerable loss occurs due to oxidation of phenol. Hence the trinitration is done indirectly. Phenol is sulphonated to the disulphonic acid stage and then nitric acid is added. Nitric acid replaces the two sulphonic acid groups with nitro groups and introduces an additional nitro group, as follows:

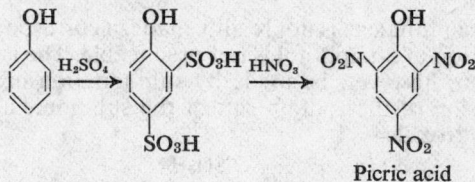

Picric acid

The electron density on the ring is maintained at a reasonably low level throughout the reaction by the presence of the electron-withdrawing sulphonic acid groups, thus keeping undesirable side oxidation at a minimum. Picric acid ($K_A = 4 \cdot 2 \times 10^{-1}$) is almost as strong as a mineral acid and liberates carbon dioxide from carbonates. Picric acid is a yellow solid with an intensely bitter taste. In fact, picric acid is named because of this property (Greek *pikros* means bitter). It is useful as a dye for silk and wool, as a drug for treatment of burns, and also as an explosive. The aminophenols may be made by reduction of nitrophenols. *m*-Aminophenol is produced in industry by heating resorcinol with ammonia in the presence of ammonium chloride in an autoclave.

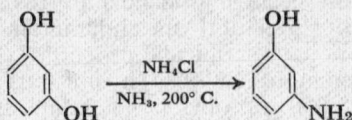

p-Aminophenol is also obtained by reduction of nitrobenzene. In industry, this reduction is carried out electrolytically. All the amino-phenols are important intermediates for dyes. The *o*- and *p*-isomers are also used as photographic developers because of their strong reducing action. Phenacetin is the acetyl derivative of *p*-phenetidine (*p*-amino-phenetole) and is used to relieve headaches and mild fever. Dulcin, another derivative of phenetidine, is remarkable because of its sweet taste. It is about 250 times sweeter than cane sugar.

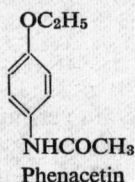

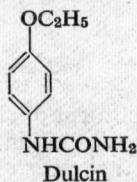

Phenacetin   Dulcin

Of the monohydric phenols derived from the polynuclear hydro-carbons, the naphthols are important. α-Naphthol can be obtained by reducing α-nitronaphthalene to α-naphthylamine and converting the amino group into the −OH group by the diazo reaction, as shown below.

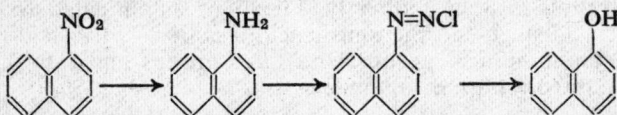

Since β-nitronaphthalene is not readily made, a corresponding method for the preparation of β-naphthol is not practicable. The latter as well as the α-isomer can, however, be made by sulphonation of naphthalene, followed by fusion of the sodium salt of the sulphonic acids obtained with sodium hydroxide.

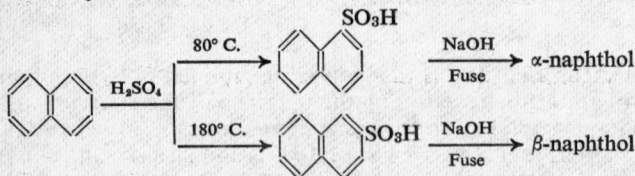

The naphthols are obtained by acidification of the sodium salts. Unlike phenol, β-naphthol gives methyl ether when heated with methanol and sulphuric acid.

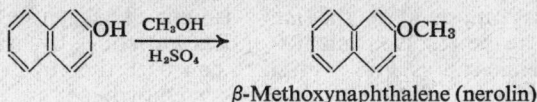

β-Methoxynaphthalene (nerolin)

β-Methoxynaphthalene has a smell resembling the oil of orange flowers (neroli oil) and is used, under the trade name **nerolin**, in perfumery. As has been mentioned previously, diazonium salts couple with α-naphthol in position 4 or if that is occupied in position 2. β-Naphthol couples only at position 1.

USES OF PHENOLS: The most important use of phenol itself is in the manufacture of Bakelite plastics. In addition, it is useful as an antiseptic and germicide. The cresols are more powerful germicides and, since they are also less toxic to man, are more often preferred. Lysol is a mixture of cresols dissolved in soap solution. Several of the phenols and the naphthols are important intermediates in the manufacture of dyes. The special uses of specific substituted phenols have already been mentioned.

POLYHYDRIC PHENOLS

Polyhydric phenols have the properties of phenol. The increased number of hydroxyl groups makes these substances more water soluble and also contributes to an increased tendency to become oxidized. The three dihydric phenols derived from benzene are the following:

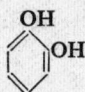   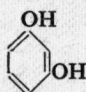   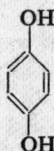

    Catechol   Resorcinol   Hydroquinone

Catechol can be obtained by alkali fusion of *o*-hydroxybenzene–sulphonic acid or by heating *o*-chlorophenol with alkali in an autoclave. The monomethyl ether of catechol, known as **guaiacol**, occurs in beechwood tar.

Adrenaline, the hormone (see Chapter Eighteen) from the adrenal glands, is a derivative of catechol. It can be synthesized from catechol, as follows:

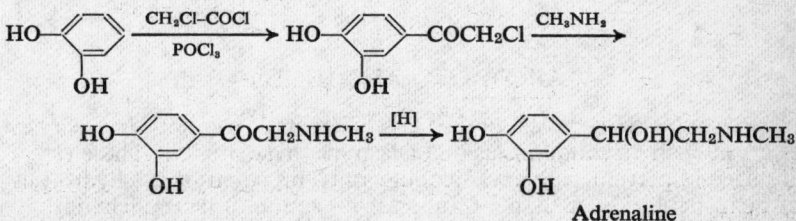

Adrenaline

The natural compound is the *laevo* isomer. The synthetic racemic compound can be resolved into its optical antipodes by using tartaric acid. Adrenaline causes blood vessels to constrict, thereby raising blood pressure. It is used in medicine to check haemorrhage.

Resorcinol can be obtained by alkali fusion of the sodium salt of *m*-benzenedisulphonic acid.

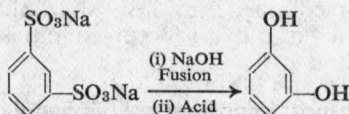

Resorcinol and its mono acetyl derivative are used in medicine as antiseptics. The powerful antiseptic 4-*n*-hexylresorcinol is prepared by condensing caproic acid with resorcinol in the presence of zinc chloride and reducing the keto group in the resulting compound by Clemmensen's method:

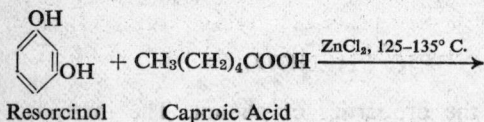

Resorcinol   Caproic Acid

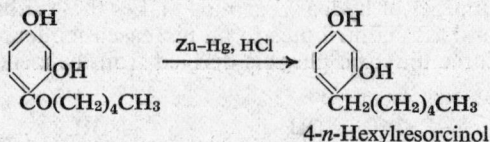

4-*n*-Hexylresorcinol

Quinol or hydroquinone is an excellent reducing agent and is commonly used as a photographic developer and as an antioxidant or a stabilizer for certain materials which polymerize in the presence of oxidizing agents. It is prepared by reducing quinone with sulphurous acid.

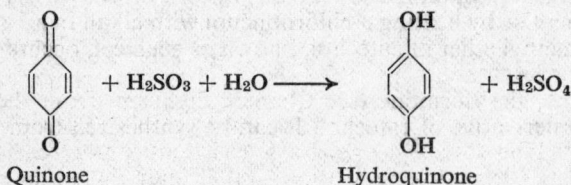

Quinone   Hydroquinone

AROMATIC ALCOHOLS

Benzyl alcohol is typical of the class of aromatic alcohols. It occurs as esters in such essential oils as oil of jasmine, hyacinth, etc. These esters are used in perfumes. Benzyl alcohol can be prepared by the hydrolysis of benzyl chloride or by the Cannizzaro reaction of benzaldehyde.

$$\text{CH}_2\text{Cl (benzene ring)} + \text{H}_2\text{O} \xrightarrow{\text{Na}_2\text{CO}_3} \text{CH}_2\text{OH (benzene ring)} + \text{NaCl} + \text{NaHCO}_3$$

$$2\,\text{CHO (benzene ring)} + \text{NaOH} \longrightarrow \text{CH}_2\text{OH (benzene ring)} + \text{COONa (benzene ring)}$$

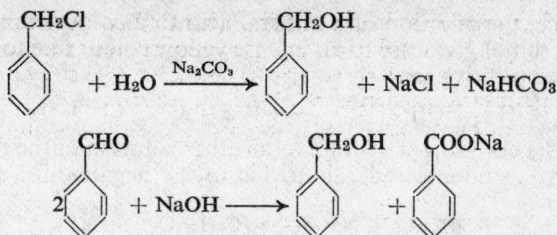

The mixture obtained from the Cannizzaro reaction is diluted with water when the sodium benzoate goes into solution. Extraction with ether furnishes benzyl alcohol. Benzyl alcohol is similar in its chemical properties to the aliphatic alcohols like methyl alcohol and ethyl alcohol. Thus it forms esters, ethers, alkoxides, and, on heating with hydrochloric or hydrobromic acid, exchanges the hydroxyl groups for chlorine or bromine. Benzyl benzoate and benzyl acetate have pleasant odours and are used in perfumes.

2-Phenylethyl alcohol occurs in rose oil in the free state as well as in the form of its esters. It has a very pleasant fragrance and is highly valued in perfumery. It can be obtained by treating ethylene oxide with phenylmagnesium bromide:

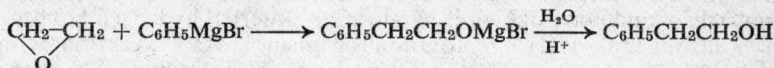

$$\text{CH}_2\text{—CH}_2 (\text{O}) + \text{C}_6\text{H}_5\text{MgBr} \longrightarrow \text{C}_6\text{H}_5\text{CH}_2\text{CH}_2\text{OMgBr} \xrightarrow[\text{H}^+]{\text{H}_2\text{O}} \text{C}_6\text{H}_5\text{CH}_2\text{CH}_2\text{OH}$$

Diphenylcarbinol or benzhydrol is obtained readily by the reduction of benzophenone.

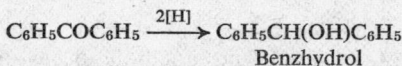

$$\text{C}_6\text{H}_5\text{COC}_6\text{H}_5 \xrightarrow{2[\text{H}]} \text{C}_6\text{H}_5\text{CH(OH)C}_6\text{H}_5$$
Benzhydrol

Diphenylcarbinol reacts even more readily than benzyl alcohol with hydrochloric acid, giving diphenylchloromethane.

Triphenylcarbinol can be obtained by treating phenylmagnesium bromide with benzophenone.

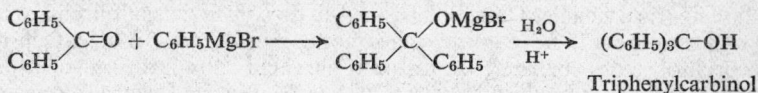

$$\begin{array}{c}\text{C}_6\text{H}_5 \\ \text{C}_6\text{H}_5\end{array}\!\!\text{C=O} + \text{C}_6\text{H}_5\text{MgBr} \longrightarrow \begin{array}{c}\text{C}_6\text{H}_5 \\ \text{C}_6\text{H}_5\end{array}\!\!\text{C}\begin{array}{c}\text{OMgBr} \\ \text{C}_6\text{H}_5\end{array} \xrightarrow[\text{H}^+]{\text{H}_2\text{O}} (\text{C}_6\text{H}_5)_3\text{C—OH}$$
Triphenylcarbinol

Triphenylcarbinol and its halide are very reactive. They take part in reactions involving the triphenylmethyl group, either as a free radical, or an ion positive or negative depending on the conditions of the reaction.

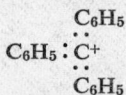

$$\begin{array}{c}\text{C}_6\text{H}_5 \\ \text{C}_6\text{H}_5 : \text{C}^+ \\ \text{C}_6\text{H}_5\end{array} \qquad \begin{array}{c}\text{C}_6\text{H}_5 \\ \text{C}_6\text{H}_5 : \text{C} :^- \\ \text{C}_6\text{H}_5\end{array}$$

Triphenyl carbonium ion   Triphenyl carbanion

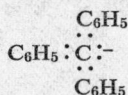

Each of these species is stabilized by several structures in which a benzene ring is quinonoid in nature, and hence is intensely coloured.

238   *Organic Chemistry Made Simple*

For instance, the addition of a mineral acid to a colourless solution of triphenylcarbinol gives rise to an intense yellow colour due to formation of the above positive ion:

$$(C_6H_5)_3C\text{-}OH + HCl \longrightarrow (C_6H_5)_3C^+ + Cl^- + H_2O$$

Similarly, the addition of sodium to an ether solution of the free radical gives rise to an intense red colour due to the negative ion mentioned above.

$$(C_6H_5)_3C^\cdot + Na^\cdot \longrightarrow (C_6H_5)_3C^{\cdot-} + Na^+$$

SUMMARY

Phenols contain an $-OH$ group attached to an aromatic ring. Phenol obtained industrially from: (1) coal tar; (2) chlorobenzene; (3) benzene-sulphonic acid. **Reactions of phenol**: (I) Weaker acid than carbonic acid. (II) Alkylation of sodium salt to give alkyl ethers. (III) **Acylation**: Phenol to phenyl esters by treatment with acid chlorides and pyridine. Difference from esterification of aliphatic alcohols. (IV) Exchange Reactions: phenol to triphenyl phosphate; phenol to thiophenol. (V) Colouration with $FeCl_3$. (VI) Catalytic reduction to cyclohexanol. Substitution Reactions include: (I) Halogenation, for example, phenol to o- and p-chlorophenol. (II) Sulphonation to o- and p-sulphonic acids. Similarly o- and p-nitrophenols by nitration. Internal hydrogen bonding in o-nitrophenol. (III) Mercuration of phenol. (IV) Reaction with formaldehyde to give Bakelite plastic. (V) Reimer–Tiemann Reaction: Phenol to o- and p-hydroxy benzaldehyde. (VI) Kolbe Reaction: Sodium phenolate to salicylic acid by heating with CO_2 at high pressure and temperature.

Other Phenols. Cresols from coal tar and also from toluidines. Thymol, carvacrol, eugenol and anethole occur naturally. Indirect nitration of phenol to 2,4,6-trinitrophenol (Picric acid). Aminophenols and their uses. Phenacetin and Dulcin, useful derivatives. Naphthols obtained by fusion of α- and β-naphthalenesulphonic acids with sodium hydroxide. α-Isomer also available from α-naphthylamine. Nerolin, used in perfumery, is β-methoxynaphthalene. Uses of phenols.

Polyhydric Phenols. Catechol from o-hydroxybenzenesulphonic acid or o-chlorophenol. Adrenaline, a hormone, is a derivative of catechol. Resorcinol from benzene m-disulphonic acid. Preparation of 4-n-hexylresorcinol, a powerful antiseptic. Hydroquinone from quinone by reduction with sulphur dioxide.

Aromatic Alcohols. Contain $-OH$ in the side chain. Benzyl alcohol from benzyl chloride or benzaldehyde. 2-Phenylethyl alcohol, benz-hydrol, and triphenylcarbinol. Free radical and ionic reactions of triphenylcarbinol.

Problem Set No. 11

1. What are the organic products obtainable by treating phenol with the following?

   (a) Phosphorus oxychloride.
   (b) Bromine water.

(c) Two molecular proportions of sulphuric acid.
(d) Dilute nitric acid.
(e) Dimethyl sulphate and alkali.
(f) Nitrous acid.

2. How are the following synthesized?

(a) Triphenylcarbinol from benzophenone.
(b) Dimethylphenylcarbinol from ethyl benzoate.
(c) Cyclohexanol from benzene.

3. How can the following mixtures be separated?

(a) Benzene, phenol, and benzoic acid.
(b) Phenol and benzyl alcohol.
(c) Phenol, cyclohexanol, and picric acid.

CHAPTER FIFTEEN

AROMATIC ALDEHYDES, KETONES, AND QUINONES

$$\underset{|}{\overset{H}{}}$$

The aromatic aldehydes contain a $-\overset{|}{C}=O$ group directly linked to an aromatic nucleus, benzaldehyde being a typical example. The aromatic aldehydes show several reactions of the aliphatic aldehydes, though in one point a distinction is to be made. The aromatic aldehydes lack a hydrogen atom in the α-position, and those reactions, depending upon the existence of the α hydrogen, are not observed in their cases. Aldehydes, like phenyl acetaldehyde, which contain the aldehyde group in the side chain, behave like aliphatic aldehydes.

PREPARATIONS OF AROMATIC ALDEHYDES: The aromatic aldehydes may be prepared by any of the general methods applied in the case of aliphatic aldehydes. Special methods applicable for the aromatic compounds are summarized below.

1. **Oxidation of Side Chain.** An alkyl group attached to the benzene nucleus can be oxidized to a $-CHO$ group under suitable conditions. Thus toluene can be oxidized by manganese dioxide and sulphuric acid to benzaldehyde.

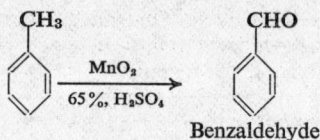

Benzaldehyde

The conditions must be carefully regulated, or else the benzaldehyde produced is further oxidized to benzoic acid. o-Nitrobenzaldehyde can be obtained by a similar oxidation of o-nitrotoluene. An alternative process used in industry consists of halogenation of toluene to benzal chloride, followed by alkaline hydrolysis.

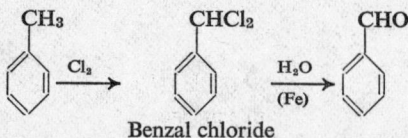

Benzal chloride

2. **Formylation.** An aromatic hydrocarbon when treated with a mixture of carbon monoxide and hydrogen chloride, in the presence of anhydrous aluminum chloride and cuprous chloride, is directly converted to an aldehyde, that is, formylated (Gattermann–Koch reaction).

Thus toluene condenses with carbon monoxide and hydrogen chloride to give *p*-tolualdehyde.

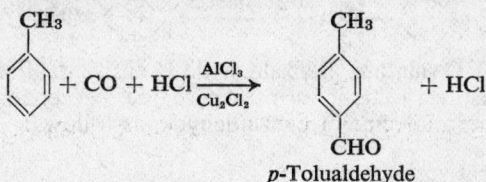

p-Tolualdehyde

The requisite equimolecular mixture of carbon monoxide and hydrogen chloride can be prepared conveniently by adding chlorosulphonic acid to formic acid.

$$HCOOH + ClSO_3H \longrightarrow CO + HCl + H_2SO_4$$

In a modification of the above reaction, particularly useful for the preparation of phenolic aldehydes (and their ethers), hydrogen cyanide is used instead of hydrogen chloride. Instead of using poisonous hydrogen cyanide as such, a mixture of zinc cyanide and hydrogen chloride is employed. Excellent yields of anisaldehyde are obtained, for example, from anisole:

Anisole   Anisaldehyde

Good yields are obtained by this method, even when it is applied to toluene and xylene, provided the reaction is carried out at 100° C.

The Reimer–Teimann reaction, leading to phenolic aldehydes like salicylaldehyde, has already been discussed.

3. **Aldehydes from Nitriles and Acids.** When anhydrous hydrogen chloride is passed into a mixture of an aromatic nitrile and anhydrous stannous chloride in ether a complex stannic salt separates which on hydrolysis gives an aldehyde.

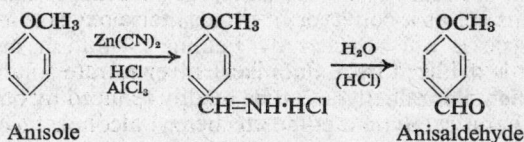

Imino chloride

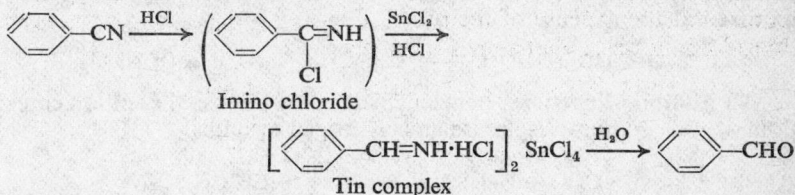

Tin complex

An imino chloride, formed by addition of hydrogen chloride to $-C\equiv N$, is believed to be an intermediate, and the reaction is known as the Stephen reaction.

A number of methods are available for the conversion of an aromatic acid to the corresponding aldehyde via the acid chloride. The

Rosenmund method of reduction, for example, can be applied to prepare benzaldehyde.

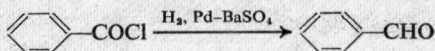

$$\text{C}_6\text{H}_5-\text{COCl} \xrightarrow{\text{H}_2,\ \text{Pd--BaSO}_4} \text{C}_6\text{H}_5-\text{CHO}$$

REACTIONS: (I) **Oxidation**. Benzaldehyde is easily oxidized to benzoic acid. On exposure to air, it forms a peroxide, perbenzoic acid, which oxidizes another molecule of benzaldehyde, as follows:

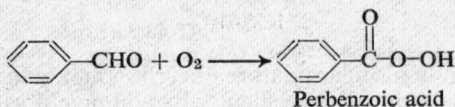

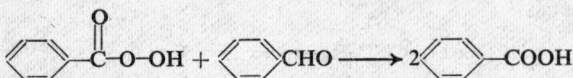

$$\text{C}_6\text{H}_5-\text{CHO} + \text{O}_2 \longrightarrow \text{C}_6\text{H}_5-\overset{\text{O}}{\overset{\|}{\text{C}}}-\text{O--OH}$$

Perbenzoic acid

$$\text{C}_6\text{H}_5-\overset{\text{O}}{\overset{\|}{\text{C}}}-\text{O--OH} + \text{C}_6\text{H}_5-\text{CHO} \longrightarrow 2\ \text{C}_6\text{H}_5-\text{COOH}$$

This type of oxidation, known as auto-oxidation, may be prevented by the addition of small amounts of hydroquinone, one part of hydroquinone in about 50,000 parts of the aldehyde acting as an effective antioxidant. Some substances, which are not oxidized by air, may be induced to become oxidized by the addition of a small quantity of benzaldehyde, probably by the formation of perbenzoic acid. Benzaldehyde may thus act as a conveyer of atmospheric oxygen.

Benzaldehyde is not affected by Fehling's solution or Benedict's solution, but is oxidized by ammoniacal silver nitrate solution.

(II) **Reduction**. Benzaldehyde can be readily reduced by both chemical (e.g. LiAlH₄) and catalytic methods to benzyl alcohol.

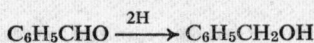

$$\text{C}_6\text{H}_5\text{CHO} \xrightarrow{2\text{H}} \text{C}_6\text{H}_5\text{CH}_2\text{OH}$$

(III) **Action of Alkali–Cannizzaro Reaction**. An aliphatic aldehyde having an α-hydrogen atom when treated with alkali gives a resin. Aromatic aldehydes, on the other hand, lack α-hydrogen atoms and undergo a reaction involving simultaneous oxidation and reduction—a reaction known as Cannizzaro reaction. Thus benzaldehyde gives benzyl alcohol and sodium benzoate when warmed with a concentrated solution of sodium hydroxide. Here one molecule of the aldehyde is oxidized at the expense of the other.

$$2\text{C}_6\text{H}_5\text{CHO} + \text{NaOH} \longrightarrow \text{C}_6\text{H}_5\text{CH}_2\text{OH} + \text{C}_6\text{H}_5\text{COONa}$$

(IV) **Addition Reactions**. Benzaldehyde gives the usual addition compounds with sodium bisulphite and hydrogen cyanide.

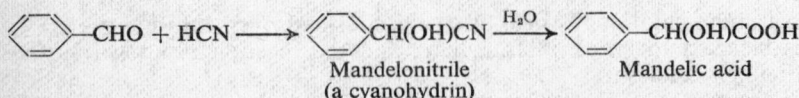

$$\text{C}_6\text{H}_5-\text{CHO} + \text{NaHSO}_3 \longrightarrow \text{C}_6\text{H}_5-\text{CH(OH)SO}_3\text{Na}$$

$$\text{C}_6\text{H}_5-\text{CHO} + \text{HCN} \longrightarrow \underset{\substack{\text{Mandelonitrile} \\ \text{(a cyanohydrin)}}}{\text{C}_6\text{H}_5-\text{CH(OH)CN}} \xrightarrow{\text{H}_2\text{O}} \underset{\text{Mandelic acid}}{\text{C}_6\text{H}_5-\text{CH(OH)COOH}}$$

Hydrolysis of the cyanohydrin gives *dl*-mandelic acid.

Like the aliphatic aldehydes, benzaldehyde gives the usual derivatives with hydroxylamine, phenylhydrazine, semicarbazide, etc.

(V) Reaction with Ammonia and Amines. With ammonia, unlike the other aldehydes which form addition compounds, benzaldehyde gives hydrobenzamide with the elimination of water.

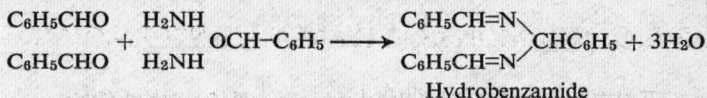

$$
\begin{array}{l}
C_6H_5CHO \\
C_6H_5CHO
\end{array}
+
\begin{array}{l}
H_2NH \\
H_2NH
\end{array}
OCH{-}C_6H_5 \longrightarrow
\begin{array}{l}
C_6H_5CH{=}N \\
C_6H_5CH{=}N
\end{array}
\Big\rangle CHC_6H_5 + 3H_2O
$$

Hydrobenzamide

Benzaldehyde condenses with aniline and other primary amines to give Schiff's Bases. Reduction of the Schiff's Base will give secondary amines, and this constitutes a route to the preparation of secondary amines uncontaminated with primary or tertiary amines.

(VI) Condensation with Aliphatic Aldehydes and Ketones. Even though benzaldehyde lacks an α-hydrogen atom, it still undergoes reactions similar to the aldol type of condensation with other carbonyl compounds having α-hydrogen atoms.

For example, with acetaldehyde in the presence of sodium ethoxide or hydroxide, benzaldehyde gives cinnamaldehyde.

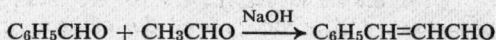

$$C_6H_5CHO + CH_3CHO \xrightarrow{\text{NaOH}} C_6H_5CH{=}CHCHO$$

Cinnamaldehyde is the chief flavouring principle of oil of cinnamon. Similarly, acetone may be condensed in the presence of alkali with one or two molecules of benzaldehyde to give benzalacetone or dibenzalacetone respectively.

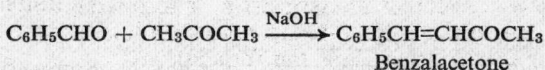

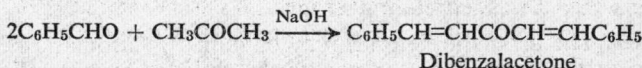

$$C_6H_5CHO + CH_3COCH_3 \xrightarrow{\text{NaOH}} C_6H_5CH{=}CHCOCH_3$$

Benzalacetone

$$2C_6H_5CHO + CH_3COCH_3 \xrightarrow{\text{NaOH}} C_6H_5CH{=}CHCOCH{=}CHC_6H_5$$

Dibenzalacetone

(VII) The Perkin Reaction. When benzaldehyde is heated with acetic anhydride and sodium acetate at 180° C. for several hours cinnamic acid is obtained.

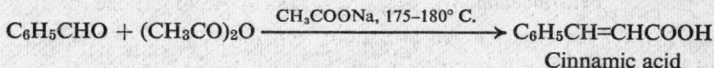

$$C_6H_5CHO + (CH_3CO)_2O \xrightarrow{\text{CH}_3\text{COONa, 175-180° C.}} C_6H_5CH{=}CHCOOH$$

Cinnamic acid

This reaction is capable of wide extension, and a variety of substituted cinnamic acids can be prepared.

(VIII) Condensation with Active Methylene Compounds. Benzaldehyde undergoes the KNOEVENAGEL reaction with malonic ester to give the benzalmalonic ester. A secondary amine like piperidine is used as a catalyst.

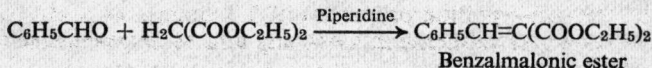

$$C_6H_5CHO + H_2C(COOC_2H_5)_2 \xrightarrow{\text{Piperidine}} C_6H_5CH{=}C(COOC_2H_5)_2$$

Benzalmalonic ester

With nitromethane, benzaldehyde gives nitrostyrene.

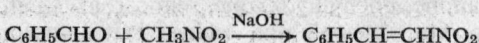

$$C_6H_5CHO + CH_3NO_2 \xrightarrow{NaOH} C_6H_5CH{=}CHNO_2$$

(IX) **Benzoin Condensation,** When boiled in an alcohol solution containing a little potassium cyanide or sodium cyanide, benzaldehyde undergoes self-condensation to give benzoin.

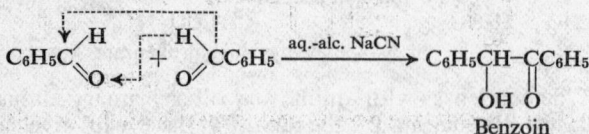

$$\text{(benzaldehyde)} \xrightarrow{\text{aq.-alc. NaCN}} \underset{\substack{| \quad \; \| \\ OH \; O}}{C_6H_5CH{-}CC_6H_5}$$
Benzoin

Oxidation of benzoin with nitric acid furnishes the diketone, benzil.

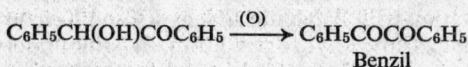

$$C_6H_5CH(OH)COC_6H_5 \xrightarrow{(O)} C_6H_5COCOC_6H_5$$
Benzil

When benzil is treated with aqueous potassium hydroxide it undergoes a peculiar rearrangement known as the benzilic acid rearrangement and gives the potassium salt of benzilic acid.

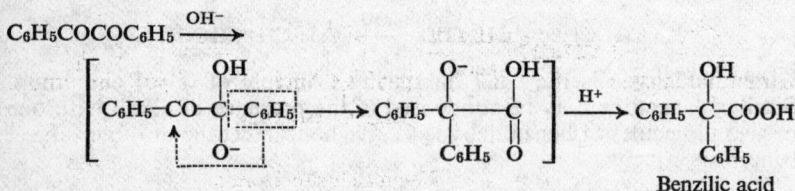

$$C_6H_5COCOC_6H_5 \xrightarrow{OH^-}$$

Benzilic acid

The reaction is believed to proceed by initial addition of OH⁻ to one of the carbonyl groups followed by migration of a phenyl group with a pair of electrons as previously shown.

Ketones. The aromatic ketones may contain two aromatic residues attached to the keto group, or one of the residues may be aliphatic. Thus we have benzophenone and acetophenone.

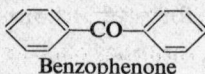

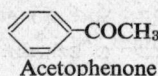

Benzophenone   Acetophenone

PREPARATION: (1) Friedel and Crafts reaction of benzene with benzoyl chloride or with phosgene gives benzophenone.

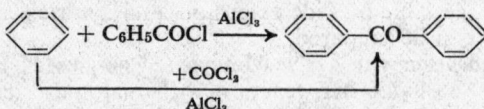

Similarly, Friedel–Crafts reaction of benzene with acetyl chloride gives acetophenone.

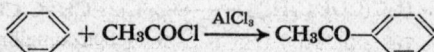

(2) Several phenolic ketones are prepared by means of the Fries rearrangement, which consists of heating an acyl derivative of a phenol with aluminum chloride in an inert solvent. The acyl group can migrate to either the *ortho* or *para* position.

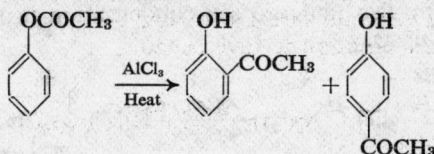

The ratio of the *ortho* to the *para* isomer depends on the condition of the reaction as the solvent, temperature, and amount of catalyst used.

(3) Another method applicable to the preparation of phenolic ketones is the Hoesch synthesis, which is a modification of the Gattermann aldehyde synthesis. In this method a phenol is condensed with acetonitrile and hydrogen chloride in the presence of zinc chloride or aluminum chloride as a catalyst in a solution of ether or chlorobenzene. Thus phloracetophenone can be prepared from phloroglucinol.

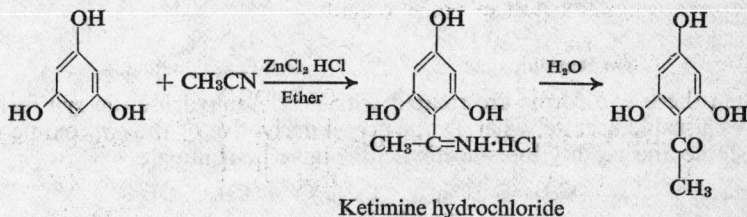

Ketimine hydrochloride

Acetophenone has been used in medicine as a hypnotic under the trade name Hypnone. On chlorination in acetic acid solution, acetophenone gives phenacyl chloride, which is a powerful lachrymator and is used as a tear gas in police work.

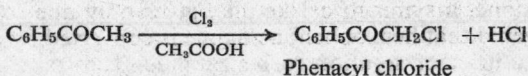

$$C_6H_5COCH_3 \xrightarrow[CH_3COOH]{Cl_2} C_6H_5COCH_2Cl + HCl$$
Phenacyl chloride

Bromination of acetophenone, similarly, furnishes phenacyl bromide.

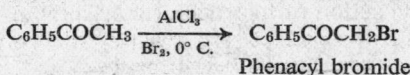

$$C_6H_5COCH_3 \xrightarrow[Br_2,\ 0°\ C.]{AlCl_3} C_6H_5COCH_2Br$$
Phenacyl bromide

Phenacyl bromide is often used to identify organic acids. It reacts with sodium salts of acids to give solid phenacyl esters which are readily purified and crystallized.

$$C_6H_5COCH_2Br + RCOONa \longrightarrow RCOOCH_2COC_6H_5 + NaBr$$
Phenacyl ester

Acetophenone reacts with hydroxylamine, hydrogen cyanide, etc., in the manner expected of a ketone. It does not form a bisulphite compound, however. The methyl group in acetophenone is adjacent to a carbonyl group and takes part in condensation reactions. For example, acetophenone readily condenses with benzaldehyde, as follows:

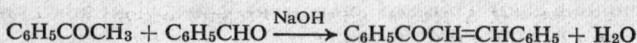

$$C_6H_5COCH_3 + C_6H_5CHO \xrightarrow{NaOH} C_6H_5COCH=CHC_6H_5 + H_2O$$

Acetophenone can also undergo self-condensation in the presence of mild alkaline catalysts.

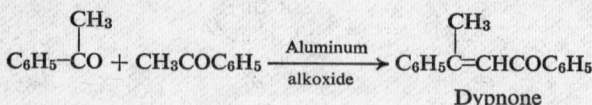

$$\underset{\overset{|}{C_6H_5-CO}}{\overset{CH_3}{}} + CH_3COC_6H_5 \xrightarrow[\text{alkoxide}]{\text{Aluminum}} \underset{\overset{|}{C_6H_5C}=CHCOC_6H_5}{\overset{CH_3}{}}$$

Dypnone

STEREOCHEMISTRY OF THE OXIMES

Oximes of aldehydes and unsymmetrical ketones (both aliphatic and aromatic) exhibit geometrical isomerism. The double bond between carbon and nitrogen in the oxime hinders free rotation, and hence the substance must exist in *syn* and *anti* forms. Thus benzaldehyde gives two stereoisomeric oximes.

$$\begin{array}{cc} C_6H_5-C-N & C_6H_5-C-H \\ \| & \| \\ N-OH & HO-N \end{array}$$

syn-Benzaldoxime   *anti*-Benzaldoxime

When these two forms are treated with acetic anhydride the *syn* form gives a stable acetate, whereas the acetyl derivative of the *anti* oxime is unstable and readily loses acetic acid to give benzonitrile.

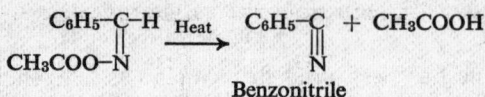

$$\begin{array}{c} C_6H_5-C-H \\ \| \\ CH_3COO-N \end{array} \xrightarrow{Heat} \begin{array}{c} C_6H_5-C \\ \| \\ N \end{array} + CH_3COOH$$

Benzonitrile

BECKMANN REARRANGEMENT OF KETOXIMES

Benzophenone, a symmetrical ketone, gives only one oxime. BECKMANN discovered that benzophenone oxime is converted by phosphorus pentachloride in ether solution to benzanilide. The transformation of an oxime into an amide is known as the Beckmann rearrangement. The rearrangement is brought about by phosphorus pentachloride in dry ether or by the action of hydrochloric or sulphuric acids.

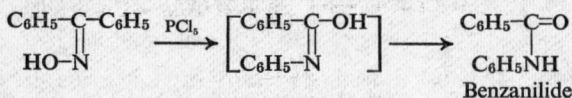

$$\begin{array}{c} C_6H_5-C-C_6H_5 \\ \| \\ HO-N \end{array} \xrightarrow{PCl_5} \left[\begin{array}{c} C_6H_5-C-OH \\ \| \\ C_6H_5-N \end{array} \right] \longrightarrow \begin{array}{c} C_6H_5-C=O \\ \\ C_6H_5NH \end{array}$$

Benzanilide

The Beckmann rearrangement is a *trans* phenomenon. During the rearrangement there is an exchange between the hydroxyl group and the group *trans* to it, followed by the ketonization of the enolic intermediate, as illustrated above. The *trans* nature of this rearrangement is so well established that it is sometimes used to fix the geometric structure of an oxime. Aldoximes may undergo the Beckmann transformation, but they leave a considerable amount of dehydration products.

QUINONES

The quinones are a special class of ketones in which the carbonyl groups form part of a ring. Some of the common types of quinones are given below:

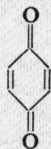 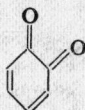 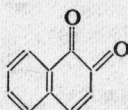 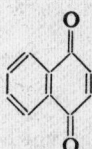

p-Quinone   *o*-Quinone   1,2-Naphthoquinone   1,4-Naphthoquinone

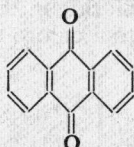

Anthraquinone

The quinone ring contains only two double bonds, and hence the quinones are not aromatic in nature. *p*-Benzoquinone and other *p*-quinones are yellow, while the *o*-quinones are generally red or orange. The oxidation of aniline with sodium dichromate and sulphuric acid provides a convenient preparation of *p*-quinone. The quinones show the reactions of both ketones and olefines and exhibit a marked tendency to become aromatic. On reduction with sulphurous acid, *p*-quinone gives hydroquinone.

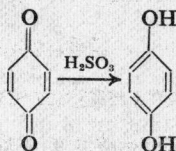

With phosphorus pentachloride, *p*-quinone is converted into *p*-dichlorobenzene.

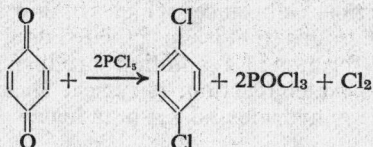

Oxidation of *p*-quinone furnishes maleic anhydride.

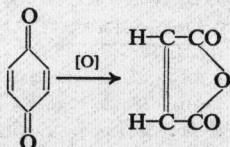

Maleic anhydride

The quinones form mono- and dioximes. The monoxime of *p*-quinone is tautomeric with *p*-nitrosophenol.

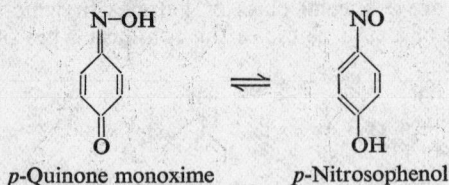

p-Quinone monoxime   *p*-Nitrosophenol

As a typical olefine, quinone adds on two and then four atoms of bromine, giving quinone dibromide and tetrabromide respectively. Quinone adds on hydrogen chloride in the 1,4 position, and the addition compound tautomerizes to a chloro derivative of quinol:

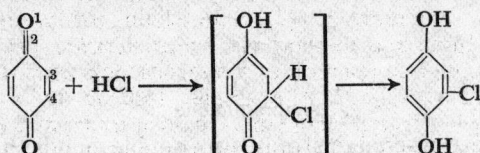

With chlorine, however, quinone gives a substitution product chloranil.

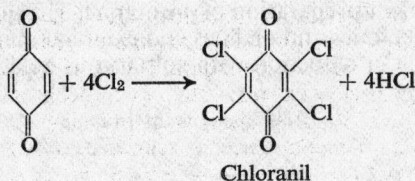

Chloranil

When a yellow alcoholic solution of quinone is added to a colourless solution of hydroquinone in alcohol, dark green crystals separate. This substance is a molecular complex of equimolecular proportions of quinone and hydroquinone, called quinhydrone.

Diels–Alder Reaction. Substances which contain a double bond between two carbonyl groups (that is, an activated double bond) add on to conjugated dienes forming 1,4 addition products. This reaction is known as the Diels–Alder reaction. A typical Diels–Alder reaction is that between maleic anhydride and 1,4-butadiene.

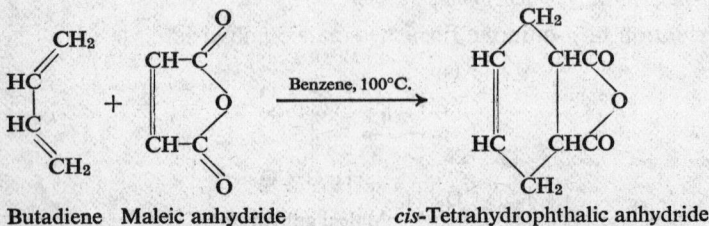

Butadiene   Maleic anhydride   *cis*-Tetrahydrophthalic anhydride

Similarly, quinones add on to butadiene, as follows:

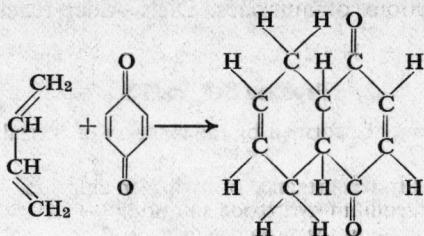

SUMMARY

Aromatic Aldehydes and Ketones. Contain C=O attached to an aromatic ring. Preparation: (1) Side chain oxidation: toluene to benzaldehyde. Industrial method via benzalchloride. (2) Gattermann–Koch reaction involves reaction of a hydrocarbon with HCl and CO in the presence of $AlCl_3$. For example, toluene to *p*-tolualdehyde. Hydroxyaldehydes from phenol ethers, zinc cyanide, hydrogen chloride, and aluminum chloride. (3) From nitriles and acids: benzonitrile to benzaldehyde (Stephen's reaction); Rosenmund reduction of benzoyl chloride to benzaldehyde.

Reactions of Benzaldehyde. (I) Air oxidation to benzoic acid. Does not reduce Fehling's or Benedict's solution. (II) Chemical and catalytic reductions give benzyl alcohol. (III) Cannizzaro reaction: benzaldehyde to benzyl alcohol and sodium benzoate. (IV) Carbonyl derivatives with hydroxylamine, phenylhydrazine, etc. (V) With ammonia, hydrobenzamide is formed. Schiff's Bases with primary amines. (VI) Aldol-type condensations: benzaldehyde to cinnamaldehyde, benzalacetone, etc. (VII) Perkin reaction: benzaldehyde to cinnamic acid. Applicable to a variety of aldehydes. (VIII) Condensation with active methylene compounds: benzaldehyde to benzalmalonic ester. (IX) Benzoin condensation: benzaldehyde converted to benzoin in the presence of cyanide ion. Oxidation of benzoin to benzil and the benzilic acid rearrangement.

Aromatic Ketones. Prepared by: (1) Friedel–Crafts reaction: benzene to benzophenone. (2) Fries rearrangement: *o*- and *p*-hydroxyacetophenone from phenylacetate. (3) Hoesch's synthesis: Phloroglucinol to phloracetophenone. Reactions of acetophenone: halogenation to phenacyl halides: formation of usual carbonyl derivatives; condensation with aldehydes. For example, acetophenone to benzalacetophenone; condensation to dypnone.

Stereoisomerism of Oximes. *Syn-* and *anti-*benzaldoxime and differing reactions with acetic anhydride. Beckmann rearrangement involves conversion of an oxime to an amide in the presence of an acidic catalyst, for example, benzophenone oxime to benzanilide. *Trans* nature of the rearrangement.

Quinones. Formulae of the common quinones. Preparation: *p*-quinone by oxidation of aniline and *o*-quinone by oxidation of catechol. Reactions: reduction to hydroquinone, formation of *p*-dichlorobenzene

250 *Organic Chemistry Made Simple*

with phosphorus pentachloride, addition of bromine, hydrogen chloride, etc. Ketonic reactions of quinones. Diels–Alder reaction with conjugated dienes.

Problem Set No. 12

1. What products may be formed by the reaction of benzaldehyde with each of the following?

    (*a*) Ethyl alcohol in the presence of hydrogen chloride.
    (*b*) Concentrated sodium hydroxide solution.
    (*c*) Diethyl malonate in the presence of piperidine.
    (*d*) Fusion with sodium propionate and propionic anhydride.
    (*e*) Acetaldehyde in the presence of dilute sodium hydroxide solution.
    (*f*) A dilute solution of potassium cyanide.

2. Suggest a method of preparing:

    (*a*) *p*-Tolualdehyde from toluene.
    (*b*) 2-Phenylethanol from benzaldehyde.
    (*c*) Ethylbenzene from benzene.
    (*d*) Benzilic acid from benzaldehyde.
    (*e*) Benzophenone from benzaldehyde.

3. Compound A is reacted with toluene to give compound B in a typical Friedel–Crafts reaction. B forms an oxime C, which when treated with phosphorus pentachloride gives compound D. Hydrolysis of D furnishes *o*-toluidine and *p*-toluic acid. Write structures for A, B, C, and D.

CHAPTER SIXTEEN

AROMATIC CARBOXYLIC ACIDS

The aromatic acids are very similar to the aliphatic acids in their chemical behaviour. The influence of an aromatic nucleus on the properties of the carboxyl group is much less pronounced than the effect on other groups like the halogen, amino, and hydroxy groups. Thus, for example, the unsubstituted benzene ring gives rise to a slightly stronger acid than an unsubstituted alkyl group—benzoic acid having a dissociation constant $K_A = 6\cdot3 \times 10^{-5}$ and acetic acid having a dissociation constant $K_A = 1\cdot8 \times 10^{-5}$. If the benzene ring is substituted the usual substituents (other than the $-NH_2$ or groups derived from it) in the *ortho* position increase the acidity, irrespective of the nature of the substituting group, thus definitely pointing to the fact that factors other than induction and resonance are operative. Substituents in the *meta* position influence acidity primarily by means of their inductive effect, since their resonance effects cannot be relayed to a *meta* carboxyl group. For instance, groups like $-OH$, $-OCH_3$, halogen, $-NO_2$, in the *meta* position exert an inductive pull on the electron pair binding the proton and thereby increasing the acidity. *Para* substituents influence acidity, depending on the net result of both their inductive and resonance effects. Thus *m*-nitrobenzoic acid has $K_A = 3\cdot2 \times 10^{-4}$ and *p*-nitrobenzoic acid $K_A = 3\cdot8 \times 10^{-4}$ as contrasted with K_A of benzoic acid, given above. *p*-Nitrobenzoic acid is the stronger acid, since the resonance effect operates in addition to the inductive effect and in the same direction, with the $-NO_2$ group in the *para* position. In the case of a halogen substituent, where the inductive and resonance effects operate in opposite directions, the *para* acid is somewhat weaker than the *meta* acid, though both are stronger than the parent unsubstituted acid. For example, *m*-chlorobenzoic acid has $K_A = 1\cdot5 \times 10^{-4}$ while *p*-chlorobenzoic acid has $K_A = 1\cdot0 \times 10^{-4}$.

PREPARATION: (1) **By Oxidation**. A side chain attached to the nucleus on oxidation gives rise to the carboxyl group. Thus toluene, ethylbenzene, or styrene on oxidation give rise to benzoic acid.

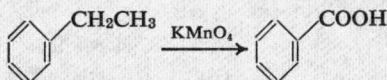

The oxidizing agent is generally aqueous potassium permanganate. The presence of an oxygen atom in the side chain increases the ease of oxidation to such an extent that benzyl alcohol, acetophenone, etc., become easily oxidized to benzoic acid.

$$C_6H_5CH_2OH \searrow [O]$$
$$\rightarrow C_6H_5COOH$$
$$C_6H_5COOCH_3 \nearrow [O]$$

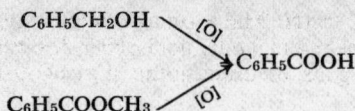

The presence of a nitro group *ortho* or *para* to the side chain also facilitates oxidation. Thus *p*-nitrotoluene is easily oxidized to *p*-nitrobenzoic acid.

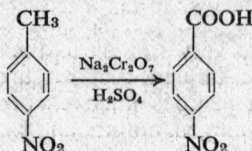

Commercially, benzoic acid is obtained by hydrolysis of benzotrichloride with lime followed by acidification to liberate the free acid.

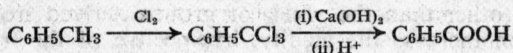

$$C_6H_5CH_3 \xrightarrow{Cl_2} C_6H_5CCl_3 \xrightarrow[\text{(ii) } H^+]{\text{(i) } Ca(OH)_2} C_6H_5COOH$$

(2) **Hydrolysis of Nitriles.** Aromatic nitriles can be hydrolysed to give acids in the usual manner.

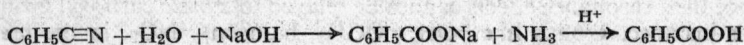

$$C_6H_5C{\equiv}N + H_2O + NaOH \longrightarrow C_6H_5COONa + NH_3 \xrightarrow{H^+} C_6H_5COOH$$

The nitriles are easily prepared from the hydrocarbon by sulphonation, followed by fusion with potassium cyanide or from the diazonium salt by Sandmeyer reaction, and hence this method of preparing aromatic acids is quite useful.

(3) **Carbonation of Grignard Reagents.** The Grignard reaction may also be used to prepare aromatic acids. An ether solution of Grignard reagent is added slowly to solid carbon dioxide, and on treatment of the product with mineral acid an aromatic acid is obtained. For instance, mesitoic acid (2,4,6-trimethylbenzoic acid) is obtained from 2-bromomesitylene as follows:

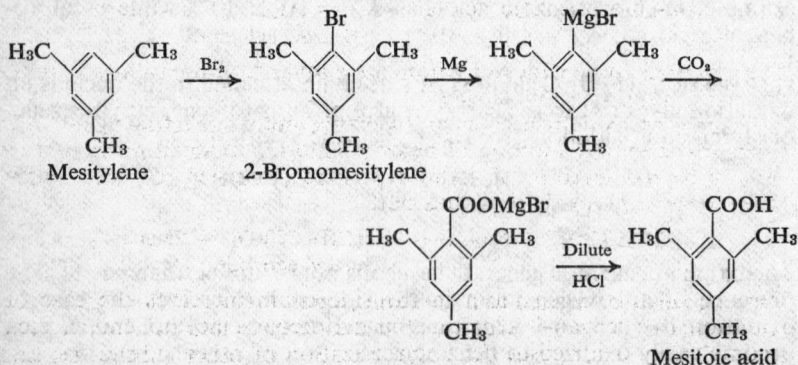

Mesitylene   2-Bromomesitylene

Mesitoic acid

PROPERTIES AND REACTIONS: The aromatic acids show the usual properties of acids and form esters, salts, acid chlorides, etc. They differ from their aliphatic analogues in undergoing typical aromatic substitution reactions.

Benzoic acid occurs in certain resins, particularly in gum benzoin and as its benzyl ester in peru and tolu balsam. It can be crystallized from water as glistening leaflets. Like other acids, the aromatic acids dissolve in sodium bicarbonate solution with liberation of carbon dioxide. Aromatic acids furnish esters when treated with alcohols in the presence of mineral acids, even though substituted benzoic acids with substituents in both *ortho* positions esterify slowly or not at all. Thus, mesitoic acid is hard to esterify in the conventional manner; this difficulty has been attributed to steric hindrance.

The action of phosphorus pentachloride on benzoic acid furnishes benzoyl chloride, a typical aromatic acid chloride.

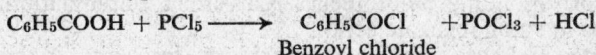

$$C_6H_5COOH + PCl_5 \longrightarrow C_6H_5COCl + POCl_3 + HCl$$
Benzoyl chloride

Industrially, it is prepared by the chlorination of benzaldehyde without a halogen carrier.

$$C_6H_5CHO + Cl_2 \longrightarrow C_6H_5COCl + HCl$$

Benzoyl chloride is used for the benzoylation of amines and phenols. It is less reactive than acetyl chloride and reacts very slowly with water or dilute sodium hydroxide solution. Advantage is taken of this fact in the Schotten–Baumann reaction. A small quantity of the hydroxy or amino compound to be benzoylated is shaken with benzoyl chloride in dilute alkali solution for a few minutes until the odour of benzoyl chloride disappears. The benzoyl derivative usually separates as an oil which solidifies on cooling and may be collected, washed, and recrystallized. Of the different acid chlorides used for acylation, acetyl chloride is the most reactive, benzoyl chloride less reactive, and benzene sulphonyl chloride the least reactive.

Benzoic anhydride can be prepared by heating benzoyl chloride with sodium benzoate, a method which is very similar to the preparation of acetic anhydride.

$$C_6H_5COCl + C_6H_5COONa \longrightarrow (C_6H_5CO)_2O + NaCl$$
Benzoic anhydride

The reactions of benzoic anhydride are similar to those of acetic anhydride, but proceed more slowly.

Benzoyl peroxide is used as an oxidizing and disinfecting agent. It is easily made by the addition of benzoyl chloride in small portions to a rapidly stirred mixture of sodium peroxide, crushed ice, and water. Solid benzoyl peroxide separates out.

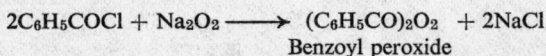

$$2C_6H_5COCl + Na_2O_2 \longrightarrow (C_6H_5CO)_2O_2 + 2NaCl$$
Benzoyl peroxide

Benzoyl peroxide is used as a bleaching agent for wheat flour under the trade name of Novadel. Because of its dissociation into two free radicals $(C_6H_5COO\cdot)$, it catalyses the polymerization of other substances, and hence is used in the plastic industry. Benzoyl peroxide is explosive, and care must be taken when handling it.

Benzamide ($C_6H_5CONH_2$) is prepared by adding benzoyl chloride to ammonium hydroxide solution. It is similar to acetamide in its properties.

OTHER SPECIFIC ACIDS

The three toluic acids (*o*, *p*, and *m*) may be obtained by the selective oxidation of one methyl group of the corresponding xylenes by nitric acid or by the hydrolysis of the corresponding tolunitriles, which in turn may be obtained from the toluidines by the Sandmeyer reaction.

Benzoic acid undergoes direct *m*-substitution to give a number of *m*-substituted products. Halogenation furnishes *m*-halobenzoic acids. The isomeric *ortho* and *para* halobenzoic acids are obtained by the oxidation of the corresponding halotoluenes. On direct sulphonation, benzoic acid gives *m*-sulphobenzoic acid.

The *ortho* and *para* sulphobenzoic acids are important in the preparation of saccharin and chloramine-T and have been dealt with previously. Nitration of benzoic acid leads to *m*-nitrobenzoic acid, the *ortho* and *para* isomers being obtained by the oxidation of the corresponding nitrotoluenes. Reduction of nitrobenzoic acids gives the aminobenzoic acids. *o*-Aminobenzoic acid, made by the reduction of *o*-nitrobenzoic acid, is known as anthranilic acid and is an intermediate in the synthesis of indigo.

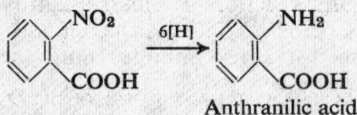

Anthranilic acid

Anthranilic acid is, however, usually prepared by the action of sodium hypohalite (Hofmann degradation) on the readily available phthalimide (see under phthalic anhydride later).

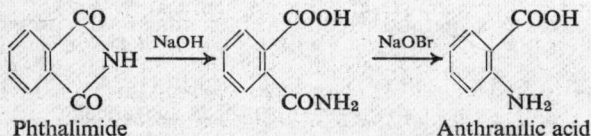

Phthalimide   Anthranilic acid

The methyl ester of anthranilic acid occurs in oil of orange blossom and in oil of jasmine. Esters of *p*-aminobenzoic acid are useful local anaesthetics. Thus the two following compounds, Anaesthesine and Butesin, are used as local anaesthetics:

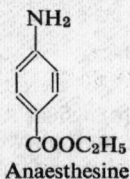

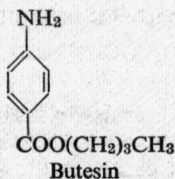

Anaesthesine   Butesin

Butesin, in the form of its picrate, is used in ointments for relief of burns, ulcers, etc. Novocaine, the best known local anaesthetic, is derived from *p*-aminobenzoic acid and has the following structure:

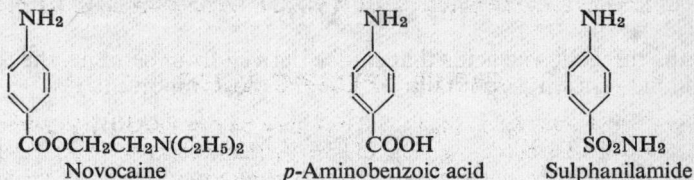

COOCH₂CH₂N(C₂H₅)₂   COOH   SO₂NH₂

Novocaine   *p*-Aminobenzoic acid   Sulphanilamide

p-Aminobenzoic acid itself is a vitamin essential to the growth of some bacteria that are responsible for certain bacterial diseases in man. It will be seen that this compound has a structural similarity to sulphanilamide. When sulphanilamide is administered to a person the pathogenic bacteria feed on the toxic sulpha drugs—apparently mistaking it for the structurally similar *p*-aminobenzoic acid—and eventually perish.

Salicylic Acid and Its Derivatives. All three hydroxybenzoic acids are known. Of them the *ortho* isomer, salicylic acid, is very important. As has already been indicated, it is made by Kolbe's method. Salicylic acid is more acidic than benzoic acid, and probably this property arises out of chelation of the salicylate ion, shown below.

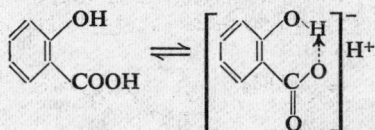

The attraction of one of the oxygen atoms of the carboxyl group for the hydrogen atom of the phenolic group, leading to the formation of a six-atom ring, may reduce the tendency of the salicylate ion to reabsorb the proton. Salicylic acid gives the reactions of both a phenol and an acid. Thus it forms esters, salts, etc., like acids. Like phenols, it gives a violet colouration with ferric chloride, couples with diazonium salts, and gives an acyl derivative of the phenolic group. Bromine water reacts with salicylic acid to give 2,4,6-tribromophenol, decarboxylation taking place during the reaction. The methyl ester of salicylic acid, methyl salicylate, is the main odoriferous constituent of the oil of wintergreen. It can be made by esterification of salicylic acid with methyl alcohol, and is used in perfumery and for medicinal purposes in the treatment of rheumatism, neuralgic pains, etc.

Acetylsalicylic acid or aspirin is produced by the acetylation of salicylic acid. Salicylic acid reacts slowly with acetic anhydride, but if a few drops of concentrated sulphuric acid are added the reaction takes place vigorously.

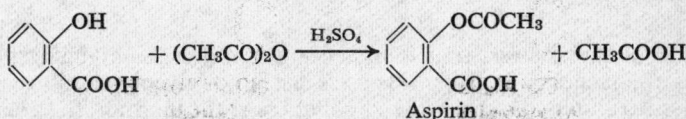

$+ (CH_3CO)_2O \xrightarrow{H_2SO_4}$ $+ CH_3COOH$

Aspirin

The uses of aspirin to reduce fever and as a remedy for headache, neuralgia, etc., are well known.

DIBASIC ACIDS

Of the three dibasic acids that can be derived from benzene, the *ortho* compound, known as phthalic acid, is the most important.

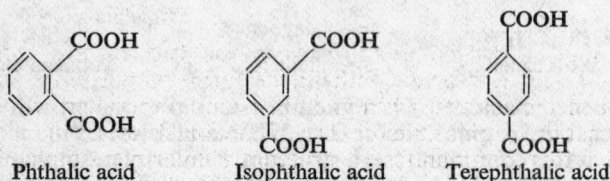

Phthalic acid   Isophthalic acid   Terephthalic acid

The above acids can be obtained by the oxidation of the corresponding xylenes with potassium permanganate or by the use of Grignard reaction or by hydrolysis of the dinitriles. The anhydride of phthalic acid is a key intermediate in the chemical industry and is obtained by oxidation of naphthalene with fuming sulphuric acid and mercury catalyst or by the vapour-phase oxidation of naphthalene with atmospheric oxygen.

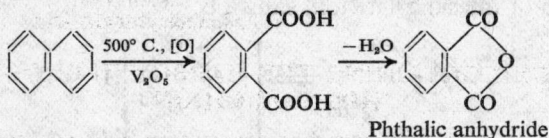

Phthalic anhydride

While phthalic anhydride is readily formed from phthalic acid, the isomeric dicarboxylic acids do not give any anhydrides.

REACTIONS OF PHTHALIC ANHYDRIDE: Esters of phthalic acid are obtained by the reaction of phthalic anhydride or the acid with alcohol and are used in diffusion pumps and in place of mercury in manometers. The diethyl and dihexyl esters are very widely used for this purpose. Dimethylphthalate is useful as an insect repellent. Phthalic anhydride is extensively used in the manufacture of the so-called alkyd resins. These resins are polyesters of acids containing two carboxyl groups and polyhydric alcohols. The polymer produced by the action of phthalic acid on ethylene glycol may be represented as follows:

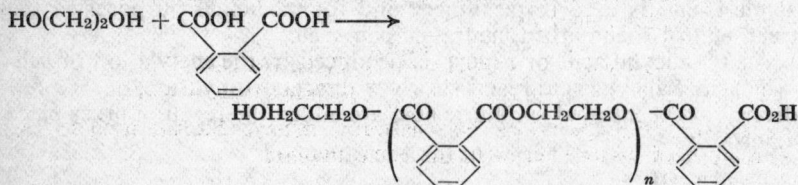

Linear polymeric esters thus obtained are useful in the manufacture of paints and lacquers. If glycerol is used instead of glycol and more of the acid is also employed, cross-linked polymers of the thermosetting variety are obtained (see Chapter Nineteen). The synthetic fibre Terylene is a polyester of terephthalic acid and ethylene glycol. Phthalimide is formed when ammonia is introduced into heated phthalic anhydride.

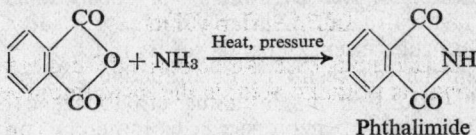

$$\underset{}{\begin{array}{c} CO \\ \diagdown \\ O \\ \diagup \\ CO \end{array}} + NH_3 \xrightarrow{Heat,\ pressure} \underset{Phthalimide}{\begin{array}{c} CO \\ \diagdown \\ NH \\ \diagup \\ CO \end{array}}$$

The imino hydrogen atom between the two carbonyl groups is distinctly acidic, and hence phthalimide forms metallic salts. Gabriel's synthesis of amines and amino acids, it may be recalled, employs the potassium salt of phthalimide. The Hofmann degradation of phthalimide, as previously mentioned, provides a convenient method of preparation of anthranilic acid. When phthalic anhydride is treated with alkaline hydrogen peroxide in the cold and the mixture then acidified monoperphthalic acid is obtained:

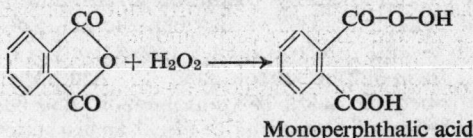

$$\underset{}{\begin{array}{c} CO \\ \diagdown \\ O \\ \diagup \\ CO \end{array}} + H_2O_2 \longrightarrow \underset{Monoperphthalic\ acid}{\begin{array}{c} CO-O-OH \\ \\ COOH \end{array}}$$

ACIDS CONTAINING THE CARBOXYL GROUP IN THE SIDE CHAIN

These acids contain the carboxyl group in side chains of varying lengths and resemble aliphatic acids more closely than true aromatic acids containing a ring carboxyl group.

Phenylacetic Acid. Phenylacetic acid may be obtained by the hydrolysis of benzyl cyanide, which is readily available from the reaction between benzyl chloride and potassium cyanide.

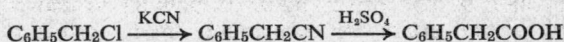

$$C_6H_5CH_2Cl \xrightarrow{KCN} C_6H_5CH_2CN \xrightarrow{H_2SO_4} C_6H_5CH_2COOH$$

The α-hydrogen atoms in phenylacetic acid and its esters are more reactive than in purely aliphatic acids.

β-Phenylpropionic Acid. β-Phenylpropionic acid is also called hydrocinnamic acid and is obtained by reduction with sodium amalgam of cinnamic acid. The latter, as mentioned before, is obtained by the Perkin reaction on benzaldehyde.

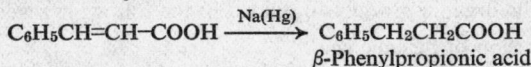

$$C_6H_5CH=CH-COOH \xrightarrow{Na(Hg)} \underset{\text{β-Phenylpropionic acid}}{C_6H_5CH_2CH_2COOH}$$

It can also be prepared by the reaction of benzyl chloride with sodium derivative of malonic ester in the usual manner.

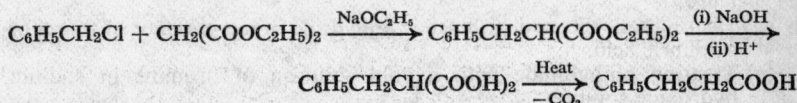

$$C_6H_5CH_2Cl + CH_2(COOC_2H_5)_2 \xrightarrow{NaOC_2H_5} C_6H_5CH_2CH(COOC_2H_5)_2 \xrightarrow[\text{(ii) H}^+]{\text{(i) NaOH}}$$

$$C_6H_5CH_2CH(COOH)_2 \xrightarrow[-CO_2]{Heat} C_6H_5CH_2CH_2COOH$$

SUMMARY

Aromatic Carboxylic Acids. Influence of substituents on acidity of benzoic acid; *o*-substituents invariably increase acidity, while the influence of a *meta* substituent depends on its inductive effect, that of a *p*-substituent on the net result of its inductive and resonance effect. Preparation: (1) Oxidation of side chains; ethylbenzene to benzoic acid; *p*-nitrotoluene to *p*-nitrobenzoic acid. Commercial process involves chlorination of toluene to benzotrichloride and then hydrolysis to benzoic acid. (2) Hydrolysis of nitriles: Benzonitrile to benzoic acid. (3) From Grignard reagents: RMgBr to RCOOH. Reactions: Usual reactions of acids such as formation of esters, acid chlorides, amides, etc. Use of benzoyl chloride in the Schotten–Baumann reaction. Benzoic anhydride and benzoyl peroxide from benzoyl chloride.

Other Specific Acids. (1) Toluic acids from toluidines or xylenes. (2) Halogenation of benzoic acid to *m*-halobenzoic acid; *o*- and *p*-halobenzoic acids obtained by oxidation of *o*- and *p*-halotoluenes. (3) Nitration of benzoic acid to *m*-nitrobenzoic acid; *o*- and *p*-isomers by oxidation of *o*- and *p*-nitrotoluenes respectively. (4) Sulphonation of benzoic acid to *m*-sulphobenzoic acid. (5) Aminobenzoic acids by reduction of nitrobenzoic acids. *o*-Aminobenzoic acid best prepared by Hofmann degradation of phthalimide. Local anaesthetics derived from *p*-aminobenzoic acid. Salicylic acid and derivatives: Salicylic acid stronger than benzoic acid. Explanation: Reactions at both the ⁻OH and ⁻COOH functions. Medicinally useful derivatives like methyl salicylate, Salol, and aspirin.

Dibasic Acids. *o*-, *m*-, *p*-Benzenedicarboxylic acids are called phthalic, isophthalic, and terephthalic acids respectively. Phthalic anhydride made commercially by vapour-phase oxidation of naphthalene. Reactions of phthalic anhydride: Esterification gives useful products. Alkyd resins are polyesters of glycols with phthalic acid. Terylene from ethylene glycol and terephthalic acid. Phthalimide from ammonia and phthalic anhydride.

Acids with Carboxyl Group in the Side Chain. (i) Phenylacetic acid by hydrolysis of phenylacetonitrile. (ii) β-Phenylpropionic acid by reduction of cinnamic acid or from diethylmalonate after alkylation with benzyl chloride and usual conversions.

Problem Set No. 13

1. What are the organic products obtained in the following reactions?

(*a*) Bromination of benzoic acid.
(*b*) Treating N-methylaniline with benzoyl chloride in the presence of sodium hydroxide.
(*c*) Treating benzene with phthalic anhydride in the presence of aluminum chloride.
(*d*) Heating salicylic acid with acetic anhydride in the presence of traces of sulphuric acid.
(*e*) Treating phthalimide with a cold solution of bromine in sodium hydroxide solution.

2. Suggest methods of preparing the following:

   (*a*) Terephthalic acid from toluene.
   (*b*) *p*-Methylcinnamic acid from *p*-tolualdehyde.
   (*c*) Methyl-*m*-nitrobenzoate from bromobenzene.

3. Two aromatic hydrocarbons, A and B, have the formula C_9H_{12}: A gives on oxidation a dicarboxylic acid C, which readily loses water to give an anhydride. B gives benzoic acid on oxidation. Write down possible structures for A, B, and C.

PART IV

HETEROCYCLIC COMPOUNDS, PHYSIOLOGICALLY ACTIVE COMPOUNDS, AND POLYMERS

CHAPTER SEVENTEEN

HETEROCYCLIC COMPOUNDS AND ALKALOIDS

Ring systems in which the rings are composed entirely of carbon atoms are called homocyclic. Examples are benzene, cyclopentane, etc. Ring compounds constituted of atoms of different elements are called heterocyclic compounds. Such ring systems usually contain several carbon atoms and one or more atoms of other elements. The usual hetero atoms are nitrogen, oxygen, and sulphur. Some of these heterocyclic rings resist opening and remain intact through vigorous reactions, as does the benzene ring, and to a greater or lesser extent have the characteristic property of benzene referred to as aromaticity. Some of the very common parent heterocyclic compounds are given below, listed in the order of increasing aromaticity. For the sake of reference, benzene is also placed in the list in its appropriate place, indicating its degree of aromaticity, though it is not heterocyclic. The different carbon atoms in the rings are numbered, starting with the hetero atom. Greek letters are also used to indicate the different carbon atoms. Furan, pyrrole, and thiophene each have two pairs of equivalent carbon atoms, namely, 2,5 and 3,4. These are also referred to as α, α', β, and β' positions.

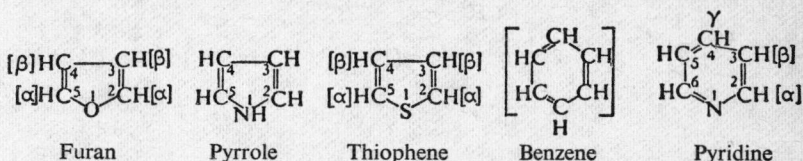

Furan Pyrrole Thiophene Benzene Pyridine

Pyridine, standing at the end of the list, is more aromatic in character and hence more stable than benzene itself. It is less easily substituted than benzene. Also, when compounds containing both a pyridine and benzene ring are subjected to oxidation the benzene ring is broken up in preference to the pyridine ring. For instance, in the oxidation of α-phenylpyridine and of quinoline the benzene ring suffers cleavage rather than the pyridine ring.

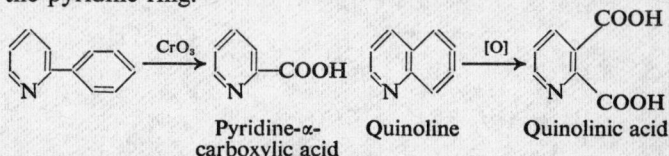

Pyridine-α-   Quinoline   Quinolinic acid
carboxylic acid

Thiophene is more subject to substitution and is less aromatic than benzene. Pyrrole and furan have still weaker aromatic character, the furan ring being something like a borderline case, since it reacts more or less like an aliphatic diene and shows only occasional glimpses of aromatic behaviour. The simple heterocyclic systems shown above may be fused to a benzene ring or to another heterocyclic ring to give bicyclic systems similar to naphthalene. Repetition of this process may lead to polynuclear heterocyclic compounds. This gives us an idea of the very large number of compounds that fall under the description of heterocyclic compounds. In fact, the number of known heterocyclic systems exceeds 4,000, and this number is constantly increasing with the discovery of new heterocyclic systems.

Heterocyclic compounds occur abundantly in plant and animal products. Those containing five or six atoms in the ring are most abundant, and a study of these systems becomes important. It is beyond the scope of this book to survey the extensive field of heterocyclic compounds. Only the simpler systems and interesting representative compounds derived from them will be treated.

PREPARATION OF FURAN, PYRROLE, THIOPHENE, AND DERIVATIVES: A method of preparation, which also throws light on the structure of furan, pyrrole, thiophene, and their derivatives, involves the use of γ-diketones. Hexane-2,5-dione, for example, will give the 2,5-dimethyl derivative of furan, pyrrole, or thiophene by treatment with phosphorus pentoxide, ammonia, or phosphorus pentasulphide respectively.

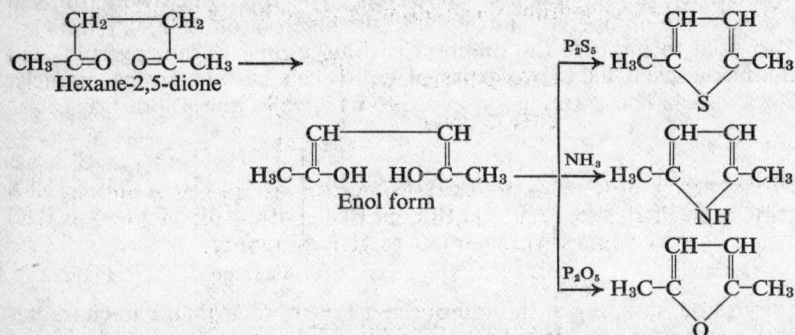

Furfural, an aldehyde derivative of furan, is a commercially important product. It is manufactured by the processing of oat hulls and corn cobs. These items contain pentoses which when heated with hydrochloric acid lose water and are converted into furfural.

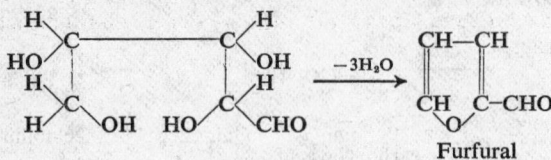

Furfural resembles benzaldehyde in its reactions (see later) and undergoes the Cannizzaro reaction to give 2-furoic acid and 2-furfuryl alcohol.

$$\underset{\text{Furfural}}{\boxed{}\!\!-\!\!\text{CHO}} \xrightarrow{33\%, \text{ NaOH}} \underset{\text{2-Furoic acid}}{\boxed{}\!\!-\!\!\text{COOH}} + \underset{\text{Furfuryl alcohol}}{\boxed{}\!\!-\!\!\text{CH}_2\text{OH}}$$

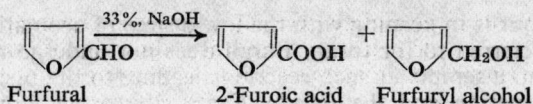

The decarboxylation of 2-furoic acid provides a convenient route to the parent furan itself:

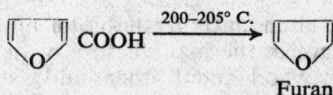

$$\boxed{}\text{COOH} \xrightarrow{200-205°\text{ C.}} \underset{\text{Furan}}{\boxed{}}$$

Pyrrole and its derivatives are the most important heterocyclic compounds and are made available by a number of synthetic processes. Pyrrole itself is found in coal tar and bone oil, which is obtained by the destructive distillation of bones. Pyrrole can be made by distillation of succinimide with zinc dust:

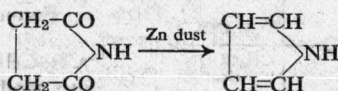

$$\begin{array}{c} \text{CH}_2\!\!-\!\!\text{CO} \\ | \qquad\quad \rangle\text{NH} \\ \text{CH}_2\!\!-\!\!\text{CO} \end{array} \xrightarrow{\text{Zn dust}} \begin{array}{c} \text{CH}\!\!=\!\!\text{CH} \\ \qquad\quad \rangle\text{NH} \\ \text{CH}\!\!=\!\!\text{CH} \end{array}$$

Thiophene occurs to the extent of 0.5% in coal-tar benzene. It can be made by treating sodium succinate with phosphorus trisulphide.

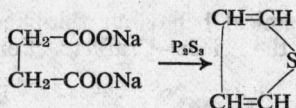

$$\begin{array}{c} \text{CH}_2\!\!-\!\!\text{COONa} \\ | \\ \text{CH}_2\!\!-\!\!\text{COONa} \end{array} \xrightarrow{\text{P}_2\text{S}_3} \begin{array}{c} \text{CH}\!\!=\!\!\text{CH} \\ \qquad\quad \rangle\text{S} \\ \text{CH}\!\!=\!\!\text{CH} \end{array}$$

Thiophene is now made industrially by the reaction of *n*-butane with sulphur at high temperature. Butane is dehydrogenated to butadiene which then combines with sulphur to give thiophene.

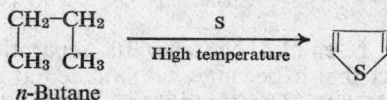

$$\underset{\text{n-Butane}}{\begin{array}{c} \text{CH}_2\!\!-\!\!\text{CH}_2 \\ | \qquad | \\ \text{CH}_3 \quad \text{CH}_3 \end{array}} \xrightarrow[\text{High temperature}]{\text{S}} \boxed{}\!\!-\!\!\text{S}$$

PROPERTIES AND REACTIONS: (*a*) **Furan and Furfural.** It will be noticed that all three heterocyclic systems described above contain a conjugated diene system. However, only furan enters readily into a Diels–Alder reaction with maleic anhydride—the reaction characteristic of conjugated dienes.

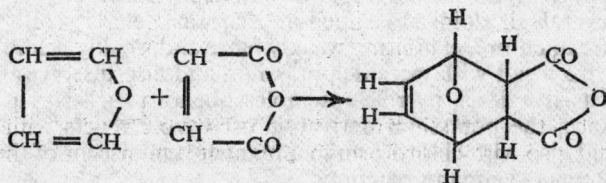

$$\begin{array}{c} \text{CH}\!=\!\text{CH} \\ | \qquad\quad \text{O} \\ \text{CH}\!=\!\text{CH} \end{array} + \begin{array}{c} \text{CH}\!-\!\text{CO} \\ \qquad\quad \text{O} \\ \text{CH}\!-\!\text{CO} \end{array} \rightarrow$$

This behaviour is in keeping with the low degree of aromatic character previously referred to for furan. Furan does not undergo nitration or sulphonation, attempts at such reactions leading to the opening of the ring. Halogenation and the Friedel–Crafts alkylation, however, take place at the 2 and 5 positions. Reduction of furan gives tetrahydrofuran, a liquid which is used increasingly as a solvent for a variety of purposes.

Furfural, the most abundantly available of all furan derivatives, is commercially important in the manufacture of furfural–phenol resins, as a preservative for wood and leather, and as a vulcanization accelerator. Its conversion to furan by means of Cannizzaro reaction has already been mentioned. Other reactions of furfural, recalling those of benzaldehyde, include the benzoin-type condensation, the Perkin reaction and the Claisen-type condensation.

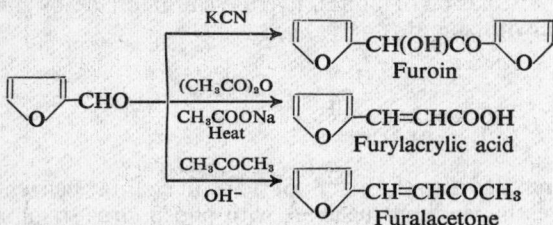

(b) **Thiophene.** In chemical behaviour thiophene is more aromatic than furan. It undergoes the Friedel–Crafts reaction and can be sulphonated and nitrated.

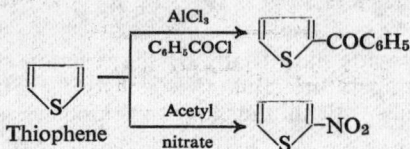

Thiophene closely resembles benzene. Its separation from benzene depends on the fact that it becomes sulphonated at room temperature while benzene does not. The thiophenesulphonic acid is soluble in sulphuric acid and hence can be separated. Alternatively, this separation can be achieved by refluxing the impure benzene with mercuric acetate, when thiophene is mercurated and the mercuric derivative precipitates out and is filtered off.

(c) **Pyrrole.** The pyrrole ring occurs in a number of important natural products, such as the haemoglobin of blood, the chlorophyll of plants, and in several alkaloids, like nicotine, cocaine, etc. Pyrrole itself is amphoteric in nature, exhibiting weakly basic and weakly acidic properties. It forms a salt with metallic potassium and also dissolves in dilute acids. In its reactions, pyrrole resembles phenol to a striking degree. For instance, the potassium derivative of pyrrole reacts with carbon dioxide and also with chloroform in a manner reminiscent of the Kolbe and the Reimer–Teimann reactions.

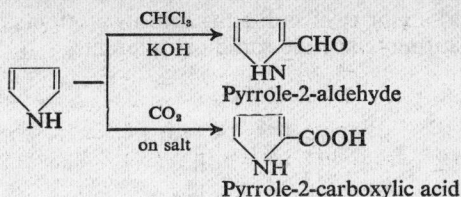

Pyrrole-2-aldehyde

Pyrrole-2-carboxylic acid

The analogy with phenol is also shown in the coupling of pyrrole with diazonium salts in the 2 position.

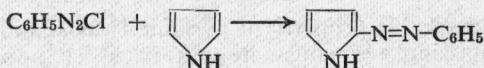

Chlorophyll, the green pigment of plants, and haemin, obtained by hydrolysis of the red pigment haemoglobin of blood, are both built of pyrrole units. Each has a central metal atom, iron in haemin and magnesium in chlorophyll, surrounded by a skeleton of four pyrrole rings connected by intermediate carbon atoms.

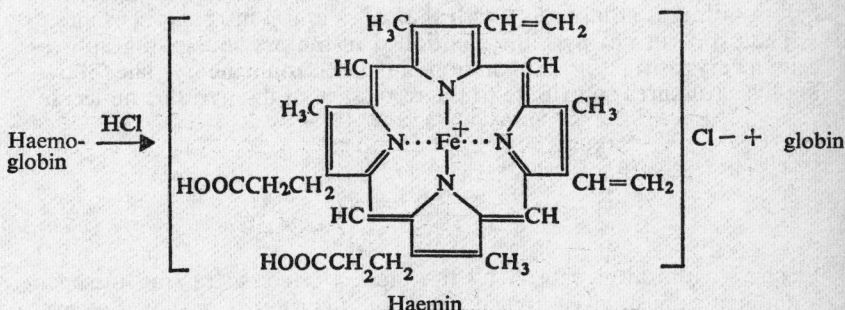

Haemin

Chlorophyll contains magnesium instead of iron and also differs in the nature of the side chains attached to the pyrrole rings.

PYRIDINE

Of the heterocyclic six-atom-ring compounds occurring in nature, the most abundant are derivatives of pyridine and piperidine, its reduction product. Pyridine and some of its homologues occur in coal tar and also in the oil obtained by dry distillation of bones from which the fat has not been removed.

The possibilities of isomerism among substitution products of pyridine are greater, since, besides the relative positions of the substituents, their positions with respect to the nitrogen atom must also be taken into account. Isomers are designated by the use of numerals or Greek letters, as indicated in the formula below. There should be three mono-substitution products and six or twelve di-substitution products, depending on whether the substituents are the same or different.

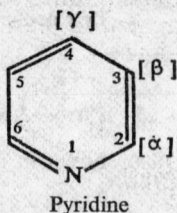

Pyridine

The most remarkable property of pyridine is its stability. Oxidizing agents, like potassium permanganate, nitric acid, etc., do not affect it. As previously mentioned, the oxidation of a system containing pyridine and benzene rings results in the degradation of the benzene ring. Furthermore, substitution reactions, like sulphonation, nitration, and halogenation, can be accomplished only under drastic conditions. Pyridine is often compared to nitrobenzene, which offers the same degree of resistance to substitution reactions. Neither undergoes the Friedel–Crafts reaction and, furthermore, the reactivity of 2- and 4-halopyridines is comparable with that of *o*- and *p*-nitrohalobenzenes.

The properties of pyridine mentioned in the previous paragraph receive a very satisfactory explanation of the electronic theory. The following five structures contribute to the resonance of the pyridine molecule.

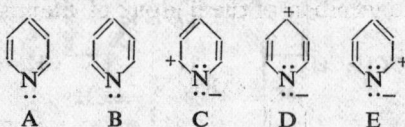

Electron-withdrawal caused by the more electron-attracting nitrogen atom creates positive centres at the 2 and 4 carbon atoms, as shown in structures C, D, and E. The usual electrophilic substitutions are, therefore, difficult, and when they occur they do so at the 3 carbon atom, which is the relatively more electron-rich centre.

Pyridine exhibits the typical properties of a tertiary amine, though it is a weaker base than aliphatic tertiary amines. It gives an amine oxide on treatment with perbenzoic acid and a quaternary ammonium salt with an alkyl halide.

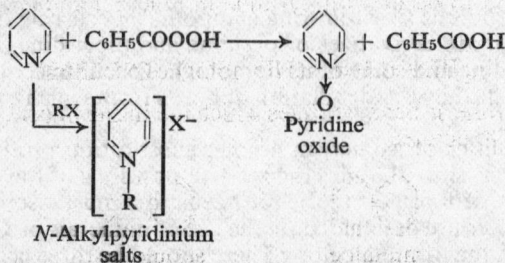

Pyridine oxide is more susceptible to electrophilic substitution than pyridine itself, substitution taking place with facility in the γ position and with less readiness at the α position. Pyridine, on reduction with sodium and alcohol or hydrogen, in the presence of nickel, gives the cyclic secondary base piperidine.

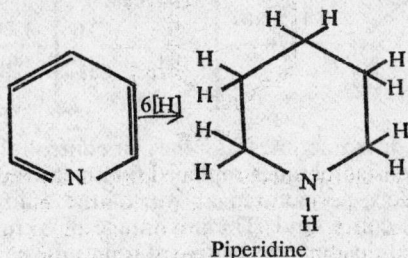

Piperidine

A characteristic reaction of pyridine is amination at the 2 position when treated with sodamide in dimethylaniline at 100° C. to give 2-amino-pyridine.

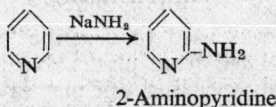

2-Aminopyridine

2-Aminopyridine is used for making sulphapyridine. Pyridoxine and nicotinic acid, both members of the B group of vitamins, are derivatives of pyridine.

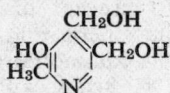

Pyridoxine
(Vitamin B₆)

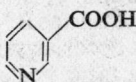

Nicotinic acid
(Niacin)

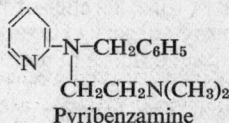

Pyribenzamine

Pyridoxine cures dermatitis in rats. Nicotinic acid and its amide are effective for curing human pellagra. Pyribenzamine used for relief of allergic disorders is also a derivative of pyridine.

The monomethyl, the dimethyl, and the trimethyl derivatives of pyridine are called picolines, lutidines, and collidines respectively. They are found to occur with pyridine in bone oil and coal tar.

QUINOLINE AND ISOQUINOLINE

These are isomeric benzpyridines which are found in coal tar and have the following structures:

Quinoline

Isoquinoline

A useful general method for the preparation of quinoline derivatives was discovered by SKRAUP. Quinoline itself is obtained by Skraup's synthesis by treating aniline with glycerol and concentrated sulphuric acid and an oxidizing agent like nitrobenzene.

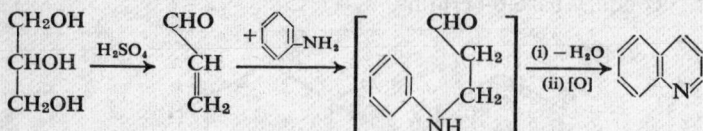

$$\begin{array}{c} CH_2OH \\ | \\ CHOH \\ | \\ CH_2OH \end{array} \xrightarrow{H_2SO_4} \begin{array}{c} CHO \\ | \\ CH \\ \| \\ CH_2 \end{array} + \langle\!\rangle\!-NH_2 \longrightarrow \left[\begin{array}{c} CHO \\ CH_2 \\ CH_2 \\ NH \end{array} \right] \xrightarrow[\text{(ii) [O]}]{\text{(i) } -H_2O} \langle\!\rangle\!N$$

The Skraup reaction sometimes assumes uncontrollable violence and hence nitrobenzene is sometimes replaced by arsenic acid. The reaction proceeds by addition, as pictured above, of the amine to an α,β unsaturated carbonyl compound. The advantage of Skraup's synthesis is its versatility, and almost any derivative of quinoline can be prepared by this method by varying the aldehyde and the amine components.

ALKALOIDS

The name alkaloid was used to describe 'alkali-like' organic substances which were isolated from plants. No precise definition of the term is possible. Roughly speaking, one may say that alkaloids are basic nitrogenous compounds of vegetable origin which have marked physiological properties. They generally have the structural features of pyrrole, pyridine, quinoline, or isoquinoline. Many alkaloids are highly toxic and have been used as poisons. In very small doses they find use in medicine. Thus, for instance, quinine is used to prevent malaria, morphine for the relief of pain, cocaine as a local anaesthetic, and atropine in eye surgery. The alkaloids occur as salts of the common organic acids, like oxalic, malic, and tartaric acids. One is struck with the extraordinary variety in the types of bases resulting from plant biosynthesis. The long and tedious labours of many chemists have laid bare the structural pattern of several alkaloids, and one is amazed at the bewildering variety of patterns. Alkaloids containing the ring systems of pyrrolidine, piperidine, pyridine, quinoline, isoquinoline, indole, and several other systems are known. The unravelling of the constitution of the several alkaloids and in some cases their synthesis, notwithstanding their complicated structure, is a tribute to the ingenuity and patient labour of chemists all over the world. We shall consider some of the simpler alkaloids in order to get some idea of the methods followed in determining alkaloidal structure and in their synthesis. Solutions containing alkaloids form precipitates with certain alkaloidal reagents. Some of these reagents are tannic acid, picric acid, phosphomolybdic acid, and phosphotungstic acid. Potassium mercury iodide, known as Meyers' reagent, is an important alkaloidal precipitant.

Coniine. Coniine, α-propylpiperidine, is the simplest alkaloid. It occurs in the seeds and other parts of the spotted hemlock. It will be remembered that this was the poison administered to Socrates. The base and its salts are toxic and cause death by paralysis. We shall illustrate the general method followed in the investigation of alkaloidal structure by using coniine as a simple example.

The molecular formula of the alkaloid was found to be $C_8H_{17}N$. Zinc dust distillation gave a compound called conyrine ($C_8H_{11}N$), which on oxidation gave a substance identified as pyridine α-carboxylic acid. Hence conyrine must be a derivative of pyridine with only one side chain at the 2 position. Of the eight carbon atoms, five are taken up by the pyridine ring, and the remaining three carbon atoms must be situated in the 2 position as a propyl or isopropyl side chain. Since conyrine differs from coniine by six hydrogen atoms, coniine must be a derivative of piperidine.

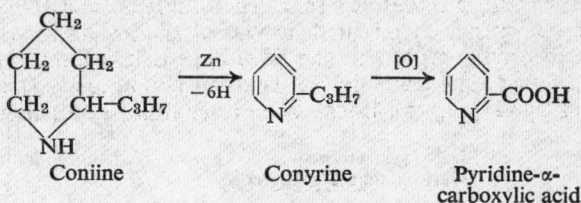

Coniine   Conyrine   Pyridine-α-carboxylic acid

The reduction of coniine by hydrogen iodide gave ammonia and normal octane, thus showing that the side chain is *n*-propyl.

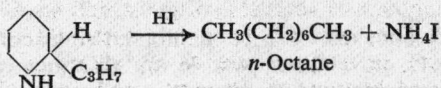

$$CH_3(CH_2)_6CH_3 + NH_4I$$
n-Octane

The synthesis of coniine was accomplished by LADENBURG in 1886 and was the first synthesis of a natural alkaloid. It is shown below.

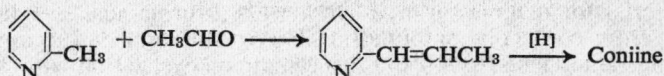

The synthetic product could be resolved to give the naturally occurring antipode.

Nicotine. Nicotine is the chief alkaloid from tobacco leaves. It is a colourless liquid and is highly toxic.

Its molecular formula was found to be $C_{10}H_{14}N_2$. On oxidation nicotine gave nicotinic acid, identified as pyridine β-carboxylic acid. Hence nicotine must contain the pyridine nucleus C_5H_4N with a $C_5H_{10}N$ side chain in the β-position, $(C_{10}H_{14}N_2 - C_5H_4N = C_5H_{10}N)$. Nicotine forms no acetyl or benzoyl derivative, and hence both the nitrogen atoms should be tertiary.

If methyl iodide is added to nicotine and the resulting quaternary ammonium salt is oxidized carefully with potassium ferricyanide a ketonic compound (*N*-methylnicotine) is obtained, which on stronger oxidation with chromic acid is converted into 1-methyl pyrrolidine-2-carboxylic acid. These observations lead us to the conclusion that nicotine contains a pyridine and a *N*-methylpyrrolidine nucleus and that the β carbon of the pyridine ring is joined to the α carbon of the pyr-rolidine. The above degradations may then be indicated as follows:

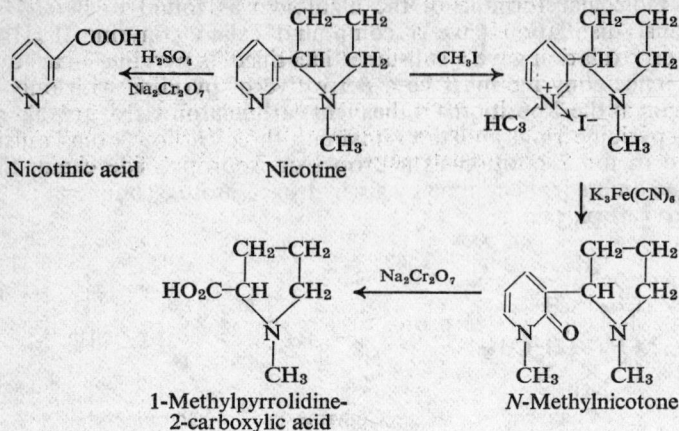

Nicotinic acid   Nicotine

1-Methylpyrrolidine-2-carboxylic acid   *N*-Methylnicotone

Several syntheses of nicotine confirm the above structure.

SUMMARY

Heterocyclic Compounds. A. Five-membered rings containing one or more atoms of N, O, or S. Furan, pyrrole, and thiophene are the simplest members and increasingly aromatic in the order mentioned. Preparation: γ-diketones react with phosphorus pentoxide or ammonia or phosphorus pentasulphide to give 2,5-di-substituted derivatives. Furfural obtained by processing of oat hulls and corn cobs serves as starting material for making furan: furfural ⟶ 2-furoic acid ⟶ furan. Pyrrole from corn cobs or bone oil or by distillation of succinimide with zinc. Thiophene from sodium succinate by cyclization with phosphorus trisulphide or industrially from *n*-butane. Reactions: (i) Furan: Diels–Alder reaction; reduction to tetrahydrofuran (solvent); furfural more important than furan and used in industry for making plastics. Reactions of furfural similar to those of benzaldehyde. (ii) Thiophene: unlike furan, undergoes aromatic substitutions, like nitration, sulphonation and acetylation. Resemblance to benzene and method of separation from it. (iii) Pyrrole: wide occurrence of pyrrole derivatives in nature. Reactions similar to those of phenols. Pyrrole units present in haemoglobin and chlorophyll.

B. Six-membered heterocyclic rings: pyridine is the most important member and is more aromatic in its reactions than benzene. Stability of the pyridine ring to oxidation. Occurrence in coal tar and bone oil. Increased number of isomers for *mono*- and *di*-derivatives. Pyridine, resistant to sulphonation, nitration, etc. Comparison with reactions of nitrobenzene. Explanation for resistance to electrophilic substitutions. Shows reactions of a tertiary amine, for example, methylation and amine oxide formation. Reduction of pyridine to piperidine. Amination with sodamide to 2-aminopyridine. Pyridoxin and niacin, members of the B vitamin group, are pyridine derivatives. Quinoline and isoquinoline are benzpyridines. Skraup's synthesis of quinoline using aniline, glycerol, sulphuric acid, and nitrobenzene.

Alkaloids. These are basic, nitrogen-containing compounds of plant origin with physiological properties. Coniine, the simplest alkaloid, is 2-*n*-propylpiperidine. Nicotine, the tobacco alkaloid, has a pyridine ring and a reduced pyrrole ring incorporated in its structure. Degradations involved in the structure determinations of the above alkaloids.

CHAPTER EIGHTEEN

DRUGS, HORMONES, VITAMINS, AND ENZYMES

It is common knowledge that the practice of medicine has been revolutionized in recent years by the introduction of a number of wonder drugs and antibiotics. Most of today's prescriptions could not have been written a quarter of a century ago, since these medicinals were unknown then. Several of them have been referred to previously. The chemistry of a few other medicinals, together with that of other complex materials, known as vitamins, hormones, and enzymes, required for the assimilation of foods both in the human and other animal organisms, will be presented here.

DRUGS

A large number of human diseases are due to micro-organisms. Even though Jenner, Pasteur, and others proved the efficacy of vaccination and serum therapy in certain contagious diseases, like smallpox, typhoid, diphtheria, etc., the other infectious diseases did not respond to such treatment. Prior to 1900, very few specific remedies for infectious diseases were known, even though the organic substances from natural sources have been part of medicinal practice since early times. The history of the development of modern drugs is intimately linked with the name of PAUL EHRLICH (1854–1915), who, in addition to discovering a number of useful drugs, laid the foundation for the branch of medicine which he called **chemotherapy**. This term refers to the treatment of bacterial diseases by chemical agents which selectively destroy the micro-organisms, without affecting the tissues of the host. The arsenicals, anti-malarials, sulpha drugs, and various antibiotics come under this classification of chemotherapeutic agents.

Arsenicals. Ehrlich's early work on the selective action of certain dyes in staining tissues led him to believe that it might be possible to find a dye which would attach itself to a parasite organism and, depending on the groups present in its molecule, would destroy the organism. This resulted in his discovery of the activity of the dye Trypan Red against trypanosomes in infected mice. The fact that this dye is an azo dye immediately suggested the study of analogous compounds containing arsenic—an element in the same group as nitrogen in the periodic table. An earlier report about the trypanosomicidal action of an arsenic compound called **Atoxyl** acted as a further incentive. Atoxyl was found to be the sodium salt of p-arsanilic acid, but was too toxic for use. Other arsenic compounds were examined, and this resulted in the discovery of the anti-syphilitic properties of 3,3'-diamino-4,4'-dihydroxy-benzene—known as *salvarsan* in Germany and as *arsphenamine* in the United States.

274

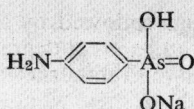

OH H_2N $NHCH_2OSO_2Na$

H_2N—⟨ ⟩—As=O HO—⟨ ⟩—As=As—⟨ ⟩—OH

ONa

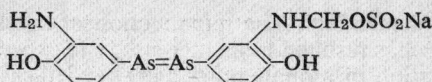

Atoxyl   Neoarsphenamine

The preparation of arsphenamine was referred to in Chapter Thirteen. Arsphenamine is commercially supplied as a hydrochloride since the free base is unstable. Prior to an injection, this hydrochloride is converted into the soluble sodium salt of arsphenamine. Neoarsphenamine, a derivative of arsphenamine developed later, is more stable and soluble and can be administered as such. Both the above compounds are still used in the treatment of syphilis, although a great number of other organic arsenic compounds have been tested.

Antimalarials. The use of quinine as a specific for malaria dates back to 1639. Several attempts have been made to find a better substitute for quinine. In 1926 SCHULEMANN of Germany reported that a drug called Plasmochin was sixty times more active than quinine in avian malaria. But it was found that plasmochin was too toxic.

Further development led to the production of atabrine in 1933. It is an acridine derivative having the same side chain as plasmochin.

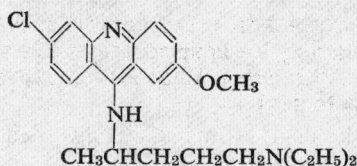

Cl
     N

            OCH_3

NH

$CH_3CHCH_2CH_2CH_2N(C_2H_5)_2$

Atabrine

Atabrine was extensively used by the armed forces during the Second World War and proved especially valuable in the treatment of certain forms of malaria (benign tertiary and malignant tertiary forms). Its main disadvantage is that it must be administered over long periods and causes a yellow coloration of the skin.

Paludrine, another anti-malarial, was developed during the Second World War. It has the following formula:

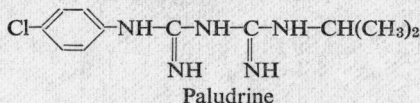

Cl—⟨ ⟩—NH—C—NH—C—NH—$CH(CH_3)_2$

              NH NH

Paludrine

Paludrine has the advantage of being effective at lower dosages and being less toxic than atabrine. Other effective anti-malarials discovered during wartime research are pentaquine and chloroquine.

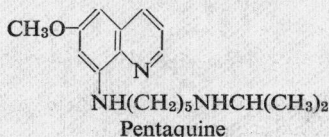

CH_3O

         N

$NH(CH_2)_5NHCH(CH_3)_2$

Pentaquine

Sulphonamides. The introduction of sulpha drugs followed by anti-biotics has resulted in an increased life expectancy and a great degree of alleviation of human suffering. Dreaded diseases like diphtheria, lock-jaw, and blood poisoning are almost obsolete, thanks to these wonderful drugs. As late as 1930 an attack of pneumonia almost always proved fatal, and today it is well under control. The sulpha drugs originated from the discovery by GERHARD DOMAGK (1934) that the dye Prontosil cured, with remarkable efficiency, experimental streptococcal and staphylococcal infections in test animals. The story of this discovery is a romantic chapter in the history of chemotherapy. About the year 1930 Domagk was testing certain dyes with respect to their chemotherapeutic activity. He was experimenting with mice. By some accident, his daughter pricked her finger with a needle and infection set in. The deadly infection began to spread with devastating effects, and the then known medicines were absolutely ineffective. Driven by desperation, Domagk gave his daughter oral doses of prontosil, one of the dyes he was testing, and she recovered miraculously. Prontosil has the following structure:

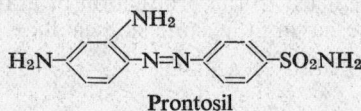

$$H_2N-\text{⟨⟩}(NH_2)-N{=}N-\text{⟨⟩}-SO_2NH_2$$

Prontosil

This was a very great discovery and, as a fitting recognition of his pioneering work in the field, Domagk was awarded the Nobel Prize in physiology and medicine in 1939.

Following Domagk's discovery, it was recognized that prontosil was degraded to sulphanilamide in the organism and that this was the therapeutically active portion of the prontosil molecule. Hundreds of derivatives of sulphanilamide were prepared and tested for their activity, but only a few were found useful. It should be noted that though there were other drugs which were active against bacteria, such as arsenicals, the sulpha drugs were found to be the most useful with the least toxicity. Further, the sulpha drugs were active against a whole array of bacterial diseases like meningitis, pneumonia, gonorrhoea, blood poisoning, scarlet fever, tonsillitis, and sinus infections.

Among the sulpha drugs that have found the widest use are sulphani-lamide, sulphapyridine, sulphathiazole, sulphapyrimidine, sulpha-guanidine, succinoylsulphathiazole, and marfanil.

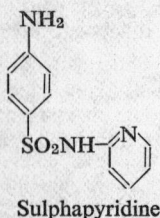

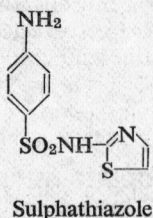

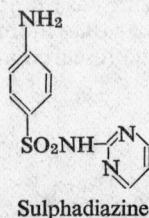

Sulphapyridine   Sulphathiazole   Sulphadiazine

NH$_2$

SO$_2$NHC$\overset{\text{NH}}{\underset{\text{NH}_2}{}}$

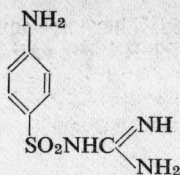

Sulphaguanidine

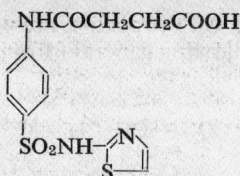

NHCOCH$_2$CH$_2$COOH

SO$_2$NH–$\overset{\text{N}}{\underset{\text{S}}{}}$

Succinoylsulphathiazole

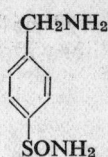

CH$_2$NH$_2$

SONH$_2$

Marfanil

A critical review of these useful drugs indicates that the grouping –SO$_2$·NH·CR=N– is essential for chemotherapeutic activity.

Antibiotics. To this point we have mainly considered synthetic compounds, like sulpha drugs and anti-malarials, which suppress the growth of the micro-organisms, and thereby the diseases caused by them. Many micro-organisms are known which produce chemical substances within themselves that inhibit the growth or destroy other micro-organisms. Such substances are called antibiotics and are effective when present even in low concentrations. Antibiotics fall, therefore, under the classification of therapeutic chemicals.

The discovery of penicillin dates back to 1928, when ALEXANDER FLEMING, in his study of bacteria, noticed that the bacteria in a dish were normal except within the range of a mould of the Penicillium species where the bacteria had all died. Fleming cultivated this mould, from which he got a broth. He found this solution, which he called penicillin, to be effective against pneumonia in mice and published his results in 1929. Fleming's work received little attention, until ten years later when FLOREY and CHAIN took up the matter again. After three years of hard work they obtained a brown powder in an attempt to get a concentrated form of Fleming's penicillin. In February 1941 the first dose of penicillin was given to a London policeman who had cut himself while shaving. His mounting fever was brought down by penicillin injection. The supply of penicillin ran out, and consequently the policeman died. Several months later penicillin was given to a boy who had an intractable streptococcal infection and the boy recovered.

During the Second World War the allies recognized the potential military value of penicillin and set up a joint research programme in Canada, England, and the United States. The aim was to find methods of mass production of penicillin and to investigate its chemical structure with a view to its ultimate synthesis. As a result of this co-operative effort, pure crystalline penicillin was isolated and its structure determined. Commercial production of the rather labile substance was started in 1943 and, although penicillin has been synthesized, it is still produced by the original method in large fermentation tanks.

Penicillin is obtained from the common mould on stale bread, cheese, or fruit. It is found to be effective against organisms like strepto-, staphylo-, pneumo-, and meningococci, but not against the bacteria causing tuberculosis. A very great advantage that penicillin has over other chemotherapeutic agents is its low toxicity. Its disadvantage is that it is not efficient when given orally, and that it is excreted rapidly. Many derivatives of penicillin have been prepared, and of these the amine salts,

especially that formed with procaine, are widely used. There are several
penicillins with the following general formula, in which R is the variable:

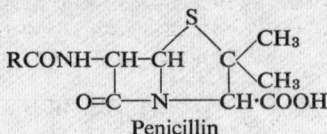

$$RCONH-CH-CH \quad C \underset{CH_3}{\overset{CH_3}{<}}$$

$$O=C-N-CH \cdot COOH$$

Penicillin

R = $C_6H_5CH_2-$ = Penicillin G
R = $-CH_2-CH=CH-CH_2CH_3$ = Penicillin F
R = $-CH_2(CH_2)_5CH_3$ = Penicillin K
R = $(p-OH-C_6H_4-CH_2-)$ = Penicillin X

Commercial samples contain mostly penicillin G. The wonderful
success that attended the use of penicillin acted as an incentive to a
search for fresh antibiotics. Streptomycin was a result of this search.
It was discovered in 1944 by WAKSMAN at the Experimental Station of
Rutgers University. The great value of this antibiotic is that it is active
against bacteria which are unaffected by penicillin, and hence is used to
supplement the action of penicillin. Streptomycin is particularly valuable
in the treatment of tuberculosis, meningitis, and pneumonia. On the
basis of degradative studies, streptomycin has been shown to be composed of two units, streptobiosamine (a nitrogen-containing disaccharide) and streptidine (a guanidine derivative of inositol). Isoniazide, the
hydrazide of isonicotinic acid, is another useful drug utilized in the treatment of tuberculosis.

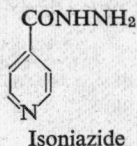

$$CONHNH_2$$

Isoniazide

Among the more useful of a large number of antibiotics that have
been isolated in recent years are aureomycin and terramycin, known as
the **tetracyclines**. They are so named because the crystals of aureomycin
have a golden colour (aureo meaning gold) and terramycin was first
isolated from a sample of earth (terra meaning earth). The two antibiotics have been shown to be derivatives of a complex substance containing four fused rings, and are therefore called tetracycline:

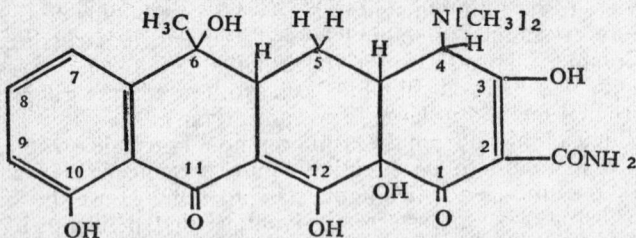

Aureomycin is the 7-chloro and terramycin the 5-hydroxy derivative of
tetracycline. Both of them exhibit similar activities and are used for the

treatment of a variety of infections, including those of the urinary and intestinal tracts and those following surgery.

Chloramphenicol, also called chloromycetin, is another important antibiotic which was isolated from *Streptomyces venezuelae*. Its structure has been shown by degradation and synthesis to be the following:

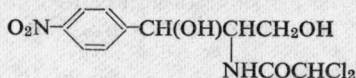

$$O_2N-\underset{\underset{\displaystyle NHCOCHCl_2}{|}}{\langle\ \rangle}-CH(OH)CHCH_2OH$$

This substance was synthesized in 1949 and is the first antibiotic for which a commercially feasible synthesis was realized. An unusual feature of this structure is the presence of the nitro group and also the dichloromethyl group—a rare occurrence in natural products. Its resemblance to the structure of ephedrin may be noted. Chloromycetin is particularly effective in the treatment of typhoid fever and intestinal infections.

It will be seen from the preceding cursory treatment that the introduction of both the sulpha drugs and the antibiotics have, in the last twenty-five years, ushered in the most amazing advance in public-health history. The picture is not all rosy, as might be imagined. There are several problems that await solution. Two of them should be mentioned:

1. The use of these wonder drugs is attended with side effects and development of drug resistance. Penicillin has a low toxicity no doubt, but some patients are allergic to it. Streptomycin produces dizziness, a ringing sound in the ear, and even impairment of hearing.

2. In several cases, after a long period of use, the drug loses its effectiveness. It is not too much to hope that these problems will be solved in the course of time. It should be noted that the antibiotics are not limited to medicinal use alone. They are added to the food of livestock to hasten their growth and are also used for food preservation.

Mechanism of Drug Action. The preceding discussion of chemotherapy is likely to give an impression that the chemists have obtained the upper hand over diseases. This is by no means true. The mechanism of drug action is far from well understood, and there are significant gaps in our knowledge. It is now believed that drug action depends upon activation of receptors located in the surface of a cell. This surface is something like a mosaic of keyholes formed by enzymes and other biological molecules. Foreign molecules of the right size, polar distribution, and steric grouping fit the keyholes selectively like a key fitting a lock. Some of them may be molecules, not required for the growth of the cells, which have the requisite conformation to fit the 'keyholes' and hence block the entry of biologically essential molecules. This will have an adverse effect on the cells, which will gradually perish. This is the principle of competitive inhibition. A simple illustration will be useful. *p*-Aminobenzoic acid, a member of the vitamin B group, is necessary for the survival of certain bacteria. Sulphanilamide, which has a certain resemblance in size to *p*-aminobenzoic acid, can fit into the enzyme surface normally reserved for *p*-aminobenzoic acid. This leads to a blocking of the metabolism of the germ, which gradually dies. In other words, this means the substance has a therapeutic value. This theory of competitive inhibition accounts for the fact that antibiotics are ineffective against viruses. The antibiotics work by inhibiting some key enzyme, but viruses have no enzymes of their own, and make use of those of the host. The development of the principle of competitive inhibition has led to the designing of several molecules, made to order, with a view to their use in chemotherapy.

The **antihistamine** drugs, Benadryl, Pyribenzamine, etc., are so designated because they neutralize the effects of histamine, a substance involved in respiratory diseases like colds, hay fever, and asthma. Interest in antihistaminic substances dates back to 1933, when it was observed that certain phenolic ethers counteracted the effects of histamine. Large-scale tests conducted to evaluate the remedial effects of different antihistamines seem to indicate that they are effective in many cases, although there is difference of opinion among medical men as to their efficacy.

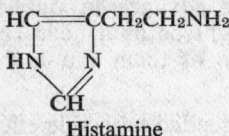

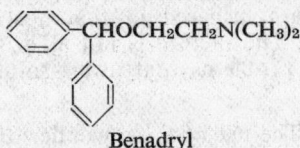

$$HC\!=\!\!=\!CCH_2CH_2NH_2$$
$$HN\quad N$$
$$CH$$
Histamine

$$C_6H_5\!\!-\!\!CHOCH_2CH_2N(CH_3)_2$$
Benadryl

HORMONES, VITAMINS, AND ENZYMES

The chemical reactions involved in the proper utilization of food by the body are mainly organic in nature and take place fairly rapidly at room temperature. Since organic reactions take place slowly as a rule, apparently powerful catalytic factors must be involved in the body processes. These factors are known as vitamins, hormones, and enzymes. The distinction between the first two is not chemical but physiological in nature. Hormones are produced in the body, while vitamins are supplied by food. They do not belong to any particular chemical category. The known vitamins are relatively simple non-colloidal substances, while the hormones may be simple or complex proteins. An enzyme is a structurally complex colloidal protein which may include a vitamin or hormone residue in it. All three are essential for our physical and mental well-being.

Hormones. Adrenaline was the first hormone to be isolated (1895) and chemically identified. The adrenal glands consist of two parts—one a central mass called the medulla, and the other an outer layer called the cortex. Adrenaline is obtained from the medulla. It is a derivative of catechol. The natural compound is the laevo isomer, which is much more potent than the dextro isomer. Adrenaline raises the blood pressure—a property which it shares with a number of related 2-phenylethylamines. Commercially, it is obtained from the adrenal medulla of cattle or by several synthetic routes, one of which was mentioned in Chapter Fourteen. Adrenaline is used for the control of haemorrhages and for the treatment of nasal congestion. The drug ephedrine, originally obtained from a Chinese herb but now available by synthesis, has similar reactions. Benzedrine, a related compound, is widely used for the treatment of hay fever, asthma, etc.

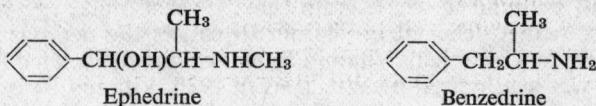

Ephedrine   Benzedrine

The hormone derived from the thyroid gland is thyroxine, which has been shown to have the following structure:

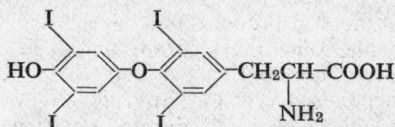

Thyroxine controls the speed of oxygen consumption in the body. Man's thyroid gland normally contains about 25 mg. of thyroxine. In case of deficiency, caused by lack of iodine in the diet or other factors, diseases like goitre (enlargement of the thyroid gland) and cretinism result.

Vitamins. The vitamins are physiologically active substances that are found to be essential for human health. It is now believed that the vitamins are structural components of enzymes. By 1912 it was evident that foodstuffs contain at least two essential substances, one of which is fat soluble (A) and the other water soluble (B). It was also clear that lack of the fat-soluble substance in the diet caused an eye disease (Xerophthalmia) and lack of the water-soluble substance caused the disease known as beriberi. These essential substances were called vitamines, since they were found to be essential or vital to life, and were believed to be amino acids. A third water-soluble substance, C, was subsequently found to be present in certain types of food which prevented the disease called scurvy. It was soon recognized that these substances were not amines, and hence it was decided to drop the letter *e* and redesignate them as vitamins. In the following survey of the known vitamins only those essential to humans will be discussed.

Vitamin A is chemically related to the pigment (yellow or red) from carrots, tomatoes, and egg yolk. It is a growth factor, prevents night blindness, and gives resistance to bacterial infection. The vitamin A molecule is one-half of the yellow pigment β-carotene. It has been synthesized and has the following structure:

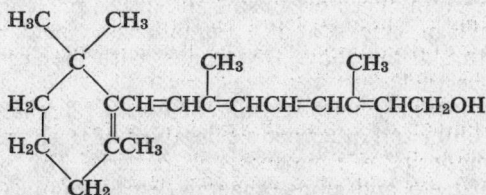

Vitamin A ($C_{20}H_{30}O$)

The original antiberiberi vitamin is now known to consist of several different vitamins and is designated as the B complex. Among its constituents are B_1 (thiamine), B_2 (riboflavin), B_{12}, and nicotinic acid. Vitamin B_1 is found in whole-grain cereals, yeast, milk, egg yolk, and asparagus. A diet deficient in vitamin B_1 causes the nervous disorder beriberi and loss of weight. Vitamin B_1 was originally isolated from rice bran and was synthesized by WILLIAMS in 1936. Vitamin B_2 is essential for perfect nutrition, since lack of it results in loss of hair, bodily weakness, and failing vision. Cheese, corn, cereals, milk, and yeast are good sources.

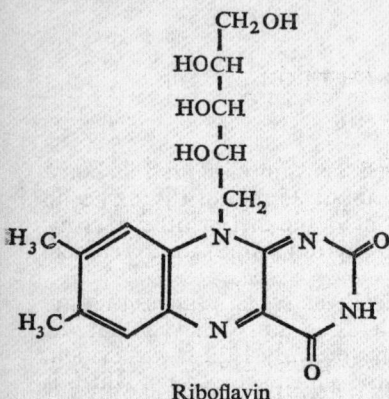

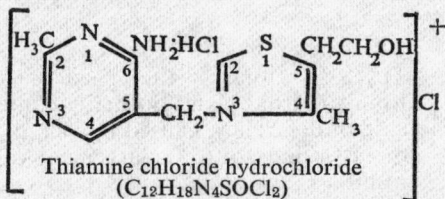

Thiamine chloride hydrochloride
($C_{12}H_{18}N_4SOCl_2$)

Riboflavin

Vitamin B_{12} was isolated from liver extract in 1948 and has anti-anaemic properties. It contains phosphorus and cobalt in addition to carbon, hydrogen, oxygen, and nitrogen. Its complete structure has been determined by X-ray methods. Nicotinic acid has been shown to be the factor which prevents pellagra among humans. Pellagra is perhaps the most widely prevalent of vitamin-deficiency diseases in the United States and manifests itself in lesions of the skin, nervous disorders, and inflammation of the digestive tract. Nicotinic acid derives its name from its original preparation by the oxidation of nicotine. Bran, liver, lean meat, fish, and wheat are good sources of this vitamin, which also stimulates the growth of certain plants. Nicotinic acid amide is also effective in the treatment of pellagra.

Among the other members of the B complex are the following: B_6 (pyridoxine), whose deficiency causes dermatitis in rats; pantothenic acid, which is an anti-dermatitis factor in chicks; folic acid, essential for the growth of certain micro-organisms and also a preventive substance against anaemia in chicks and *p*-aminobenzoic acid, essential for bacterial growth. All the above B vitamins, with the exception of B_{12}, have been synthesized.

Vitamin C is the anti-scurvy vitamin. Scurvy is a disease which at one time plagued sailors and other people deprived of fresh fruits and vegetables for a long period of time and is characterized by haemorrhages, degeneration of teeth and gums, anaemia, etc. Vitamin C is found in citrus fruits, tomatoes, cabbage, and other fresh vegetables. It is a lactone related to hexose sugars and is also called ascorbic acid. As in the case of adrenaline, one of the optically active forms of ascorbic acid (L-) is more active than the other.

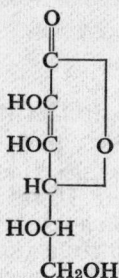

$$
\begin{array}{l}
O \\
\parallel \\
C \\
\mid \\
HOC \\
\mid \\
HOC \quad O \\
\mid \\
HC \\
\mid \\
HOCH \\
\mid \\
CH_2OH
\end{array}
$$

Vitamin C
(L-Ascorbic acid)

SUMMARY

Drugs. Chemotherapy is the treatment of bacterial diseases by chemical reagents. Ehrlich's pioneering work in chemotherapy. Arsenicals like atoxyl, arsphenamine, and neoarsphenamine for treatment of syphilis. Anti-malarials like plasmochin, atabrine, pentaquine, and paludrine. Spectacular effects of sulphonamides and the different types of sulpha drugs. Mechanism of drug action. Antibiotics are chemical substances produced by micro-organisms which destroy or prevent growth of other micro-organisms. Penicillin, its discovery and usefulness. Other antibiotics like streptomycin, aureomycin, terramycin, and chloramphenicol—each of value in the treatment of specific ailments. Anti-histamines, like Benadryl and Pyribenzamine, are useful in the treatment of respiratory diseases.

Hormones. From adrenal glands. Adrenaline from the medulla, its structure and effects.

Vitamins. Those essential to humans include vitamins A, B_1, B_2, B_{12}, and C.

CHAPTER NINETEEN

GIANT MOLECULES OR POLYMERS

In our study of organic chemistry we met with certain giant molecules, such as starch, cellulose, proteins, and rubber. These molecules are built up of smaller units joined together and form a repeating structure. The repeating structure or the recurring unit in a given molecule is called the monomer, and the entire molecule composed of several such repeating units is called a polymer. Thus we can say that rubber is a polymer of isoprene, starch of α-glucose, and cellulose of β-glucose. We can distinguish between two types of polymerization. In the case of rubber, the polymer and the monomer have the same empirical formula. The polymer is built by simple addition of several monomers without loss of any fragment. This type of polymerization is called **addition polymerization** or **polymerization by association**. In the other kind of polymerization the empirical formula of the monomer and the polymer are not the same. During such polymerizations molecules of the monomer are joined together with the elimination of a molecule of water. The repeating units in the polymer do not have the same chemical composition as the polymer. For example, the recurring units of starch and cellulose contain one less molecule of water than the corresponding monomers, and the terminal units of the polymer chain may be different from the units inside. This type of polymerization is referred to as **condensation polymerization**. Several synthetic polymers have come into industrial use. They not only compete with natural products but also, in several cases, they are useful where no natural products exist.

ADDITION POLYMERIZATION

Addition polymerization involves the combination of unsaturated monomers—either of the same kind or of two different kinds. An example of simple polymerization is the interaction of many molecules of ethylene to form a polymer known as polyethylene or polythene.

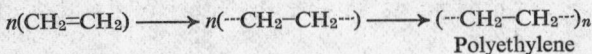

$$n(CH_2=CH_2) \longrightarrow n(\cdots CH_2-CH_2\cdots) \longrightarrow (\cdots CH_2-CH_2\cdots)_n$$
Polyethylene

A catalyst, in this case a little oxygen, is used to activate the double bond whereby the molecule seeks to combine at both ends as indicated by the dotted bonds above. When several molecules are activated in the same way they combine with one another to form long chains. This chain reaction terminates when the growing ends are saturated by some foreign groups or atoms which prevent further lengthening. Since the end groups are generally different from the monomer, it is apparent that the polymer will have an empirical formula slightly different from that of the monomer. Also, the polymer chains thus formed need not all be of the same length.

If polymerization is between two different kinds of monomers the process is referred to as **copolymerization**. Vinyl chloride can be copolymerized, for example, with vinyl acetate:

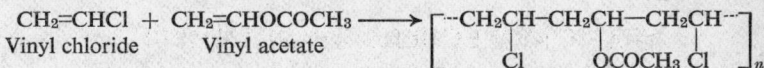

$$CH_2{=}CHCl + CH_2{=}CHOCOCH_3 \longrightarrow \left[{\cdots}CH_2CH{-}CH_2CH{-}CH_2CH{\cdots} \atop \qquad Cl \qquad\quad OCOCH_3\ Cl \right]_n$$

Vinyl chloride   Vinyl acetate

The reactions in copolymerization need not be in the same proportions, nor need they combine in a rational pattern.

CONDENSATION POLYMERIZATION

Two substances, each containing two functional groups, can form long polymer chains by splitting the elements of a simple molecule like water. If a dibasic acid, for example, is esterified with a glycol like ethylene glycol, a polyester can be formed by repeated esterification of the end groups, as follows:

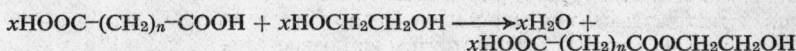

$$x\text{HOOC}{-}(CH_2)_n{-}\text{COOH} + x\text{HOCH}_2CH_2OH \longrightarrow x\text{H}_2O +$$
$$x\text{HOOC}{-}(CH_2)_n\text{COOCH}_2CH_2OH$$

followed by

$$\text{HOOC}(CH_2)_n\text{COOCH}_2CH_2O\overset{\displaystyle O}{\overset{\|}{C}}(CH_2)_n{-}\text{COOCH}_2CH_2OH$$

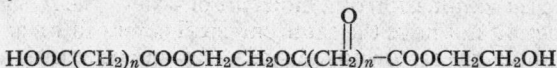

and then progressively to

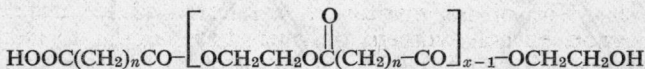

$$\text{HOOC}(CH_2)_n\text{CO}{-}\left[\text{OCH}_2CH_2O\overset{\displaystyle O}{\overset{\|}{C}}(CH_2)_n{-}\text{CO} \right]_{x-1}{-}\text{OCH}_2CH_2OH$$

The alkyd resins, referred to in Chapter Sixteen, are polymeric esters in which the acid component is phthalic acid and the alcohol component is ethylene glycol or glycerol. Glycerol has three hydroxyl groups, which makes possible the cross-linking of the individual polymer chains, to give rise to a three-dimensional polymer (see page 286).

The amount of cross-linking, and thereby the physical characteristics of the polymer, can be regulated by changing the proportions of the reactants.

The linear chain polymers obtained by either of the above types of polymerization are solid at room temperature. They become plastic when heated and can be moulded into any desired shape by application of pressure. For this reason they are called thermoplastic polymers. After cooling, the moulded object loses its plasticity and retains its new form. The process can be repeated and the plastic moulded into any other shape. The cross-linked polymers are of the thermosetting type. They have a certain amount of plasticity in the early stages of moulding, but once cooled they cannot be softened again. A large number of polymeric products with a variety of properties and uses are made

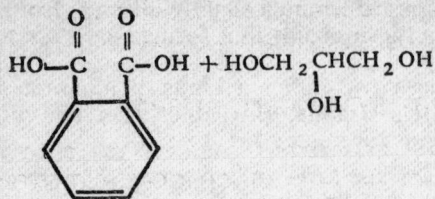

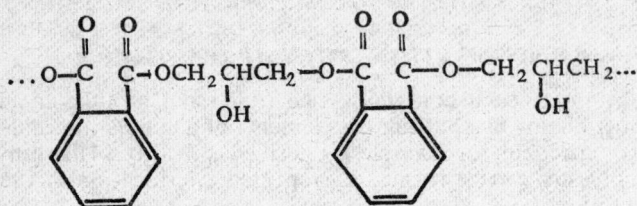

Phthalic anhydride

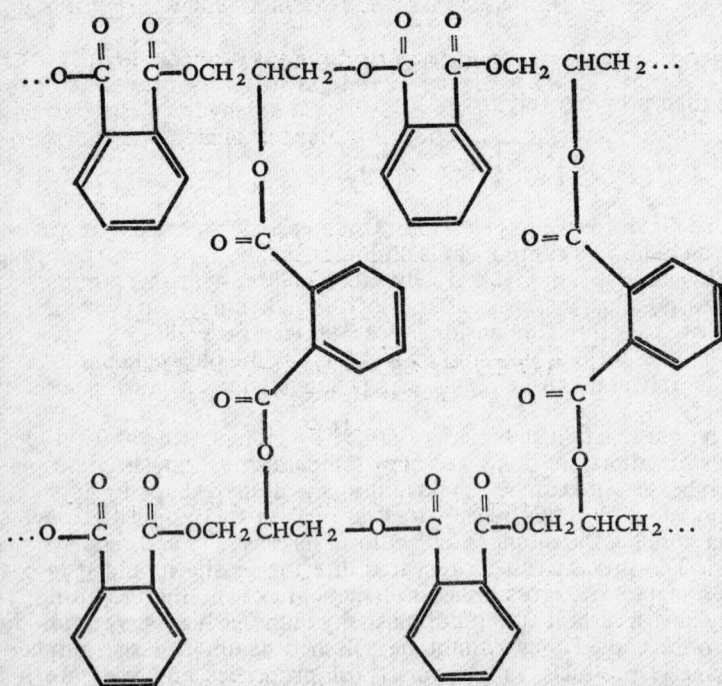

available by industry and are in everyday use. These include natural and synthetic rubbers, synthetic fibres, synthetic plastics, and the silicones.

NATURAL AND SYNTHETIC RUBBER

Natural rubber, which has served man for more than a hundred and fifty years, is obtained from the latex of the *Hevea brasiliensis* tree grown in tropical countries. This latex is a colloidal dispersion in water which is coagulated by the addition of weak acids to give raw rubber. This raw rubber is sticky and soft and is elastic only over a small range of temperature. However, on heating it with sulphur (usually 5–8%) or a compound containing sulphur to a temperature of 100–150° C., a product is obtained whose elasticity and consistency are unaffected by temperature variations. This is the well-known vulcanization process. The time for vulcanization is shortened by the addition of accelerators— usually nitrogen- or sulphur-containing compounds. An improvement in the wearing qualities and strength of the vulcanized product is also effected by the addition of filler ingredients like zinc oxide or carbon black. Increasing the sulphur content to 30% or more adds to the hardness and tensile strength to give the product known as ebonite. The nature of the vulcanization process is obscure, but there is no doubt that cross-linking of the polymer chains occurs through sulphur atoms. Vulcanized rubber also contains some anti-oxidants like phenyl-β-naphthylamine or other aromatic amines which prevent oxidative degradations.

The destructive distillation of rubber gives isoprene, which can also be polymerized in the presence of sodium to give a rubber-like product. Ozonization of rubber gives laevulinic aldehyde (CH_3COCH_2CHO). These facts indicate that rubber is a polymer of isoprene with a head-to-tail linking of the monomer units.

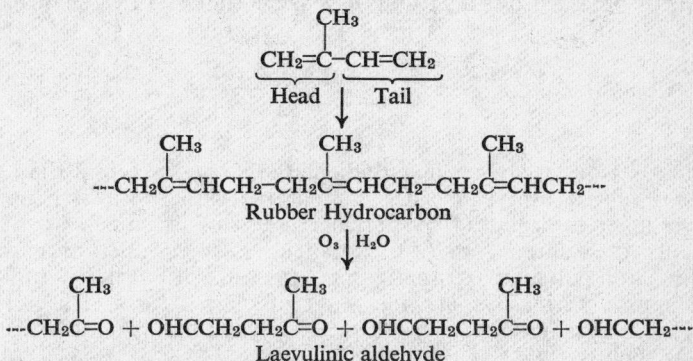

$$CH_2=\underset{\underset{\text{Head}}{|}}{\overset{\overset{\displaystyle CH_3}{|}}{C}}-CH=CH_2$$
$$\underbrace{}_{\text{Head}} \quad \underbrace{}_{\text{Tail}}$$

$$\downarrow$$

$$---CH_2\overset{\overset{\displaystyle CH_3}{|}}{C}=CHCH_2-CH_2\overset{\overset{\displaystyle CH_3}{|}}{C}=CHCH_2-CH_2\overset{\overset{\displaystyle CH_3}{|}}{C}=CHCH_2---$$

Rubber Hydrocarbon

$$O_3 \downarrow H_2O$$

$$---CH_2\overset{\overset{\displaystyle CH_3}{|}}{C}=O + OHCCH_2CH_2\overset{\overset{\displaystyle CH_3}{|}}{C}=O + OHCCH_2CH_2\overset{\overset{\displaystyle CH_3}{|}}{C}=O + OHCCH_2---$$

Laevulinic aldehyde

With rising demand for rubber products, it was natural that chemists should turn their attention to making synthetic substitutes for rubber. Before 1930 very little synthetic rubber was made. In 1941, Japan obtained control of the source of around 90% of the supply of natural rubber. This acted as a further stimulus for finding substitutes. Several types of synthetic rubbers, which duplicate and sometimes excel the properties of natural rubber, have been produced since 1920. Synthetic rubbers, like the natural product, can be vulcanized.

Neoprene. Neoprene, introduced in 1931, was the first synthetic rubber produced in the United States. It is made by polymerization of chloroprene obtained, as follows, from acetylene.

$$CH \equiv CH \xrightarrow[NH_4Cl]{Cu_2Cl_2} \overset{4}{C}H_2 = \overset{3}{C}H\overset{2}{C} \equiv \overset{1}{C}H \xrightarrow[(1,4\text{-addition})]{HCl} \left[\underset{Cl}{CH_2CH = C = CH_2} \right] \xrightarrow{Isomerization}$$

Vinylacetylene

$$\underset{\underset{Cl}{|}}{CH_2 = CH - C = CH_2} \xrightarrow{Polymerization} \cdots CH_2 - CH = \underset{\underset{Cl}{|}}{C} - CH_2 - CH_2 - CH = \underset{\underset{Cl}{|}}{C} - CH_2 \cdots$$

Chloroprene   Neoprene

Neoprene is superior to natural rubber in resistance to chemicals, oils, and the action of sunlight.

Butadiene Rubbers. The synthetic rubber produced in the largest tonnage is GR-S (Government Rubber-Styrene type) also called Buna-S. It is a copolymer consisting of 80% butadiene and 20% styrene. The name buna owes its origin to the fact that the polymerization of butadiene was originally effected with sodium (*bu*tadiene-*na*trium for sodium). Buna-S is widely used in the manufacture of tyres. Butadiene is commercially prepared by the dehydrogenation of *n*-butane or from ethanol. A German process developed during the Second World War involves the use of acetylene.

$$CH \equiv CH \xrightarrow{CH_2O} HOH_2C - C \equiv C - CH_2OH \xrightarrow{4H}$$

$$HOH_2CCH_2CH_2CH_2OH \xrightarrow{-2H_2O} CH_2 = CH - CH = CH_2$$

1,4-Butadiene

Styrene is obtained by alkylation of benzene followed by dehydrogenation.

$$CH_2 = CH_2 + \bigcirc \xrightarrow{AlCl_3} \bigcirc - CH_2CH_3 \xrightarrow[700° C.]{Al_2O_3} \bigcirc - CH = CH_2$$

Styrene

Buna-N is a copolymer of butadiene (75%) and acrylonitrile (25%) and has oil-resistant properties comparable with those of neoprene. Butyl rubber is produced by polymerization of isobutene with small amounts of isoprene (1–2%). Isobutene is a product of catalytic cracking and polymerizes rapidly in the presence of boron trifluoride or aluminium chloride over a wide range of temperatures. The polymer polybutene thus obtained is a saturated polymer and cannot be vulcanized as such.

$$n\underset{\underset{CH_3}{|}}{\overset{\overset{CH_3}{|}}{C}} = CH_2 \longrightarrow \left[\cdots \underset{\underset{CH_3}{|}}{\overset{\overset{CH_3}{|}}{C}} - CH_2 - \underset{\underset{CH_3}{|}}{\overset{\overset{CH_3}{|}}{C}} - CH_2 \cdots \right]_n$$

Polybutene

However, isobutene in the presence of small amounts of isoprene gives a copolymer containing enough unsaturation to permit vulcanization. The vulcanized product is butyl rubber and has excellent air-retentive properties. It has displaced natural rubber in the manufacture of inner tubes.

SYNTHETIC FIBRES

No discussion of organic chemistry would be complete without a reference to synthetic fibres and plastics which have found such a wide variety of application. When the average woman steps out of her house she unconsciously pays a tribute to the organic chemist—for, from head to foot, she is adorned with products of his making. Her plastic shoes, nylon stockings, the rayon or Terylene fabrics she wears, her plastic handbag, and the host of other things she uses or wears are all products that stem from organic chemistry.

The major natural textile fibres are cotton, a polysaccharide, and wool and silk, which are proteins. Synthetic fibres or artificial silks are made by suitable processes from synthetic polymers or derivatives of cellulose. In one process solutions of these polymers are forced through the capillary holes of a spinneret (a mechanical spinning device) into a bath containing a chemical which coagulates the polymer in the form of continuous fibres. A variation of the same process consists of removing the solvent as the fibre emerges from the spinneret in a stream of warm air. Another process, referred to as melt spinning, consists of melting the polymer and then forcing it through the spinneret. Synthetic fibres produced from the naturally occurring cellulose or derivatives of cellulose are called rayons. Among purely synthetic fibres produced by the polymerization of simple molecules are Nylon, Orlon, Vinyon, and Terylene.

Nylon. Nylon, the first completely synthetic fibre introduced, is a polyamide obtained by heating adipic acid with hexamethylene diamine in an autoclave. Nylon threads are produced by extruding the molten polyamide as fine filaments and then stretching them. Nylon is similar to silk in physical and chemical properties, but has superior tensile strength and elasticity. Because of their strength, nylon cords are preferred for use in car, bus, and transport-plane tyres.

$$HO-\overset{\overset{O}{\|}}{C}(CH_2)_4\overset{\overset{O}{\|}}{C}-OH + \quad H_2N(CH_2)_6NH_2 \quad \xrightarrow{-H_2O}$$

Adipic acid   Hexamethylene diamine

$$\cdots\overset{\overset{O}{\|}}{C}(CH_2)_4\overset{\overset{O}{\|}}{C}NH(CH_2)_6NH\overset{\overset{O}{\|}}{C}(CH_2)_4\overset{\overset{O}{\|}}{C}NH(CH_2)_6NH\overset{\overset{O}{\|}}{C}(CH_2)_4\overset{\overset{O}{\|}}{C}NH(CH_2)_6NH\cdots$$

Nylon

Orlon. Orlon, one of the newer synthetic fibres, is woven into wool-like fabrics produced from a polymer of acrylonitrile.

$$nCH_2=CH-CN \longrightarrow \left[\cdots CH_2-\underset{CN}{CH}-CH_2-\underset{CN}{CH}\cdots\right]_{\frac{n}{2}}$$

Polyacrylonitrile

Orlon threads are produced by spinning solutions in special solvents in a dry spinning process. Orlon threads are unaffected by moisture, micro-organisms, and a number of chemicals.

Vinyon. Vinyon is a fibre made from a copolymer of vinyl chloride (90%) and vinyl acetate (10%). Vinyon is stable to the action of chemicals, but is not too widely used because of its relatively low softening point.

Terylene. Terylene is made from a polyester obtained by esterification of terephthalic acid with ethylene glycol. Clothes made of Terylene fibres have remarkable crease-resistant properties.

There are other synthetic fibres in use and newer ones are bound to be developed.

SYNTHETIC PLASTICS AND RESINS

The terms plastics and resins are used interchangeably by the polymer chemist, and there is no need to differentiate between them here. They are generally products that can be moulded, extruded, or cast under the effect of temperature and pressure. Frequently, chemicals called plastic-izers are added to attain the plasticity needed before moulding. Some of the more commonly used plastics will be discussed in the following section.

Cellulose Plastics. The first item to be introduced among the synthetic plastics was pyroxylin or cellulose dinitrate. The product obtained by plasticizing pyroxylin with camphor is known as celluloid. It is used in the manufacture of photographic films. Celluloid has the advantage of being both light in weight and tough, but its use is hazardous because of its inflammability. Esters of cellulose, like the diacetate plasticized with dialkylphthalate (or triphenyl phosphate), give rise to plastics with more desirable properties than the cellulose nitrate plastics. Pyroxylin, in a modified form, dissolves to a greater extent in organic solvents. The lacquers thus obtained can be applied to a hard surface such as the body of a car. On evaporation of the solvent, a smooth and stable surface coating is obtained. The solvent usually used for cellulose nitrate lacquers is n-butyl acetate.

Formaldehyde Plastics. The condensation of formaldehyde with phenol or substituted phenols gives rise to phenolic plastics which belong to the thermosetting type. Bakelite, the best known of these, was dis-cussed in Chapter Fourteen and constitutes almost 50% of the total production of plastics. It is moulded into any number of articles of everyday use, including toilet articles, umbrella handles, car parts, insulators, radio cabinets, etc., and also several decorative articles. It is used in the fabrication of laminated sheets and also as an adhesive for plywood. The use of furfural instead of formaldehyde has also furnished useful plastics. The reaction of formaldehyde with urea in the presence of amines or ammonia provides another class of plastics, also of the thermosetting type. These urea plastics have found several applications similar to those of Bakelite.

Alkyd Resins. Alkyd resins have been referred to earlier. They are used in the preparation of synthetic enamels for refrigerators and other similar household appliances.

Vinyl Plastics. Compounds containing the vinyl radical $CH_2=CH-$ polymerize to give thermoplastic resins. Thus, vinyl chloride and vinyl acetate polymerize in the presence of ultra-violet light or peroxide or other catalysts to give polyvinyl chloride and polyvinyl acetate respectively.

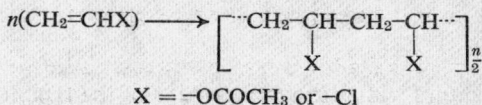

$$n(CH_2=CHX) \longrightarrow \left[\begin{array}{c} \cdots CH_2-CH-CH_2-CH\cdots \\ | \quad\quad | \\ X \quad\quad X \end{array} \right]_{\frac{n}{2}}$$

$$X = -OCOCH_3 \text{ or } -Cl$$

A polymer of tetrafluoroethylene, Teflon, is a remarkable stable plastic which is unaffected by drastic treatment with hot acids, and hence is used in the manufacture of containers and pipes for acids. Polyethylene, obtained by polymerization of ethylene at high temperatures and pressures, has excellent insulating properties. Vinyl resins are used increasingly in the manufacture of films, dishes, phonograph records, automobile seat covers, raincoats, etc.

Acrylate Resins. Acrylic ($CH_2=CH-CO_2R$) and methacrylic

$$CH_3$$
$$|$$

($CH_2=C-CO_2R$) esters polymerize in the same way as the above vinyl monomers. Methyl methacrylate furnishes polymethylmethacrylate, which is sold under the trade name perspex. The optical properties of high transmission of light and high internal reflection displayed by polymethylmethacrylate, coupled with its great strength, are responsible for its use in surgical instruments, roadway reflectors, aeroplane windows, industrial goggles, etc. The preparation of methyl methacrylate has already been referred to.

Polystyrene Resins. Under the influence of peroxide or other catalysts, styrene gives polystyrene, with a high molecular weight and a head-to-tail union of the monomeric units.

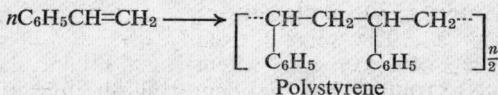

$$nC_6H_5CH=CH_2 \longrightarrow \left[\begin{array}{c} \cdots CH-CH_2-CH-CH_2\cdots \\ | \quad\quad\quad | \\ C_6H_5 \quad\quad C_6H_5 \end{array} \right]_{\frac{n}{2}}$$

Polystyrene

Polystyrene is used to produce electric insulators, toys, shoe soles, and dishes.

SILICONES

The silicones form a relatively new group of synthetic plastics. Silicon is an element below carbon in the periodic table; it has a valence of four. It can be substituted for carbon in organic compounds as follows:

Methane	CH_4	Silane	SiH_4
Ethane	C_2H_6	Disilane	Si_2H_6

The silicon bonds are more polar than those of carbon. When an alkyl dichlorosilane is treated with cold water it is hydrolysed to the dihydroxy compound; this rapidly condenses, splitting off water

between two −OH groups of two consecutive molecules to give silicone
polymers of the linear type.

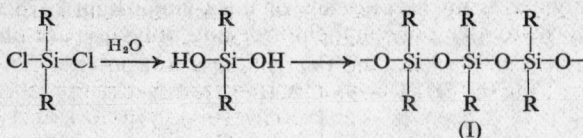

$$\underset{\underset{R}{|}}{\overset{\overset{R}{|}}{Cl-Si-Cl}} \xrightarrow{H_2O} \underset{\underset{R}{|}}{\overset{\overset{R}{|}}{HO-Si-OH}} \longrightarrow \underset{\underset{R}{|}}{\overset{\overset{R}{|}}{-O-Si-O-}}\underset{\underset{R}{|}}{\overset{\overset{R}{|}}{Si-O-}}\underset{\underset{R}{|}}{\overset{\overset{R}{|}}{Si-O-}}$$
$$(I)$$

Copolymerization with a trihydroxysilane of the type $RSi(OH)_3$ gives
rise to complex cross-linked polymers of the type shown below:

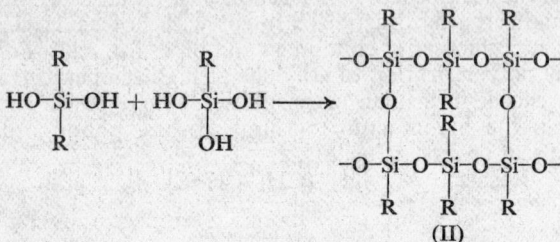

$$\underset{\underset{R}{|}}{\overset{\overset{R}{|}}{HO-Si-OH}} + \underset{\underset{OH}{|}}{\overset{\overset{R}{|}}{HO-Si-OH}} \longrightarrow$$

$$-O-Si-O-Si-O-Si-O-$$

(II)

A variety of silicones can be prepared by variations in the organic part
of the molecule. The polychloro derivatives of silane are obtained by a
Grignard reaction with silicon tetrachloride, as follows:

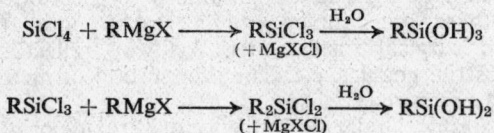

$$SiCl_4 + RMgX \longrightarrow \underset{(+MgXCl)}{RSiCl_3} \xrightarrow{H_2O} RSi(OH)_3$$

$$RSiCl_3 + RMgX \longrightarrow \underset{(+MgXCl)}{R_2SiCl_2} \xrightarrow{H_2O} RSi(OH)_2$$

The alkyl halides can react in the vapour phase with silicon at very high
temperatures in the presence of copper and silver as catalysts.

$$Si + 2RCl \longrightarrow R_2SiCl_2$$

Linear polymers of the type (I), where R is CH_3 and the end groups
are trimethylsilyl groups $[-Si(CH_3)_3]$, are high-boiling oils which are
valuable lubricants. Unlike hydrocarbon lubricants, they retain their
viscosity over a wide range of temperature. With increasing complexity
of the polymers, the silicones change to a rubber-like consistency and
then to a resin. Silicone rubbers of varying properties result from introducing cross-links into type (I) (R=CH_3). They retain their flexibility
over a wide range of temperature, stick to different types of surfaces,
and have good insulating properties. They are increasingly used to make
gaskets, insulating sheaths for electric wires and cables, and hydraulic
seals. The silicone oils are used as heating baths, as brake fluids for cars,
and for waterproofing surfaces and clothing.

The field of polymers is a vast and ever-growing area. In the space
available it is not possible to do anything more than present an elementary survey of the field. It is hoped, however, that the reader has
been able to get some idea of a branch of chemistry that has provided
him with so many useful products in everyday life.

SUMMARY

Polymerization is the process whereby giant molecules are built up, consisting of repeating units called monomers; it consists of two types: (i) addition polymerization, and (ii) condensation polymerization. The former refers to the combination of monomers either of the same kind or different kinds by a process of addition involving no loss of fragments, for example, ethylene $\longrightarrow$ polythene. Copolymerization refers to polymerization involving two different types of monomers, for example, vinyl chloride with vinyl acetate. Condensation polymerization refers to the combination of monomers by a process involving loss of a simple fragment, for example, polyester from a dialcohol and a diacid. The alkyd resins are such polymers obtained from phthalic acid and glycol or glycerol. Linear polymers are thermoplastic, and cross-linked polymers are thermosetting.

Rubber. Natural rubber from the tree *Hevea brasiliensis*; vulcanization consists of heating raw rubber with sulphur to improve its properties. Use of accelerators, anti-oxidants, etc. Structure of rubber as a polyisoprene with a head-to-tail linking of the monomers. Synthetic rubber substitutes: Neoprene is obtained by polymerization of chloroprene, which in turn is obtained from acetylene. Butadiene rubbers include Buna-S and Buna-N. Buna-S is a copolymer of 80% butadiene with 20% styrene, and is produced in the largest tonnage. Buna-N is a copolymer of 75% butadiene with 25% acrylonitrile. Butyl rubber is obtained by polymerization of isobutene with small amounts of isoprene. Properties of the above substitutes compared with those of natural rubber.

Synthetic Fibres. Processes used for making fibres from polymers. Rayons from modified cellulose polymers. Purely synthetic fibres include Nylon (from adipic acid and hexamethylene diamine), Orlon (polymer of acrylonitrile), Vinyon (a copolymer of vinyl chloride with vinyl acetate), and Terylene (polyester of terephthalic acid with ethylene glycol).

Synthetic Plastics and Resins. Plastics and resins are interchangeable terms and refer to polymers that can be moulded and cast. Plastics from cellulose dinitrate (pyroxylin) and cellulose diacetate. Lacquers from modified pyroxylin used in surface coatings. Bakelite plastics (from phenol and formaldehyde) in the fabrication of articles of everyday use. Plastics from formaldehyde and urea, furfural–phenol. Polymerization of vinyl-type compounds like vinyl chloride, vinyl acetate, acrylic and methacrylic esters, and styrene gives useful polymers. Teflon, a fluorine-containing polymer. Silicones a new class of synthetic plastics from dialkyl dichlorosilanes (R_2SiCl_2). Nature of silicone oils and silicone rubbers.

APPENDIXES

The appendixes contain the following material which will be helpful to you when working in organic chemistry. A list of items appears below:

A—EXPERIMENTS AND PREPARATIONS

B—ESSENTIAL ITEMS FOR EXPERIMENTS

C—SUGGESTED FURTHER READING

D—ANSWERS TO PROBLEM SETS

E—GLOSSARY OF SOME COMMON ORGANIC
     COMPOUNDS

APPENDIX A

EXPERIMENTS AND PREPARATIONS

Organic chemistry is far more than the abstract discussion of formulae, reactions, structures, etc. The experiments in this section will emphasize the practical nature of organic chemistry. Practical experiments provide a better appreciation of the principles discussed and a first-hand knowledge of some of the techniques used in the laboratory. The available space excludes any systematic treatment of the practical aspects of organic chemistry.

The toxic properties and high inflammability of many organic compounds make the systematic treatment of practical organic chemistry suitable only for specially equipped laboratories under expert supervision. In such a place you are unlikely to have a serious accident, because your mishaps can be dealt with promptly. However, if you were to pursue such a course on your own a small mistake could easily turn into injury, illness, or a major fire.

The object of this section is to introduce you as safely as possible to some of the simple techniques used in the organic chemistry laboratory and to enable you to prepare, separate and test a few compounds.

REACTIONS REQUIRING HEAT

Organic reactions involve breaking some covalent bonds and forming others. These tend to be slower than the reactions between ions, common in inorganic chemistry, so heat is often needed to break bonds and to speed things along generally. A wide range of reactions in which heat plays an essential part can be performed elegantly by the 'wet asbestos' method: for example, the bond between carbon and iodine in methyl iodide can be broken and two methyl groups joined together to form ethane as follows.

EXPERIMENT NO. 1: Pour methyl iodide into a test-tube to a depth of 1 in. Push asbestos wool into the liquid with a glass rod until the liquid is soaked up. Set up the test-tube in a clamp and introduce about 1 g. of copper turnings half-way along the tube so that it forms a small heap. Fit the test-tube with a cork and delivery tube, as shown in Fig. 37, and heat the copper turnings with a gentle flame (the flame should be just yellow at the tip before placing the bunsen burner under the tube). A few test-tubes of ethane can be collected by displacements of water:

$$2CH_3I + 2Cu = CH_3-CH_3 + 2CuI$$

Similarly, butane may be made by replacing the methyl iodide with ethyl iodide.

297

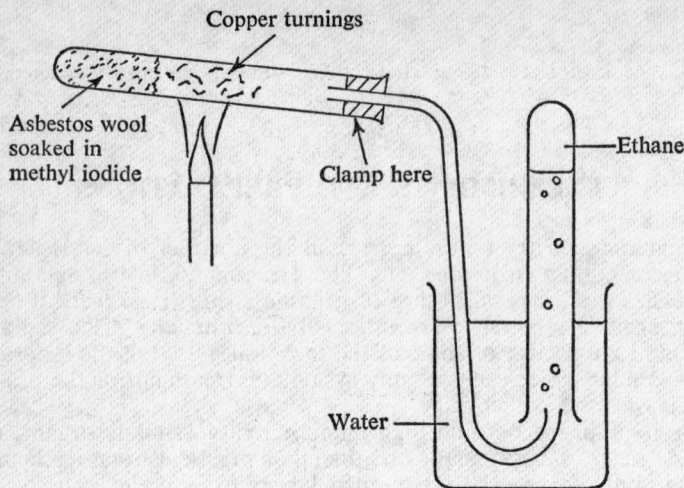

Fig. 37

The elements of water may be removed catalytically from ethanol (rectified spirit) soaked in the asbestos wool, to form ethylene, by using alumina as the solid reagent:

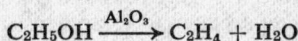

$$C_2H_5OH \xrightarrow{Al_2O_3} C_2H_4 + H_2O$$

Paraffin oil soaked in the asbestos can be 'cracked' catalytically by using chips of porous pot, packing 3 in. of the tube as the solid reagent.

If fresh soda-lime is used as the solid reagent glacial acetic acid can be converted to methane and ethylene dibromide to acetylene:

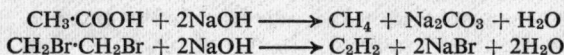

$$CH_3 \cdot COOH + 2NaOH \longrightarrow CH_4 + Na_2CO_3 + H_2O$$
$$CH_2Br \cdot CH_2Br + 2NaOH \longrightarrow C_2H_2 + 2NaBr + 2H_2O$$

Several of these liquid reagents are volatile and highly inflammable, so no flame should be burning anywhere near where they are being poured out. Glacial acetic acid is corrosive to the skin, so any splashes should be washed off immediately.

TESTS which can be applied to all the gases produced:

(i) Apply a lighted splint to the mouth of the tube in which the gas was collected. Note the appearance of any flame. Test the residual gas for carbon dioxide by pouring in lime-water and shaking. Do not try to light the gas emerging from the delivery tube: it may have air mixed with it, and the apparatus would then blow up.

(ii) Tests for unsaturation: (*a*) shake the gas with dilute (pale pink) potassium permanganate solution and sodium carbonate solution; (*b*) shake the gas with bromine water or a solution of bromine in carbon tetra-chloride. Note any colour changes (cf. pages 42 and 43). N.B. Pure bromine is a pungent, corrosive, dangerous liquid. Avoid the pure liquid if you are working on your own and avoid contact with its solutions in water or carbon tetrachloride.

REFLUXING

A special technique is needed for heating volatile reaction mixtures for long periods if the reactants are not to be lost altogether: this is the process of refluxing, whereby the vapour is continually condensed and run back into the reaction vessel. As an example try the preparation of soap from an animal fat or a vegetable oil (cf. page 98).

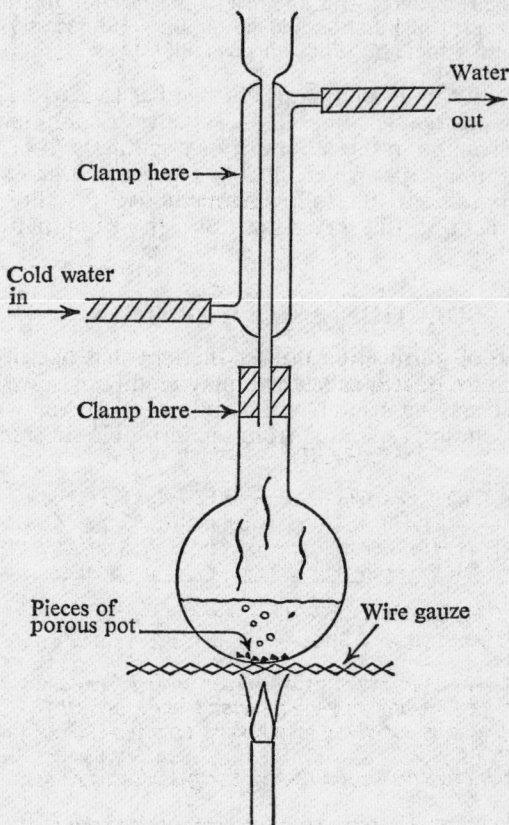

Fig. 38

EXPERIMENT NO. 2: Into a 250-ml. round-bottomed flask pour 20 g. olive oil, 50 ml. of 10% aqueous sodium hydroxide solution, and about 40–50 ml. industrial methylated spirit (**CAUTION—INFLAMMABLE**). Drop in two or three pieces of porous pot to facilitate smooth boiling, then cork in a water-cooled condenser as shown in Fig. 38. Run cold water through the condenser cooling jacket and boil the liquid in the flask gently at first, over a small flame, being careful to avoid undue frothing, until all the oil has disappeared and a golden, homogeneous liquid remains.

The flask now contains a soap and glycerol in aqueous alcoholic caustic soda:

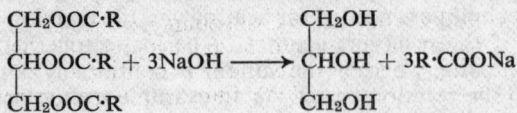

$$\begin{array}{l} CH_2OOC \cdot R \\ | \\ CHOOC \cdot R \\ | \\ CH_2OOC \cdot R \end{array} + 3NaOH \longrightarrow \begin{array}{l} CH_2OH \\ | \\ CHOH \\ | \\ CH_2OH \end{array} + 3R \cdot COONa$$

Separation of the soap from this is one example of the general problem of separating wanted products from the rest of the reaction mixture. This time we use two processes: distillation and salting out, and necessary instructions will be found under these headings on page 306.

One of the first things an organic chemist has to learn is the methods of purification of organic compounds. Whether a compound is derived from nature or made in the laboratory by synthesis, it is usually contaminated with impurities which must be removed. Organic compounds are generally purified by the following methods: (i) solution and filtration; (ii) extraction; (iii) crystallization; (iv) sublimation; and (v) distillation.

SOLUTION AND FILTRATION

This method of purification utilizes the fact that one of the components of a mixture of two substances may readily dissolve in a solvent while the other may be insoluble. The mixture is shaken with a solvent and the clear solution separated from the insoluble substance either by

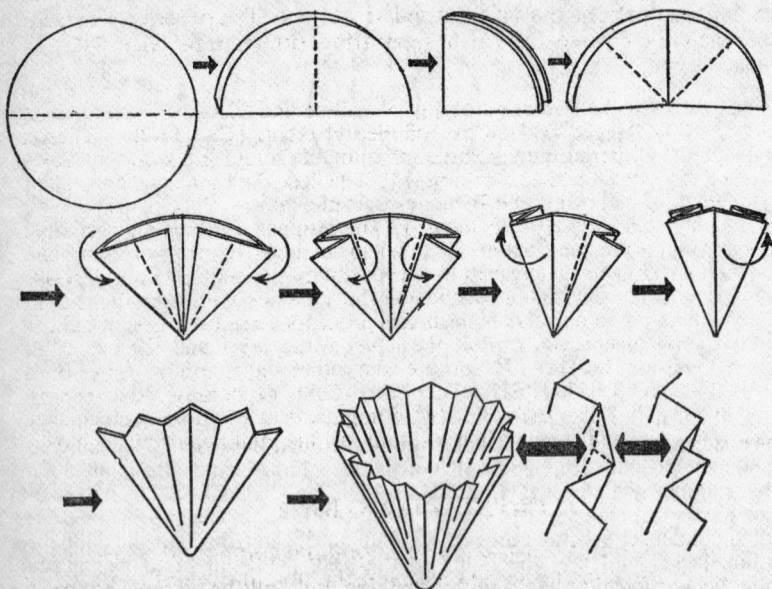

Fig. 39. Steps in Making Fluted Filter-paper.

pouring off, that is, decanting the supernatant layer, or by filtration. Evaporation of the solution will then furnish the soluble substance.

EXPERIMENT NO. 3: To a mixture of 0·5 g. of common salt and 0·5 g. of benzoic acid add about 15 ml. of water. Agitate thoroughly and filter through a glass funnel with a filter paper folded in quarters (9 cm. diameter), receiving the filtrate in a small clean beaker. Evaporate filtrate to dryness either by setting aside at room temperature for a couple of days or by heating over a water-bath. Heat a pinch of the insoluble substance on a spatula over a small flame. The substance first melts, then burns with a smoky flame and disappears leaving no residue, which indicates that it is the organic compound benzoic acid. Repeat the experiment with the substance recovered by evaporation. You will observe that it cannot be melted or ignited, which indicates that it is salt.

EXTRACTION

The process of extraction is based on the principle that a substance distributes itself between a pair of solvents which are immiscible or only slightly miscible with each other in such a way that the ratio of its concentrations in the two solvents is a constant at constant temperature. Usually water is one of the solvents, and the other is an organic solvent like diethyl ether, benzene, chloroform, etc. The choice of the solvent used to extract an organic substance from an aqueous suspension or solution depends upon the solvent's immiscibility with water, its ability to dissolve the organic substance and its ease of removal from the extracted organic material, that is its boiling point. The extraction will be more efficient if a given volume of solvent is used in lots of two or more smaller portions than if it is used all at once. The process of extraction is usually carried out in a separating funnel (see page 312 for illustration).

EXPERIMENT NO. 4: To a mixture of 5 g. of aniline and 50 ml. of water, taken in a separating funnel, add 30 ml. of diethyl ether. [CAUTION: Diethyl ether (b.p. 35° C.) is a highly volatile inflammable liquid. Be sure that there is no free flame around when working with it.] Close the funnel with the glass stopper and shake thoroughly. Be careful when shaking, since pressure may develop within the funnel and blow off the stopper, spilling the contents. To avoid this, hold the stoppered funnel in an inverted position with one hand holding the stopper in position and the stem inclined upwards. Release any pressure inside the funnel by opening the stopcock with the other hand. Close the stopcock and shake. Repeat this procedure several times and allow the two layers to separate. Collect the lower water layer and the top ether layer in separate beakers. Evaporate the ether layer on a water-bath (CAUTION: NO FLAME SHOULD BE PRESENT because ether is very volatile and highly inflammable) or by setting aside at room temperature for a few hours. You will observe that a liquid residue, aniline, is left behind.

CRYSTALLIZATION

The best method of purifying a solid organic compound is by means of crystallization. The process consists of the preparation of a saturated solution of a compound in a suitable solvent at an elevated temperature, filtering while hot to remove undissolved impurities, and allowing the compound to crystallize out on cooling. A suitable solvent for crystallizing a compound should not react with the substance to be purified, should dissolve the substance easily at high temperatures but only poorly at a low temperature, and should be sufficiently volatile to permit its easy removal from the crystallized substance. Several organic solids are coloured due to contamination with coloured impurities. These are usually removed by treatment of a solution of the hot substance with animal charcoal.

The filtration of a solution before crystallization can be accomplished through a glass funnel with a fluted filter-paper or better by suction. A fluted filter-paper is made from a circular filter-paper by folding it as indicated in Fig. 39 and creasing along the fold lines.

Suction filtration is rapid and is carried out with the aid of a filter pump connected by rubber tubing to a filter flask. The latter is attached to a Buchner funnel by means of a one-holed rubber stopper, as in the diagram (Fig. 40). Two filter papers of the right diameter to just fit in the Buchner funnel are used.

In boring a rubber stopper both the stopper and the borer of the required size are lubricated with some glycerol. The stopper is held firmly on a bench with one hand, while the borer is driven into the stopper from a vertical position with the other hand. No attempt should be made to apply too much pressure, since this results in an uneven hole with a small diameter. To bore a bark cork, first wet it with water and soften it in a cork press or roll it between your hands. The borer is driven halfway through the cork and then removed. The cut pieces inside the borer are pushed out, and the rest of the hole is bored from the other side by properly centring the borer.

EXPERIMENT NO. 5: Weigh about 4 g. of impure acetanilide in a 250-ml. beaker, add 125 ml. of water and heat to boiling, while stirring with a glass rod when a clear solution is obtained. Add 0·5 g. of animal charcoal, continue boiling for another 5 minutes and filter while hot through either a funnel fitted with a fluted paper or better by suction filtration. In the latter case, wet the filter-papers with some hot water, apply suction to keep the filter-

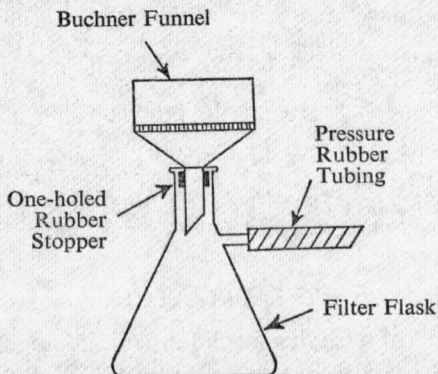

Fig. 40. Apparatus for Suction Filtration.

paper in position, reject this water, reassemble the set-up, turn on suction, and filter the hot solution of acetanilide. If any charcoal seeps into the filtrate, heat the filtrate to dissolve any substance that may have crystallized out and filter again. Transfer equal portions of the filtrate to two separate beakers. Set one aside to cool slowly by itself to room temperature and cool the other beaker in an ice-bath. You will observe that slow cooling gives bigger crystals than rapid cooling.

After crystallization is complete, collect crystals from both the beakers on a Buchner funnel by suction filtration. Wash with 10–20 ml. of cold water and press crystals down on the filter-paper with the flat side of a glass stopper to remove adhering water as much as possible. Transfer the crystals to a filter-paper, spread out the crystals, and leave for drying overnight. When completely dry, find the weight and calculate the percentage of purified substance recovered from impure material.

FRACTIONAL CRYSTALLIZATION

A mixture of two solid substances can very often be separated into pure components by fractional crystallization from a solvent in which they dissolve, but to unequal extents. The difference in solubilities must be appreciable; otherwise the process of separation will be tedious. The mixture is treated with the solvent either at room temperature or at a higher temperature and filtered. The undissolved portion is mostly the more insoluble component. The filtrate on cooling deposits, depending on the difference in solubilities of the components, more of the insoluble substance, or the soluble substance, or a mixture of both. At any rate, the filtrate from this crop will be richer in the soluble component, which can be obtained by concentration and cooling. The more insoluble and the more soluble portions thus obtained are subjected separately and repeatedly to the above process of solution and crystallization until pure fractions are obtained. The purity of each crop of crystals may be tested by a melting-point determination (see Experiment No. 11) or by examination of the crystals under a microscope to see if the two substances have different crystalline forms.

EXPERIMENT NO. 6: Mix 3 g. of cinnamic acid intimately with 3 g. of benzoic acid and place the mixture in a 250-ml. beaker. Add 75 ml. of water heated to 50° C., heat on a water-bath, and stir for 10 minutes while maintaining the temperature at 50° C. Allow the undissolved portion to settle down and filter by suction while the solution is still warm. Repeat extraction of the undissolved portion similarly five times. Crystallize the insoluble portion remaining after the last extraction from boiling water and collect crystals. Combine filtrates from the various extractions and evaporate on a water-bath to about 80 ml. Cool and collect the crystals thus obtained. The melting points of the soluble and the insoluble substances to be determined as described in Experiment No. 11 will be 121° C. (benzoic acid) and 133° C. (cinnamic acid) respectively.

SUBLIMATION

Some solids, when heated, pass directly into the vapour state without melting to a liquid and, by suitable arrangement, the vapours can be condensed to give back the original solid in a purer condition. This process is known as sublimation.

304 Organic Chemistry Made Simple

EXPERIMENT NO. 7: Place 4 g. of crude *p*-quinone on the bottom of a beaker (250 ml.), cover the beaker with a watch-glass and heat on a water-bath for a few hours. Beautiful yellow needles of pure *p*-quinone collect on the bottom of the watch-glass. Repeat the experiment with camphor or naphthalene.

DISTILLATION

The foregoing methods of purification refer to substances which are solids at room temperature. For liquids, the universal method of purification is by distillation. This process involves conversion of a liquid to the vapour state by the application of heat followed by condensation of the vapours to a liquid which is collected in a vessel other than the one used for vaporization. The temperature at which a liquid distils is a constant at a given pressure, known as its boiling point (b.p.), and serves as a criterion of purity.

The distillation of a simple homogeneous liquid is usually carried out in the set-up shown in Fig. 41.

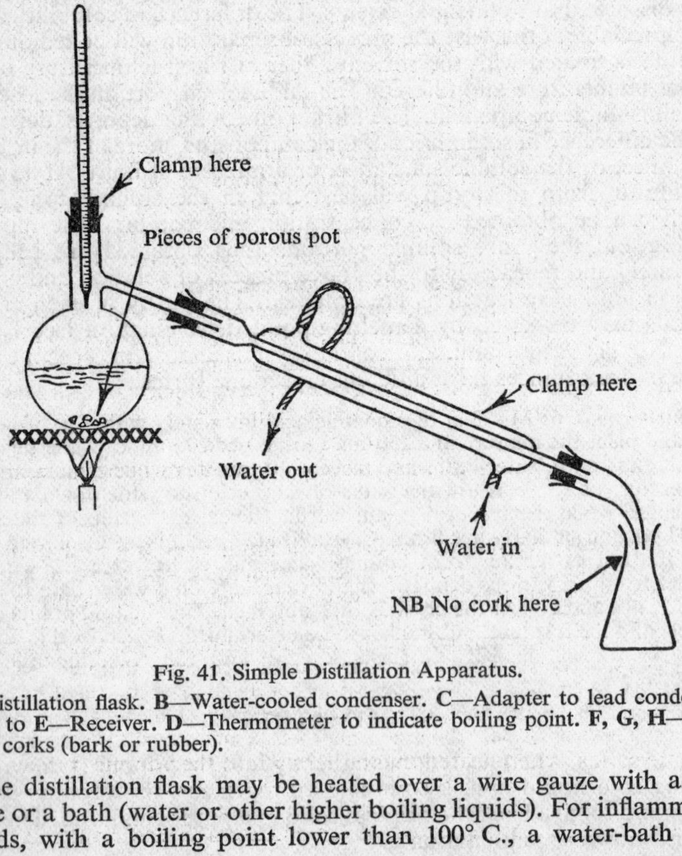

Fig. 41. Simple Distillation Apparatus.

A—Distillation flask. B—Water-cooled condenser. C—Adapter to lead condensed liquid to E—Receiver. D—Thermometer to indicate boiling point. F, G, H—One-holed corks (bark or rubber).

The distillation flask may be heated over a wire gauze with a free flame or a bath (water or other higher boiling liquids). For inflammable liquids, with a boiling point lower than 100° C., a water-bath or a steam cone or electrical heating is employed. It is common practice to add one or two porcelain bits to the liquid to be distilled. This prevents bumping of the liquid during distillation. For liquids with a boiling point above 150° C. there is no need to circulate water through the condenser, since air cooling is effective enough to condense the vapours of such liquids.

In the laboratory the need often arises to separate a mixture of two or more liquids. This can be done satisfactorily enough in the set-up described above if the components have boiling points widely different from one another. For instance, a mixture of benzene (b.p. 80°). [CAUTION—HIGHLY INFLAMMABLE] and nitrobenzene (b.p. 211°) can be separated by distilling the mixture slowly in the above set-up and collecting the distillates separately. If the components have similar boiling points a device known as a fractionating column is used, and the distillation is then referred to as fractional distillation. The principle underlying fractional distillation is essentially as follows:

When a liquid mixture is volatilized a mixture of vapours is obtained which is richer in the more volatile component, that is the substance with the lower boiling point. This vapour mixture on partial condensation gives a liquid mixture richer in the less-volatile component. Partial vaporization and partial condensation then, if it can be effected successively, will give distillates richer in the more-volatile components in the earlier fractions and richer in the less-volatile component in the later fractions. A fractionating column usually consists of a vertical tube containing a large number of indentations or bulbs or filled with glass beads or helices. It, therefore, offers a large cooling surface to the ascending vapours, which are at the same time brought into intimate contact with the condensing liquid flowing downwards. This intimate contact results in exchange between the ascending vapours and the descending liquid whereby the vapour is enriched in the volatile component and the returning liquid in the less-volatile component. Different types of fractionating columns are known; a simple type is known as a Vigreaux column. It consists of a vertical tube with pairs of indentations made from opposite sides, consecutive pairs being in planes at right angles to each other. The indentations are so arranged that the ascending vapours cannot pass directly through without meeting at least one indentation.

EXPERIMENT NO. 8: Because of the danger of fire if there is a mishap during distillation the apparatus shown in Fig. 41 should be assembled in a metal tray containing sand sufficient to retain all the liquid being distilled. Place 15 ml benzene and 20 ml nitrobenzene along with two or three pieces of unglazed porcelain (porous pot) in the 250 ml distilling flask A. Replace the thermometer in position and start circulating tap water through the condenser. Heat the flask over a wire gauze with a free flame until the distillation commences. Adjust the heating so that the distillate collects at the rate of 2 drops per second. Reject the first 5 ml. of distillate and collect the fraction coming over at a constant temperature. This will be the benzene (b.p. 80° C.). When the temperature again begins to rise rapidly, stop heating and extinguish the burner. Replace the receiver E containing benzene with a clean one. Drain the water from the cooling jacket of the condenser B and resume heating.

Reject any distillate coming over below 207° C. and collect the fraction coming over between 207 and 211° C. This will be the nitrobenzene. Do not distil to dryness but leave some residual liquid (5–10 ml.) in the distilling flask.

EXPERIMENT NO. 9: Use the same apparatus and fire precautions as in experiment no. 8 to try and separate a mixture of 10 ml. petrol and 10 ml. paraffin into fractions. The first few drops of distillate will probably appear at about 54° C. With this simple apparatus the boiling point will rise steadily as distillation proceeds and one can only divide the distillate arbitrarily into fractions boiling over short ranges, e.g. 50–70°, 70–90°, 90–110°, etc. For better separation a fractionating column should be fitted and well lagged to reduce heat losses.

EXPERIMENT NO. 2 CONTINUED: Transfer the liquid containing the soap to a 250-ml. distillation flask and add two or three pieces of porous pot. Fit with a thermometer, condenser, and rubber stoppers as shown in Fig. 41. Run cold water slowly through the cooling jacket and distil off the industrial spirit at 74–81° C. (*N.B.* The distillate you collect is volatile and inflammable.) Pour the residual liquid from the distillation flask into a beaker. On cooling the soap appears as a colloidal brownish gel on the surface.

SALTING OUT

EXPERIMENT NO. 2 COMPLETED: On adding a saturated solution of common salt, with stirring, the soap separates as a whitish solid. (These curds can be filtered through a Buchner funnel, washed with distilled water, and put aside to set.)

STEAM DISTILLATION

Many organic substances which have high boiling points or boil only with decomposition can be volatilized in a current of steam. The boiling point of any liquid is the temperature at which the vapour pressure equals the atmospheric pressure. Now the vapour pressure of a mixture of two immiscible liquids is the sum of the separate vapour pressures of the components, and the boiling point of the mixture will consequently be lower than that of either component. For example, aniline and water are a pair of immiscible liquids whose vapour pressures are 43 mm. and 717 mm. respectively, at about 98·5°. The sum of the vapour pressures is 760 mm., which is the atmospheric pressure. Aniline will distil over with water when a mixture of both is heated to 98·5° C. According to elementary physics, the partial vapour pressures are proportional to the number of moles of each component in the vapour state, that is at 98·5°. See below.

$$\frac{\text{Moles of aniline}}{\text{Moles of water}} = \frac{43}{717}$$

The ratio of the actual weights of aniline to the water distilling with it =

$$\frac{\text{Moles of aniline} \times \text{molecular weight of aniline}}{\text{Moles of water} \times \text{molecular weight of water}} = \frac{43}{717} \times \frac{93}{18} = 0.3099$$

In other words, about 31 g. of aniline will distil over with 100 g. of water. Thus the higher molecular weight of aniline as compared with that of water makes possible the distillation of larger amounts of aniline than the partial pressures would indicate. Steam distillation is employed in the perfumery industry for isolation of the volatile essences of plants.

EXPERIMENT NO. 10: Set up apparatus shown below:

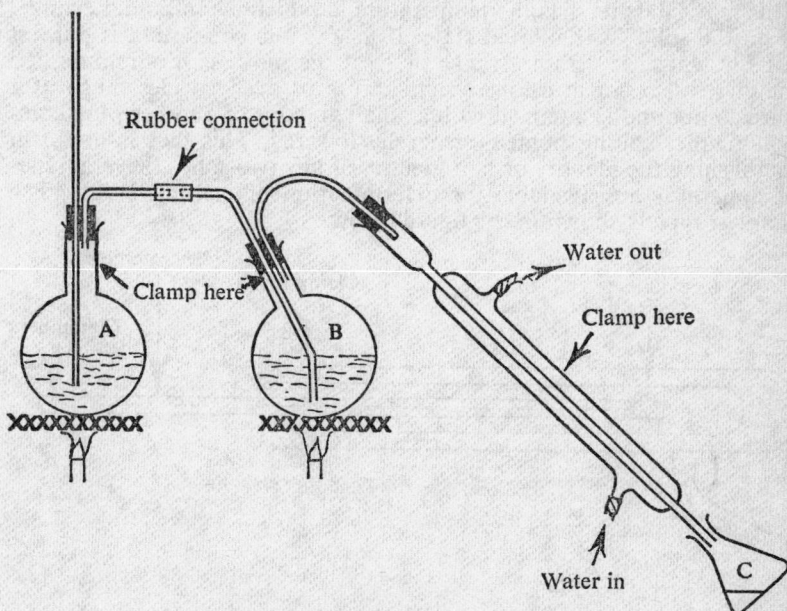

Fig. 42. Steam Distillation Apparatus.

A is a 500-ml. round-bottomed flask fitted with a two-holed rubber stopper carrying a safety tube and a right-angled bent tube which serves as a steam generator. B is another 500-ml. round-bottomed flask fitted with a two-holed stopper carrying a long bent tube for delivery of steam and a shorter one for leading the vapours to the condenser. The longer tube is connected to the steam generator by means of a rubber tubing, and the shorter one is connected to the condenser by means of a rubber stopper. C is the receiver for the distillate and is a 250-ml. conical flask. Place 10 g. of crude aniline and 50 ml. of water in flask B and heat to just near the boiling point and then lead in steam from the steam generator A. Preliminary heating of B avoids unnecessary condensation of steam and an increase in bulk of contents. When no more oily drops collect in the distillate, disconnect the steam generator with care and stop the current of steam. The oil which separates at the bottom of C is aniline, which you may isolate and free from water, if you so desire. For this, extract the distillate with 50 ml. of ether using a separating funnel as described in Experiment No. 4, and then dry the extract by agitating with 5 g. of anhydrous magnesium sulphate for 15 minutes. Fllter and evaporate the filtrate on a previously heated water-bath (**CAUTION: NO FLAME SHOULD BE PRESENT** because ether is very volatile and highly inflammable). Distil residual liquid as in Experiment No. 8. Aniline distils at 182° C. and, therefore, there is no need to pass water through the condenser.

DETERMINATION OF MELTING POINTS

The criterion of purity of a solid organic substance is its melting point. This may be defined as the temperature at which a compound changes from the solid to the liquid state. If an organic compound is pure, it usually will melt within a range of 1° at the most. If it is impure, the melting will occur at a lower temperature or over a wider range. If a pure compound is admixed with a small amount of another pure compound, the melting point is generally lowered. This fact is useful in establishing the identity or non-identity of any two solids. Several types of apparatus are available for determination of melting points. The simplest type is shown in the figure below.

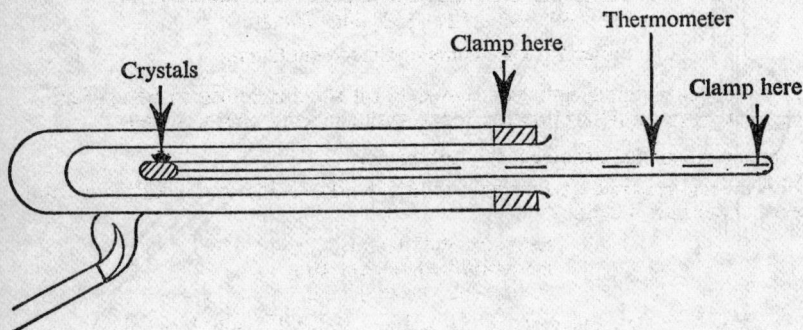

Fig. 43

A $\frac{1}{2}$ in. diameter test tube is held in place inside a 1 in. diameter boiling tube by either a ring of cork or two or three turns of asbestos cord. Put a few crystals of the dry solid onto the bulb of a thermometer with a suitable temperature range, mounted horizontally so that the melting point apparatus can be slid over it (see Fig. 43). 'Stroke' the boiling tube gently and regularly with a bunsen flame about 1 in. tall so that the temperature of the thermometer rises slowly. When approaching the melting point of the solid, the temperature should only rise by about two degrees per minute—the motion of the mercury thread will hardly be apparent. The commonest mistake made with this apparatus is to heat too quickly. A slightly impure solid nearly always melts at a lower temperature than the pure material.

EXPERIMENT NO. 11: Determine the melting points of pure samples of naphthalene (80·2° C.), benzoic acid (122° C.), salicylic acid (157° C.), and phthalimide (238° C.). From the observed melting points and those given within the parentheses, calculate the thermometer corrections and plot a correction curve for your thermometer, as follows:

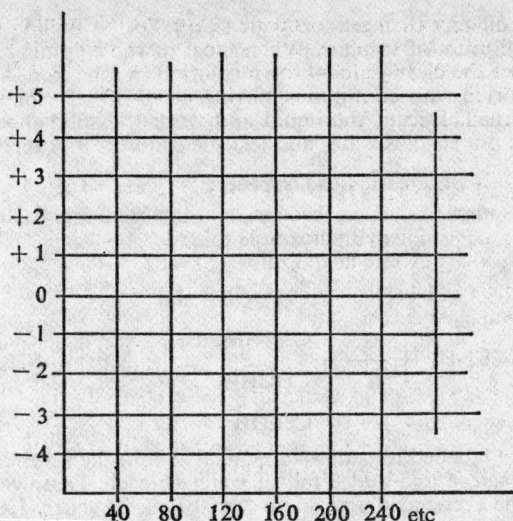

Fig. 44. Thermometer Correction Curve.

With the aid of this graph, you can read off the correction to be applied to any melting-point determination made with the same thermometer.

EXPERIMENT NO. 12: **Iodoform** (an antiseptic):

Chemicals required (common chemicals, like acids and alkalis, are not listed under this heading):

Ethyl alcohol 10 ml.
Potassium carbonate 10 g.
Iodine (pulverized) 10 g.

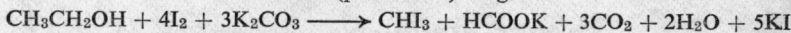

$$CH_3CH_2OH + 4I_2 + 3K_2CO_3 \longrightarrow CHI_3 + HCOOK + 3CO_2 + 2H_2O + 5KI$$

Dissolve the potassium carbonate in 40 ml. of water taken in a 125-ml. conical flask and add the ethyl alcohol. Heat the mixture gradually to 75–80° C. on a water-bath, since ethyl alcohol is inflammable, add the iodine in small portions at a time and stir. When the colour of the iodine is discharged, cool and collect the yellow crystals by suction filtration. Wash the crystals with water. Redissolve by heating with the minimum amount of ethyl alcohol, (about 25–30 ml.), again using a water-bath and add water dropwise until the solution becomes turbid—but avoid an excess and let the solution cool. Filter the crystals and dry them by pressing them between filter-papers. The crystals (2–3 g.) are lemon yellow in colour and melt at 119° C.

EXPERIMENT NO. 13: **Thiokol** (a synthetic rubber):

Chemicals required: Sulphur 5 g.
Sodium hydroxide 2 g.
Ethylene dichloride 10 ml.

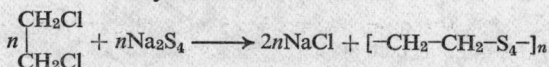

$$n \begin{matrix} CH_2Cl \\ | \\ CH_2Cl \end{matrix} + nNa_2S_4 \longrightarrow 2nNaCl + [-CH_2-CH_2-S_4-]_n$$

Dissolve the sodium hydroxide pellets in 50 ml. boiling water in a beaker. (*N.B.* Sodium hydroxide is caustic: do not touch the pellets with your fingers. Wash off any that gets on your skin.) To the boiling solution add carefully 5 g. flowers of sulphur. When most of the sulphur has dissolved, filter and collect the clear yellow-brown solution in a beaker. Add 10 ml. of ethylene dichloride with continuous stirring at 70–80° C. Spongy lumps of rubber are formed. Decant the liquid and work the solid thoroughly with water. Squeeze out surplus water and test the solid for its rubber properties.

EXPERIMENT NO. 14: **Formation of Osazones:**

Chemicals required: Glucose, fructose, and lactose 0·5 g. each
Phenylhydrazine 3 ml.
Acetic acid 3 ml.

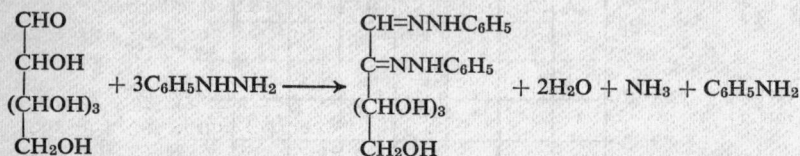

$$
\begin{matrix}
CHO \\
| \\
CHOH \\
| \\
(CHOH)_3 \\
| \\
CH_2OH
\end{matrix}
\; + 3C_6H_5NHNH_2 \longrightarrow \;
\begin{matrix}
CH{=}NNHC_6H_5 \\
| \\
C{=}NNHC_6H_5 \\
| \\
(CHOH)_3 \\
| \\
CH_2OH
\end{matrix}
\; + 2H_2O + NH_3 + C_6H_5NH_2
$$

Place in three labelled test-tubes (20 cm. × 2 cm.) the samples of glucose, fructose, and lactose and add 6 ml. of water to each. Dissolve the phenyl-hydrazine in the acetic acid and dilute with 10 ml. of water. Heat a 600-ml. beaker half full of water to boiling; while the water is boiling add 3 ml. of the phenylhydrazine reagent to each of the test-tubes, cork them loosely, and place them immediately in the boiling water. Note the time required from this instant for the appearance of each of the osazones. Continue heating for another half an hour. Allow solutions to cool, filter the yellow crystals, and find their melting points. Glucose and fructose give the same osazone (m.p. 205° C.), while the osazone of lactose melts at 200° C. with decomposition. Examine the osazones for their crystalline forms under a microscope, if available. Glucosazone crystallizes in the form of sheaves of corn, while lactosazone crystallizes in clusters of thin needles. Fructose gives the osazone in 2 minutes, glucose in 4–5 minutes, and lactose takes as much as 2 hours to give the osazone.

ESSENTIAL ITEMS FOR EXPERIMENTS

The list below includes some of the items which you will need to carry out experiments in organic chemistry. Some of the items are illustrated on page 312.

1 Beaker, 100 ml.
1 Beaker, 250 ml.
1 Beaker, 400 ml.
3 Bottles, glass stoppered, 250 ml.
2 Bunsen burners.
3 Clamps, iron.
1 Condenser, Liebig.
12 Corks (bark), different sizes.
1 Cork-borer set.
1 Cylinder, graduated, 100 ml.
1 Cylinder, graduated, 10 ml.
1 Dish (porcelain), 7·5 cm. diameter.
1 Distilling adapter.
2 Droppers, medicine.
1 File, triangular, iron.
2 Packages filter-paper, circular (one to fit Buchner funnel).
1 Filter pump (glass or metal).
1 Flask, conical, 50 ml.
1 Flask, conical, 120 ml.
1 Flask, conical, 250 ml.
1 Flask, conical, 500 ml.
1 Flask, distilling, 250 ml.
1 Flask, distilling, 50 ml.
1 Flask, filter, 500 ml.
1 Flask, flat bottom, 250 ml.
1 Flask, flat bottom, 100 ml.
1 Flask, round bottom, 250 ml.
2 Flasks, round bottom, 500 ml.

1 Fractionating column, 25 cm.
1 Funnel, Buchner, 8 cm. diameter.
1 Funnel, glass, 7·5 cm. diameter.
1 Funnel, separating, 250 ml.
2 Litmus (red and blue) books.
Matches.
1 Pestle and mortar.
2 Rings, iron, 10 cm. diameter.
2-ft of Rod, glass, 8 mm. diameter.
1 Spatula (metal).
1 Packet of wooden splints.
3 Stands, iron.
12 Stoppers, rubber (different sizes).
12 Test-tubes, soft glass, 15 cm. long.
3 Boiling tubes.
1 Test-tube brush.
1 Test-tube holder.
1 Test-tube stand.
6 Sample tubes, 1·5 cm × 4 cm.
1 Thermometer, 0–360° C.
1 Thermometer, 0–110° C.
2-ft of Tubing, glass, 8 mm.
9-ft of Tubing, rubber, 7 mm.
3-ft of Tubing, rubber, pressure.
2 Watch glasses, 7·5 cm. diameter.
1 Water-bath.
1 Wire gauze.

Flat Bottomed Flask

Filter Flask

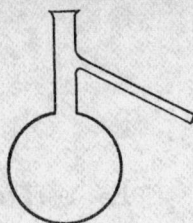
Distilling Flask

Round Bottomed Flask

Conical or Erlenmeyer Flask

Filter Funnel

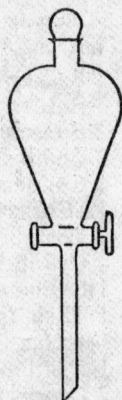
Separating Funnel

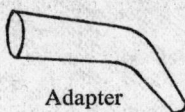
Adapter

Buchner Funnel

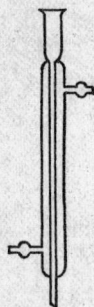
Water Condenser

Porcelain Dish

Beaker

Measuring Cylinder

Spatula

Fig. 45. Some Common Laboratory Items.

APPENDIX C

SUGGESTED FURTHER READING

Introductory Texts

Bezzant, *Basic Organic Chemistry*. McGraw-Hill: London, 1967.

Conrow and McDonald, *Deductive Organic Chemistry*. Addison-Wesley: London, 1966.

Cremlyn, *A Concise Organic Chemistry*. Pitman: London, 1966.

English and Cassidy, *Principles of Organic Chemistry*. McGraw-Hill: London, 1961.

Fieser and Fieser, *Basic Organic Chemistry*. Harrap: London, 1959.

Hart and Schuetz, *Organic Chemistry*. IUE—TABS-Educational: New York, 1967.

Thomas, *A Mechanistic Approach to Organic Chemistry*. Arnold: London, 1967.

Advanced Texts

Brewster and McEwen, *Organic Chemistry*. Prentice-Hall: London, 1961.

Conant and Blatt, *Chemistry of Organic Compounds*. Collier-Macmillan: London, 1966.

Corwin and Bursey, *Introductory Organic Chemistry*. Addison-Wesley: London, 1966.

Cram and Hammond, *Organic Chemistry*. McGraw-Hill: London, 1967.

Ferguson, *Textbook of Organic Chemistry*. Van Nostrand: London, 1965.

Fieser and Fieser, *Organic Chemistry*. Reinhold: London, 1956.

Finar, *Organic Chemistry*. Vols. I and II. Longmans: London, 1964.

Noller, *Chemistry of Organic Compounds*. W. B. Saunders: London, 1965.

Packer and Vaughan, *Organic Chemistry*. Oxford: London, 1958.

Schmidt, *Organic Chemistry*. (Ed. Campbell.) Oliver and Boyd: London, 1964.

Turner and Harris, *Organic Chemistry*. Longmans: London, 1960.

Wheland, *Advanced Organic Chemistry*. Wiley: London, 1960.

Applied Organic Chemistry

Astle, *Chemistry of Petrochemicals*. Reinhold: London, 1956.

J. A. Kent, *Riegel's Industrial Chemistry*. Reinhold: London, 1962.

Laboratory Manuals

Adams and Johnson, *Elementary Laboratory Experiments in Organic Chemistry*. Macmillan: New York.

Adams and Johnson, *Laboratory Experiments in Organic Chemistry*. Collier-Macmillan: London, 1963.

Cason and Rapoport, *Laboratory Text in Organic Chemistry*. Prentice-Hall: London, 1962.

Fieser, *Experiments in Organic Chemistry*. Harrap: London, 1964.

Mann and Saunders, *Introduction to Practical Organic Chemistry*. Longmans: London, 1964.

Mann and Saunders, *Practical Organic Chemistry*. Longmans: London, 1960 (4th edn.).

Sabel, *Basic Techniques of Preparative Organic Chemistry*. Pergamon Press: London.

Waddington and Finlay, *Organic Chemistry through Experiment*. Mills and Boon: London, 1965.

APPENDIX D

ANSWERS TO PROBLEM SETS

Problem Set No. 1

1. (a) The molecular weight is $4 \times 12 + 8 \times 1 + 2 \times 16 = 88$.

% Carbon $= \dfrac{48}{88} \times 100$ or $54 \cdot 5$; % Hydrogen $= \dfrac{8}{88} \times 100$ or $9 \cdot 1$;

% Oxygen $= \dfrac{32}{88} \times 100$

or $36 \cdot 4$ (values corrected to the first decimal place).

(b) Molecular weight is $4 \times 12 + 9 \times 1 + 1 \times 35 \cdot 5 = 92 \cdot 5$.

% Carbon $= \dfrac{48}{92 \cdot 5} \times 100$ or $51 \cdot 9$; % Hydrogen $= \dfrac{9}{92 \cdot 5} \times 100$ or $9 \cdot 7$;

% Chlorine $= \dfrac{35 \cdot 5}{92 \cdot 5} \times 100$ or $38 \cdot 4$.

2. The percentages add up to 100 and hence there is no oxygen in the compound. Ratio of the number of atoms of carbon, hydrogen and nitrogen in the molecule $= \dfrac{53 \cdot 34}{12} : \dfrac{15 \cdot 56}{1} : \dfrac{31 \cdot 1}{14} = 4 \cdot 445 : 15 \cdot 56 : 2 \cdot 2214 = 2 : 7 : 1$ (corrected to the nearest whole numbers). Hence the empirical formula is C_2H_7N.

3. Molecular weight of the gas $=$ weight of 22,400 ml. at 760 mm. and $0°$ C. $= \dfrac{0 \cdot 1977}{100} \times 22{,}400 = 44 \cdot 3$.

4. Percentage of oxygen in the compound $= 100 - (32 \cdot 9 + 4 \cdot 1 + 19 \cdot 25)$ or $43 \cdot 75$. Ratio of number of atoms of carbon, hydrogen, nitrogen and oxygen $= \dfrac{32 \cdot 9}{12} : \dfrac{4 \cdot 1}{1} : \dfrac{19 \cdot 25}{14} : \dfrac{43 \cdot 75}{16} = 2 : 3 : 1 : 2$ (correct to nearest whole numbers). $\therefore$ Empirical formula $= C_2H_3NO_2$; let molecular formula be $(C_2H_3NO_2)_x$. $\therefore$ Molecular weight $= (73)_x = 146$. $\therefore x = 2$. Molecular formula is, therefore, $C_4H_6N_2O_4$.

Problem Set No. 2

1. (a) CH₃–CH₂–CH₂–CH₂–CH₂–CH₃   *n*-Hexane.

(b) CH₃–CH₂–CH₂–CH–CH₃   2-Methylpentane.
                    |
                    CH₃

(c) CH₃–CH₂–CH–CH₂–CH₃   3-Methylpentane.
               |
               CH₃

(d) CH₃–CH–CH–CH₃   2,3-Dimethylbutane.
         | |
       CH₃ CH₃

       CH₃
       |
(e) CH₃–C–CH₂–CH₃   2,2-Dimethylbutane.
      |
      CH₃

2. (a) Correct name is 2,2,3-trimethylhexane. (b) Correct name is 2-methylbutane. (c) Correct name is 2-methylpentane. (d) Correct name is 2,2,3-trimethylbutane.

3. A mixture of butane, propane, and ethane.

4. (*a*) 1,2-Dibromoethene.　　　　(*b*) 2,2-Dichloropropane.

5. (*a*) Ethanol $\xrightarrow{\text{H}_2\text{SO}_4}$ ethylene $\xrightarrow{\text{Br}_2}$ ethylene dibromide $\xrightarrow[\text{KOH}]{\text{Alcoholic}}$ acetylene.

　　(*b*) Ethanol $\xrightarrow[\text{(as in (a))}]{}$ acetylene $\xrightarrow[\text{(i) SbCl}_5]{\text{2HBr}}$ 1,1-dibromoethane.

　　(*c*) Ethylene $\xrightarrow[\text{(as in (a))}]{}$ acetylene $\xrightarrow[\text{(ii) distil.}]{}$ 1,1,2,2-tetrachloroethane.

Problem Set No. 3

1. (*a*) ClCH$_2$–CH$_2$–CH$_2$–CH$_2$–CH$_2$Cl.

$$
\begin{array}{c}
\quad\quad\quad\quad\text{Cl} \\
\text{(b) CH}_3\text{–CH}_2\text{–C–CH}_3. \\
\quad\quad\quad\quad\text{CH}_3
\end{array}
$$

　　(*c*) CH$_3$–CH$_2$–CH$_2$–CH$_2$–CH=CH–CH$_2$Br.

$$
\begin{array}{c}
\quad\quad\quad\text{CH}_3\ \text{CH}_3 \\
\text{(d) CH}_3\text{CH–C–CH}_2\text{Br.} \\
\quad\quad\quad\quad\text{CH}_2\text{CH}_3
\end{array}
$$

2. (*a*) 1-Propanol $\xrightarrow[-\text{H}_2\text{O}]{\text{H}_2\text{SO}_4}$ propene $\xrightarrow{\text{HBr}}$ 2-bromopropane.

　　(*b*) 2-Bromopropane $\xrightarrow[\text{(as in (a))}]{\text{AgNO}_2}$ 2-nitropropane.

　　(*c*) Ethyl alcohol $\xrightarrow{\text{P, I}_2}$ ethyl iodide $\xrightarrow{\text{Sodium}}$ *n*-butane.

　　(*d*) Methyl alcohol $\xrightarrow{\text{P, I}_2}$ methyl iodide; ethyl alcohol $\xrightarrow{\text{Na}}$ sodium ethoxide;

　　　methyl iodide $\xrightarrow{\text{NaOC}_2\text{H}_5}$ methyl ethyl ether.

　　(*e*) 1-Bromobutane $\xrightarrow[\text{KOH}]{\text{Alcoholic}}$ butene-1 $\xrightarrow{\text{Br}_2}$ 1,2-dibromobutane $\xrightarrow[\text{KOH}]{\text{Alcoholic}}$ butyne-1.

3. No. Butene-1 $\xrightarrow{\text{HCl}}$ 2-chlorobutane (refer to Markownikoff's rule).

$$
\begin{array}{c}
\quad\quad\quad\quad\quad\quad\quad\quad\quad\quad\quad\text{CH}_3 \\
\text{4. (a) CH}_3\text{–CH}_2\text{–CH}_2\text{–CN + NaBr.}\quad\quad\quad\text{(b) CH}_3\text{–CH–MgBr.}
\end{array}
$$

$$
\begin{array}{c}
\quad\text{CH}_3 \\
\text{(c) CH}_3\text{–C=CH}_2\text{ + KCl + H}_2\text{O (elimination reaction).}
\end{array}
$$

　　(*d*) CH$_3$–CH$_2$–CH$_2$–CH$_2$OH + KCl (substitution occurs predominantly).

Problem Set No. 4

1. (i) CH$_3$–CH$_2$–CH$_2$–CH$_2$OH (Butanol-1).

　(ii) CH$_3$–CH$_2$–CHOH–CH$_3$ (Butanol-2).

　(iii) CH$_3$–CH–CH$_2$OH (2-Methylpropanol-1).

$$
\quad\quad\quad\quad\quad\quad\text{CH}_3
$$

$$
\begin{array}{c}
\quad\quad\quad\quad\quad\text{CH}_3 \\
\text{(iv) CH}_3\text{–C–OH (2-Methylpropanol-2).} \\
\quad\quad\quad\quad\quad\text{CH}_3
\end{array}
$$

　(v) CH$_3$–CH$_2$–O–CH$_2$–CH$_3$ (Diethyl ether).

(vi) CH₃O–CH₂–CH₂–CH₃ (Methyl propyl ether).

                  CH₃

(vii) CH₃O–CH   (Methyl isopropyl ether).

                  CH₃

2. Butanol $\xrightarrow{[O]}$ butanal $\xrightarrow{[O]}$ butyric acid (having the same number of carbon atoms as the starting alcohol); butanol-2 $\xrightarrow{[O]}$ butanone-2 $\xrightarrow{[O]}$ acetic acid + propionic acid (acids having fewer carbon atoms than the starting alcohol).

3. (a) Methanol $\xrightarrow{P, I_2}$ methyl iodide $\xrightarrow{Mg}$ methylmagnesium iodide $\xrightarrow{CH_3CHO}$ propanol-2.

   (b) Ethanol $\xrightarrow{[O]}$ acetaldehyde; ethanol $\xrightarrow{P, Br_2}$ ethylbromide $\xrightarrow{Mg}$ ethylmagnesium bromide $\xrightarrow{CH_3CHO}$ butanol-2.

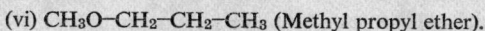

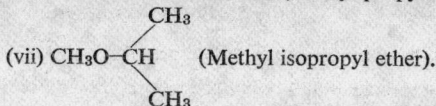

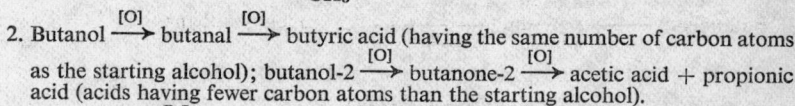

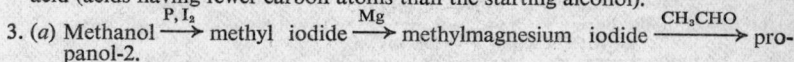

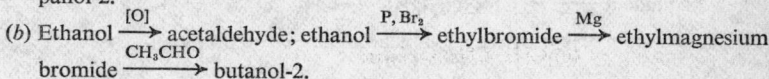

4. Three hydroxyl groups.

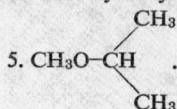

           CH₃

5. CH₃O–CH   .

           CH₃

Problem Set No. 5

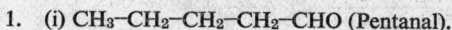

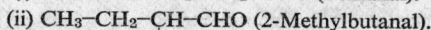

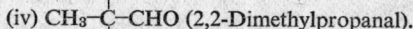

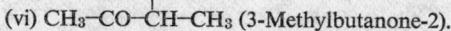

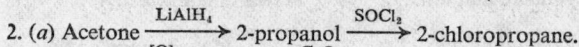

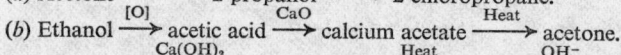

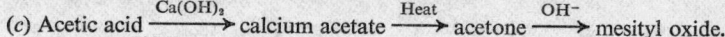

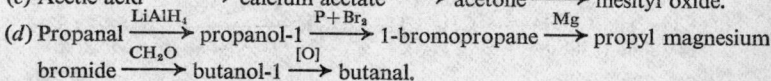

1. (i) CH₃–CH₂–CH₂–CH₂–CHO (Pentanal).

  (ii) CH₃–CH₂–CH–CHO (2-Methylbutanal).

               CH₃

  (iii) CH₃–CH–CH₂–CHO (3-Methylbutanal).

          CH₃

          CH₃

  (iv) CH₃–C–CHO (2,2-Dimethylpropanal).

          CH₃

  (v) CH₃–CO–CH₂–CH₂–CH₃ (Pentanone-2).

             CH₃

  (vi) CH₃–CO–CH–CH₃ (3-Methylbutanone-2).

  (vii) CH₃–CH₂–CO–CH₂–CH₃ (Pentanone-3).

2. (a) Acetone $\xrightarrow{LiAlH_4}$ 2-propanol $\xrightarrow{SOCl_2}$ 2-chloropropane.

   (b) Ethanol $\xrightarrow{[O]}$ acetic acid $\xrightarrow{CaO}$ calcium acetate $\xrightarrow{Heat}$ acetone.

   (c) Acetic acid $\xrightarrow{Ca(OH)_2}$ calcium acetate $\xrightarrow{Heat}$ acetone $\xrightarrow{OH^-}$ mesityl oxide.

   (d) Propanal $\xrightarrow{LiAlH_4}$ propanol-1 $\xrightarrow{P+Br_2}$ 1-bromopropane $\xrightarrow{Mg}$ propyl magnesium bromide $\xrightarrow{CH_2O}$ butanol-1 $\xrightarrow{[O]}$ butanal.

3. Since A reacts with hydroxyl amine, it must have a $-\overset{|}{C}=O$ group and B must be an oxime, $-C=N-OH$. A can be one of the following possibilities: CH₃–CH₂–CH₂–CHO (I), CH₃–CH–CHO (II), and CH₃–CO–CH₂–CH₃ (III)

                                    CH₃

Failure to reduce Tollen's or Fehling's solution indicates that A is (III). So B is

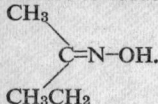

$$CH_3$$
$$\underset{CH_3CH_2}{\overset{\displaystyle}{C}}=N-OH.$$

4. Butanone-2 $\xrightarrow{\text{LiAlH}_4}$ butanol-2 $\xrightarrow[-\text{H}_2\text{O}]{\text{H}_2\text{SO}_4}$ butene-2.

5. (I) Propanol-1 $\xrightarrow{[O]}$ propanal $\xrightarrow[\text{HCl}]{\text{C}_3\text{H}_7\text{OH}}$ $CH_3CH_2CH(OC_3H_7)_2$.

(II) Acetaldehyde $\xrightarrow{\text{LiAlH}_4}$ ethanol $\xrightarrow{\text{P}+\text{Br}_2}$ ethyl bromide $\xrightarrow{\text{Mg}}$ ethylmagnesium bromide $\xrightarrow{\text{CH}_3\text{CHO}}$ $CH_3-CHOHCH_2-CH_3$.

(III) Ethanol $\xrightarrow{}$ ethylmagnesium bromide $\xrightarrow{\text{CH}_2\text{O}}$ propanol-1.

(as in (II))

Problem Set No. 6

1. (a) 1-Chloropropane $\xrightarrow{\text{NaCN}}$ $CH_3-CH_2-CH_2-CN$ $\xrightarrow[\text{(ii) acidify}]{\text{(i) KOH hydrolysis}}$ *n*-butyric acid.

(b) Ethyl propionate $\xrightarrow{\text{LiAlH}_4}$ $CH_3-CH_2-CH_2-OH$ $\xrightarrow{\text{P}+\text{Br}_2}$ $CH_3-CH_2-CH_2Br$ $\xrightarrow{\text{NaCN}}$ nitrile $\xrightarrow[\text{(ii) acidify}]{\text{(i) aq. KOH}}$ *n*-butyric acid.

2. (a) Formic acid (not acetic acid) is a reducing agent and reduces Fehling's solution.

(b) The chlorine in acetyl chloride is so reactive that hydrolysis occurs even on contact with atmospheric moisture to give fumes of hydrogen chloride. The chlorine in chloroacetic acid is relatively less reactive.

3. Only *n*-hexanoic acid will react with sodium bicarbonate to give CO_2. As between the other two, only hexanol-1 will react with sodium giving off hydrogen.

4. (a) Acetylene $\xrightarrow{\text{HCN}}$ acrylonitrile $\xrightarrow[\text{H}_2\text{O}]{\text{KOH}}$ acrylic acid.

(b) Ethanol $\xrightarrow{[O]}$ acetic acid $-\begin{cases}\xrightarrow{\text{NaOH}} \text{sodium acetate} \\ \xrightarrow{\text{SOCl}_2} \text{acetyl chloride}\end{cases}$ sodium acetate + acetyl chloride $\xrightarrow{}$ acetic anhydride.

(c) Ethanol $\xrightarrow{[O]}$ acetic acid $\xrightarrow[\text{(I}_2)]{\text{Cl}_2}$ α-chloroacetic acid $\xrightarrow{\text{NaCN}}$ $NC-CH_2COONa$ $\xrightarrow{\text{C}_2\text{H}_5\text{OH}}{\text{HCl}}$ diethyl malonate; ethanol $\xrightarrow{\text{P}+\text{Br}_3}$ ethylbromide; diethyl malonate $\xrightarrow[\text{(ii) C}_2\text{H}_5\text{Br}]{\text{(i) NaOC}_2\text{H}_5}$ $\underset{\displaystyle CO_2C_2H_5}{CH_3CH_2-CH-CO_2C_2H_5}$ $\xrightarrow[\text{(ii) H}^+]{\text{(i) KOH}}$ $\xrightarrow[-\text{CO}_2]{\text{Heat}}$ butanoic acid.

(d) CH_3CH_2COOH $\xrightarrow{\text{Br}_2,\text{P}}$ $CH_3-CHBrCOOH$ $\xrightarrow[\text{H}_2\text{O}]{\text{Na}_2\text{CO}_3}$ $\xrightarrow{[O]}$ $CH_3COCOOH$
propanoic acid.   pyruvic acid

(e) CH_3-CH_2-OH $\xrightarrow{[O]}$ CH_3COOH $\xrightarrow[\text{H}^+]{\text{CH}_3\text{CH}_2\text{OH}}$ $CH_3-COOC_2H_5$ $\xrightarrow{\text{NaOC}_2\text{H}_5}$ $CH_3COCH_2COOC_2H_5$ (I); CH_3CH_2OH $\xrightarrow{\text{P}+\text{Br}_2}$ CH_3CH_2Br (II);

(I + II) $\xrightarrow{\text{NaOC}_2\text{H}_5}$ $\underset{\displaystyle CO_2C_2H_5}{CH_3-CH_2-CH-COCH_3}$ $\xrightarrow[\text{NaOH}]{\text{dilute}}$ $CH_3-CH_2-CH_2-CO-CH_3$.

(*f*) (i) Diethyl malonate $\xrightarrow[C_2H_5Br]{NaOC_2H_5}$ $CH_2-CH_2-CH(CO_2C_2H_5)_2$ $\xrightarrow[\substack{\text{(ii) H}^+\\\text{(iii) Heat}(-CO_2)}]{\text{(i) KOH}}$

$CH_3-CH_2-CH_2-CO_2H.$

(ii) Diethyl malonate $\xrightarrow[CH_3CH_2CH_2Br]{NaOC_2H_5}$ $CH_3-CH_2-CH_2-CH(CO_2C_2H_5)_2$ $\xrightarrow[CH_3I]{NaOC_2H_5}$

$CH_3-CH_2-CH_2-\underset{\underset{CH_3}{|}}{C}(CO_2C_2H_5)_2$ $\xrightarrow[\substack{\text{(ii) H}^+\\\text{(iii) Heat}(-CO_2)}]{\text{(i) KOH}}$

$CH_3-CH_2-CH_2-\underset{\underset{CH_3}{|}}{C}H-CO_2H$

(iii) Diethyl malonate $\xrightarrow[\text{Base}]{CH_3CH_2CHO}$ $CH_3CH_2CH=C(CO_2C_2H_5)_2$ $\xrightarrow[\substack{\text{(ii) H}^+\\\text{(iii) Heat}(-CO_2)}]{\text{(i) KOH}}$

$CH_3-CH_2-CH=CH-CO_2H.$

(*g*) (i) $CH_3-CO-CH_2-CO_2C_2H_5$ $\xrightarrow[CH_3CH_2Br]{NaOC_2H_5}$

$CH_3CO-\underset{\underset{CH_2CH_3}{|}}{C}H-CO_2C_2H_5$ $\xrightarrow[\text{(ii) H}^+]{\text{(i) Concentrated NaOH}}$ $CH_3-CH_2-CH_2-CO_2H.$

(ii) $CH_3COCH_2CO_2C_2H_5$ $\xrightarrow[CH_3CH_2Br]{NaOC_2H_5}$

$CH_3-CO-\underset{\underset{CH_2CH_3}{|}}{C}H-CO_2C_2H_5$ $\xrightarrow[\text{NaOH}]{\text{Dilute}}$ $CH_3COCH_2-CH_2-CH_3.$

(iii) $CH_3-COCH_2-CO_2C_2H_5$ $\xrightarrow[BrCH_2CH_2CO_2C_2H_5]{NaOC_2H_5}$

$CH_3-CO\underset{\underset{CH_2CH_2CO_2C_2H_5}{|}}{C}HCO_2C_2H_5$ $\xrightarrow[\text{NaOH}]{\text{Dilute}}$ $CH_3COCH_2CH_2CH_2CO_2H.$

(*h*) $CH_3CONH-CH(CO_2C_2H_5)_2$ $\xrightarrow[CH_3I]{NaOC_2H_5}$ $CH_3CONH-\underset{\underset{CH_3}{|}}{\overset{\overset{CH_3}{|}}{C}}(CO_2C_2H_5)_2$ $\xrightarrow{NaOH}$

$CH_3CONH-\underset{\underset{CH_3}{|}}{C}(CO_2H)_2$ $\xrightarrow{\text{Heat}}$ $CH_3-\underset{\underset{NHCOCH_3}{|}}{C}H-CO_2H$ $\xrightarrow[\text{NaOH}]{\text{Concentrated}}$

$CH_3-\underset{\underset{NH_2}{|}}{C}H-CO_2H.$

α-Alanine

5. (*a*) $CH_3CH_2-CH_2CN.$ (*b*) $\underset{CH_2CO}{\overset{CH_2CO}{|}}\!\!\diagdown\!\!O.$

(*c*) $CH_3-CO-N(C_2H_5)_2.$
(*d*) $(CO_2C_2H_5)_2CH-CH_2CH_2CH_2CH_2-CH(CO_2C_2H_5)_2.$
(*e*) $(CH_2)_6\underset{CO_2H}{\overset{CO_2H}{\diagup}}$

(*f*) $CH_3CH_2COCH-CO_2C_2H_5.$
$\underset{\;|}{\;CH_3}$

(*g*) $\underset{O\diagdown\!\!\!\underset{CO}{CH_2}}{CH_2-CH_2}$

(*h*) $CH_3CH_2CH_2CH_2\underset{\underset{Br}{|}}{C}HCH=CH_2.$

(*i*) $CH_3-CH_2COONa.$

6. A vegetable oil is an ester, whereas a mineral oil is a hydrocarbon. So the former can be treated with sodium hydroxide to produce glycerol and the sodium salt of a fatty acid (soap). The mineral oil will be unaffected by treatment with sodium hydroxide.

7. Of the eight theoretically possible pentanols, the following will be optically active.

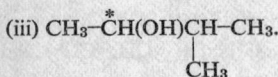

(i) $CH_3-CH_2CH_2\overset{*}{C}H(OH)CH_3$.   (ii) $CH_3-CH_2-\overset{*}{C}H-CH_2OH$.
$$CH_3$$

(iii) $CH_3-\overset{*}{C}H(OH)CH-CH_3$.
$$CH_3$$

The asymmetric carbon atoms are indicated by asterisks.

8. (a) 1 *d*, 1 *l*, and 1 *dl* forms since there is only one asymmetric carbon atom.
   (b) 1 *d*, 1 *l*, and 1 *dl* forms for the same reason as in (a).
   (c) 2 similar asymmetric carbon atoms and hence 1 *d*, 1 *l*, 1 *meso*, and 1 *dl* forms.
   (d) 1 *cis* and 1 *trans* forms:

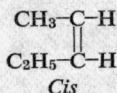

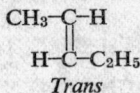

$$CH_3-C-H \qquad CH_3-C-H$$
$$C_2H_5-C-H \qquad H-C-C_2H_5$$
       Cis   *Trans*

   (e) No isomers, either optical or geometrical.
   (f) One *cis* and one *trans* forms:

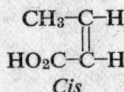

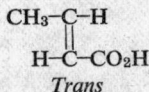

$$CH_3-C-H \qquad CH_3-C-H$$
$$HO_2C-C-H \qquad H-C-CO_2H$$
       Cis   *Trans*

   (g) Two different asymmetric carbon atoms and hence 2^2 or 4 optical isomers and 2 *dl* forms.

9. 15 ml. *N*/5-NaOH neutralizes 0·2220 g. Hence 1,000 ml. NaOH (1*N*) will neutralize $\dfrac{0\cdot2220 \times 5 \times 1,000}{15} = 74$ g. It is a monobasic acid and hence the molecular weight is 74. The $-COOH$ group contributes 45 to the molecular weight and hence the alkyl group (R) contributes $74 - 45$ or 29 to the molecular weight. The acid is therefore $CH_3CH_2CO_2H$. On heating the sodium salt with sodium hydroxide, ethane (C_2H_6) will be obtained.

10. On acetylation, each $-OH$ group is converted into $-OCOCH_3$ group. This involves an increase in molecular weight of 42 (corresponding to $COCH_2$ per OH group acetylated). Increase in molecular weight in the present case is $434 - 182$ (molecular weight of the compound $C_6H_{14}O_6$) or 252. The compound has $\dfrac{252}{42}$ or 6 hydroxyl groups.

Problem Set No. 7

1. (a) $CH_3-CH_2-CH_2NH_2 + 3NaOH + CHCl_3 \longrightarrow$
$$CH_3CH_2CH_2N{\equiv}C + 3NaCl + 3H_2O.$$
   (b) $CH_3CH_2-CONH_2 + Br_2 + 4NaOH \longrightarrow$
$$Na_2CO_3 + 2NaBr + 2H_2O + CH_3CH_2NH_2.$$
   (c) $C_6H_5SO_2Cl + NH(C_2H_5)_2 \xrightarrow{NaOH} C_6H_5SO_2N(C_2H_5)_2 + NaCl + H_2O.$
   (d) $CH_3CH_2NH_2 + HONO \longrightarrow CH_3CH_2OH + N_2 + H_2O.$
   (e) $N(C_2H_5)_3 + C_2H_5I \longrightarrow (C_2H_5)_4\overset{+-}{N}I.$
   (f) $(C_2H_5)_4\overset{+-}{N}I + AgOH \longrightarrow (C_2H_5)_4\overset{+}{N}\overset{-}{OH} + AgI.$

(g) $CH_3CH_2-CONH_2 + P_2O_5 \longrightarrow CH_3CH_2-CN + 2HPO_3.$

(h) $CH_3CH_2CH_2I + AgNO_2 \longrightarrow$
$$CH_3CH_2CH_2-NO_2(+ \text{ some } CH_3CH_2CH_2-O-NO) + AgI.$$

(i) $CH_3CH_2CH_2CONH_2 \xrightarrow{\text{LiAlH}_4} CH_3CH_2CH_2CH_2NH_2.$

(j) $N(C_2H_5)_3 + HNO_2 \longrightarrow$ No Reaction.

(k) $CH_3CH_2CH_2NO_2 + NaOCH_3 \longrightarrow CH_3CH_2CH=N{\overset{\displaystyle ONa}{\underset{\displaystyle O}{}}} + CH_3OH.$

2. (a) $CH_3CH_2Br \xrightarrow{\text{NH}_3} CH_3CH_2NH_2 +$ other products. (Or)

$\downarrow$ NaCN

$CH_3CH_2C\equiv N \xrightarrow{\text{NaOH}} CH_3CH_2COOH \xrightarrow{\text{SOCl}_2} CH_3CH_2COCl \xrightarrow{\text{NH}_3}$

$ CH_3CH_2CONH_2 \xrightarrow[\text{+NaOH}]{\text{Br}_2} CH_3CH_2NH_2.$

(b) $CH_3Br \xrightarrow{\text{NaCN}} CH_3CN \xrightarrow[\text{reduction}]{\text{catalytic}} CH_3CH_2NH_2.$

(c) $CH_3CH_2CH_2OH \xrightarrow{\text{P}_2\text{S}_5} CH_3CH_2CH_2SH \xrightarrow{\text{NaOC}_2\text{H}_5}$

$Br_2 \downarrow (P)$

$ CH_3CH_2CH_2SNa \xrightarrow{\text{CH}_3\text{CH}_2\text{CH}_2\text{Br}} (CH_3CH_2CH_2)_2S.$

$CH_3CH_2CH_2Br$

3.
$$\begin{array}{cc} CH_2-CH_2 \\ CH_2 \quad CH_2 \\ NH \end{array} \xrightarrow{\text{2CH}_3\text{I}} \begin{array}{cc} CH_2-CH_2 \\ CH_2 \quad CH_2 \\ \overset{+}{N} \quad I^- \\ CH_3 \quad CH_3 \end{array} \xrightarrow{\text{AgOH}} \begin{array}{cc} CH_2-CH_2 \\ CH_2 \quad CH_2 \\ \overset{+}{N} \quad OH^- \\ CH_3 \quad CH_3 \end{array} \xrightarrow{\text{Heat}}$$

$$\begin{array}{cc} CH_2-CH \\ CH_2 \quad CH_2 \\ N(CH_3)_2 \end{array} + H_2O \xrightarrow[\text{(excess)}]{\text{CH}_3\text{I}} \begin{array}{cc} CH_2-CH \\ CH_2 \quad CH_2 \\ ^+N(CH_3)_3 \quad I^- \end{array} \xrightarrow[\text{(ii) Heat}]{\text{(i) AgOH}}$$

$$CH_2=CH-CH=CH_2 + N(CH_3)_3 + H_2O.$$

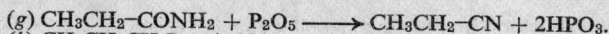

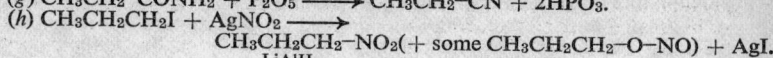

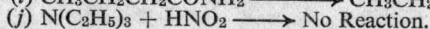

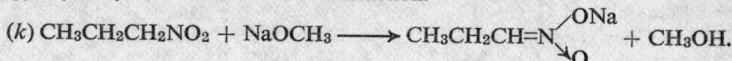

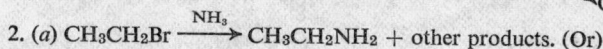

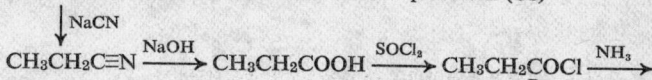

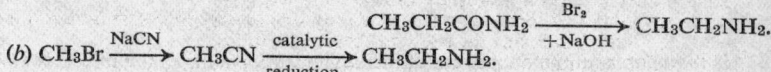

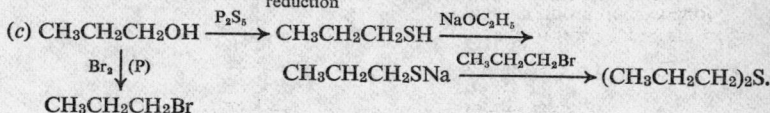

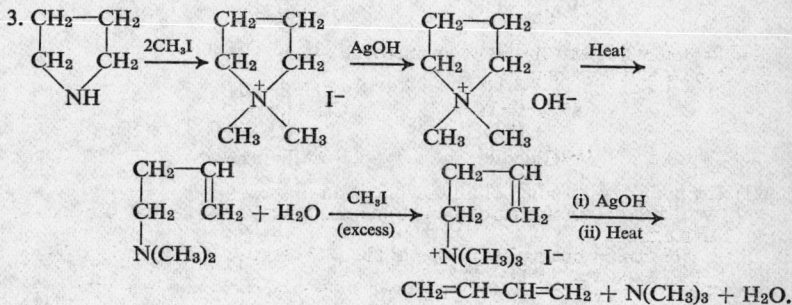

Problem Set No. 8

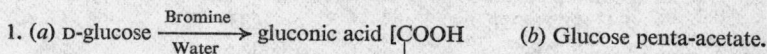

1. (a) D-glucose $\xrightarrow[\text{Water}]{\text{Bromine}}$ gluconic acid [COOH   (b) Glucose penta-acetate.

 (CHOH)$_4$

 CH$_2$OH]

(c) D-fructose gives an osazone identical with that from D-glucose.

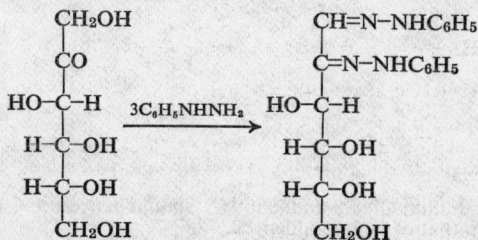

$$\begin{array}{c} CH_2OH \\ CO \\ HO-C-H \\ H-C-OH \\ H-C-OH \\ CH_2OH \end{array} \xrightarrow{\text{3C}_6\text{H}_5\text{NHNH}_2} \begin{array}{c} CH=N-NHC_6H_5 \\ C=N-NHC_6H_5 \\ HO-C-H \\ H-C-OH \\ H-C-OH \\ CH_2OH \end{array}$$

(*d*) A mixture of the epimeric cyanohydrins I and II.

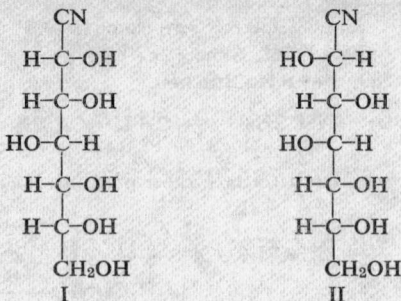

```
        CN                          CN
    H–C–OH                     HO–C–H
    H–C–OH                      H–C–OH
   HO–C–H                      HO–C–H
    H–C–OH                      H–C–OH
    H–C–OH                      H–C–OH
      CH₂OH                        CH₂OH
        I                           II
```

2. (I) D-glucose and D-mannose are epimers. They differ in the configuration of only one carbon atom, namely C_2.

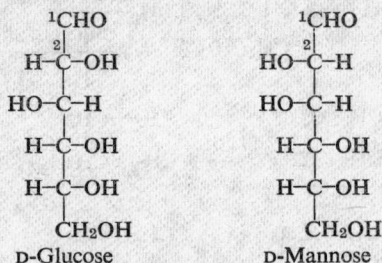

```
      ¹CHO                        ¹CHO
  H–²C–OH                    HO–²C–H
   HO–C–H                      HO–C–H
    H–C–OH                      H–C–OH
    H–C–OH                      H–C–OH
      CH₂OH                        CH₂OH
    D-Glucose                   D-Mannose
```

(II) The hydrolysis product of an osazone is an osone.

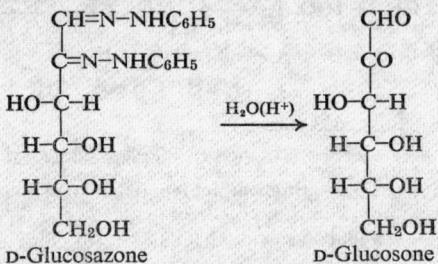

```
   CH=N–NHC₆H₅                         CHO
    C=N–NHC₆H₅                          CO
  HO–C–H              H₂O(H⁺)      HO–C–H
   H–C–OH            ─────────►     H–C–OH
   H–C–OH                           H–C–OH
     CH₂OH                            CH₂OH
  D-Glucosazone                    D-Glucosone
```

(III) Freshly prepared solutions of either the α or β form of glucose, having different specific rotations each, undergo a change in their values to the same equilibrium value on standing. This phenomenon is referred to as muta-rotation and is due to tautomeric changes.

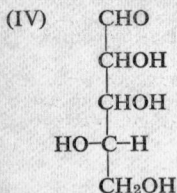

(IV)
```
      CHO
      CHOH
      CHOH
   HO–C–H
      CH₂OH
```

The configuration of C_4 and only C_4 should correspond as given in the structure to that of L-glyceraldehyde.

(V) A reducing sugar is one which reduces Fehling's or Benedict's solution and contains a potential −CHO group. For example, maltose and lactose are reducing sugars while sucrose is a non-reducing sugar.

(VI) Glucosides are ethers of the type shown below:

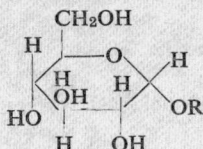

They are formed by the reaction of a hydroxy compound (ROH) with glucose with elimination of water.

3. (*a*) D-glucose $\xrightarrow{\text{NH}_2\text{OH}}$ oxime $\xrightarrow{(\text{CH}_3\text{CO})_2\text{O}}$

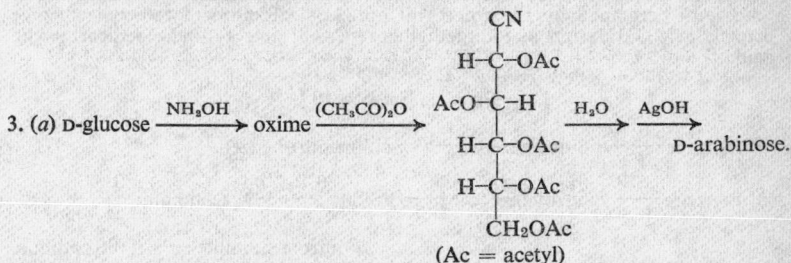

$\xrightarrow{\text{H}_2\text{O}}$ $\xrightarrow{\text{AgOH}}$ D-arabinose.

(Ac = acetyl)

(*b*) D-arabinose $\xrightarrow{\text{HCN}}$

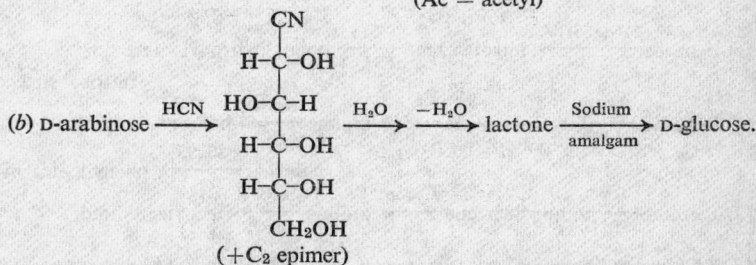

$\xrightarrow{\text{H}_2\text{O}}$ $\xrightarrow{-\text{H}_2\text{O}}$ lactone $\xrightarrow[\text{amalgam}]{\text{Sodium}}$ D-glucose.

(+C₂ epimer)

(*c*) Sucrose $\longrightarrow$ D-glucose + D-fructose. The mixture of D-glucose and D-fructose obtained on hydrolysis of sucrose is called invert sugar, since the specific rotation of a solution of sucrose is positive before and negative after hydrolysis.
+66·5° C.   +52·2° C.   −92·0° C.

(*d*) Starch $\xrightarrow[\text{(an enzyme)}]{\text{Amylase}}$ maltose.

Problem Set No. 9

1. (*a*) *m*-Dimethylbenzene (usually called *m*-xylene). (*b*) *p*-Bromoisopropylbenzene. (*c*) *m*-Nitrotoluene. (*d*) 1,2,3-Trimethylbenzene. (*e*) *p*-Nitroethylbenzene. (*f*) 1-Nitro-2-chloronaphthalene.
2. (*a*) Benzyl bromide (C₆H₅CH₂Br). (*b*) *m*-Nitrobenzenesulphonic acid. (*c*) A mixture of *o*- and *p*-nitrobromobenzenes. (*d*) *p*-Cresol. (*e*) Benzoic acid. (*f*) *o*-Methylacetophenone and *p*-methylacetophenone. (*g*) 2,4-Dinitrobenzonitrile. (*h*) 2,2′-Dinitrodiphenyl.
3. (*a*) Benzene $\xrightarrow{\text{HNO}_3}_{\text{H}_2\text{SO}_4}$ nitrobenzene $\xrightarrow{\text{Br}_2}$ *m*-nitrobromobenzene.

   (*b*) Benzene $\xrightarrow[\text{CH}_3\text{I}]{\text{AlCl}_3}$ toluene $\xrightarrow{\text{HNO}_3}_{\text{H}_2\text{SO}_4}$ *p*-nitrotoluene $\xrightarrow{\text{KMnO}_4}$ *p*-nitrobenzoic acid.

(c) Benzene $\xrightarrow[H_2SO_4]{HNO_3}$ nitrobenzene $\xrightarrow[HNO_3]{Fuming}$ m-dinitrobenzene $\xrightarrow[+H_2S]{NH_3}$ m-nitroaniline.

(d) Benzene $\xrightarrow[Fe]{Cl_2}$ chlorobenzene $\xrightarrow{H_2SO_4}$ p-chlorobenzenesulphonic acid.

(e) Benzene $\xrightarrow[AlCl_3]{CH_3CH_3Br}$ ethylbenzene.

(f) Benzene $\xrightarrow{H_2SO_4}$ benzenesulphonic acid $\xrightarrow[\substack{with \\ NaOH}]{Fusion}$ phenol.

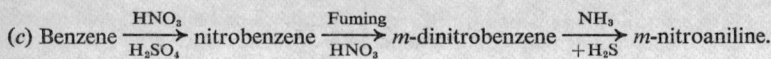

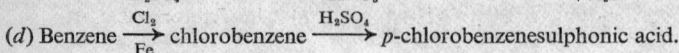

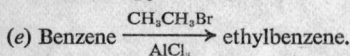

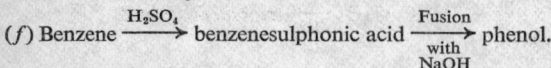

4. (a) Nuclear halogen in p-bromotoluene is not removed by alkaline hydrolysis under mild conditions. Benzyl bromide after such treatment gives ionic bromine, which can be detected by the silver nitrate test. (b) Hexene-1 will decolourize potassium permanganate; benzene will not. (c) Oxidation of toluene will give benzoic acid; oxidation of p-dimethylbenzene will give 1,4-benzenedicarboxylic acid.

Problem Set No. 10

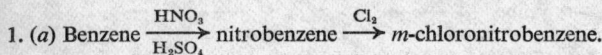

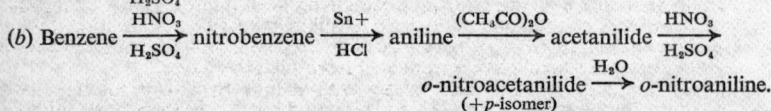

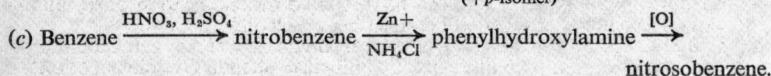

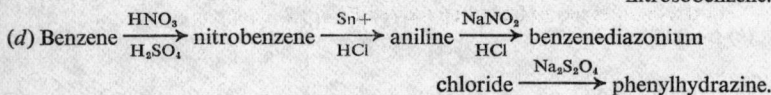

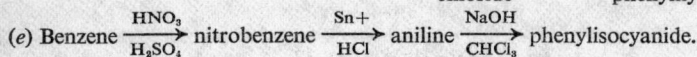

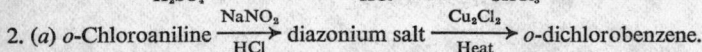

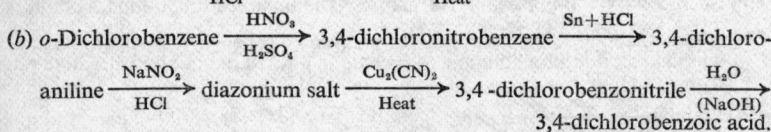

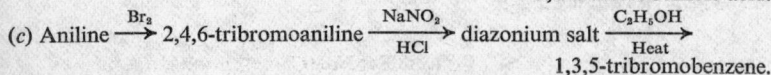

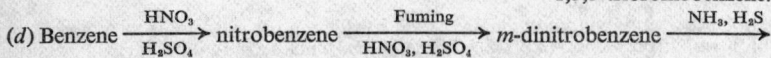

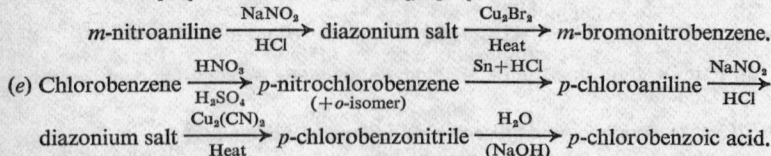

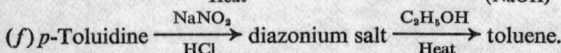

1. (a) Benzene $\xrightarrow[H_2SO_4]{HNO_3}$ nitrobenzene $\xrightarrow{Cl_2}$ m-chloronitrobenzene.

(b) Benzene $\xrightarrow[H_2SO_4]{HNO_3}$ nitrobenzene $\xrightarrow[HCl]{Sn+}$ aniline $\xrightarrow{(CH_3CO)_2O}$ acetanilide $\xrightarrow{HNO_3}$
o-nitroacetanilide $\xrightarrow{H_2O}$ o-nitroaniline.
(+p-isomer)

(c) Benzene $\xrightarrow{HNO_3, H_2SO_4}$ nitrobenzene $\xrightarrow[NH_4Cl]{Zn+}$ phenylhydroxylamine $\xrightarrow{[O]}$
nitrosobenzene.

(d) Benzene $\xrightarrow[H_2SO_4]{HNO_3}$ nitrobenzene $\xrightarrow[HCl]{Sn+}$ aniline $\xrightarrow[HCl]{NaNO_2}$ benzenediazonium
chloride $\xrightarrow{Na_2S_2O_4}$ phenylhydrazine.

(e) Benzene $\xrightarrow[H_2SO_4]{HNO_3}$ nitrobenzene $\xrightarrow[HCl]{Sn+}$ aniline $\xrightarrow[CHCl_3]{NaOH}$ phenylisocyanide.

2. (a) o-Chloroaniline $\xrightarrow[HCl]{NaNO_3}$ diazonium salt $\xrightarrow[Heat]{Cu_2Cl_2}$ o-dichlorobenzene.

(b) o-Dichlorobenzene $\xrightarrow[H_2SO_4]{HNO_3}$ 3,4-dichloronitrobenzene $\xrightarrow{Sn+HCl}$ 3,4-dichloro-
aniline $\xrightarrow[HCl]{NaNO_2}$ diazonium salt $\xrightarrow[Heat]{Cu_2(CN)_2}$ 3,4 -dichlorobenzonitrile $\xrightarrow[(NaOH)]{H_2O}$
3,4-dichlorobenzoic acid.

(c) Aniline $\xrightarrow{Br_2}$ 2,4,6-tribromoaniline $\xrightarrow[HCl]{NaNO_2}$ diazonium salt $\xrightarrow[Heat]{C_2H_5OH}$
1,3,5-tribromobenzene.

(d) Benzene $\xrightarrow[H_2SO_4]{HNO_3}$ nitrobenzene $\xrightarrow[HNO_3, H_2SO_4]{Fuming}$ m-dinitrobenzene $\xrightarrow{NH_3, H_2S}$
m-nitroaniline $\xrightarrow[HCl]{NaNO_2}$ diazonium salt $\xrightarrow[Heat]{Cu_2Br_2}$ m-bromonitrobenzene.

(e) Chlorobenzene $\xrightarrow[H_2SO_4]{HNO_3}$ p-nitrochlorobenzene $\xrightarrow{Sn+HCl}$ p-chloroaniline $\xrightarrow[HCl]{NaNO_2}$
(+o-isomer)
diazonium salt $\xrightarrow[Heat]{Cu_2(CN)_2}$ p-chlorobenzonitrile $\xrightarrow[(NaOH)]{H_2O}$ p-chlorobenzoic acid.

(f) p-Toluidine $\xrightarrow[HCl]{NaNO_2}$ diazonium salt $\xrightarrow[Heat]{C_2H_5OH}$ toluene.

Problem Set No. 11

1. (*a*) Triphenyl phosphate [PO(OC$_6$H$_5$)$_3$].   (*b*) 2,4,6-Tribromophenol.
  (*c*) 4-Hydroxybenzene-1,3-disulphonic acid.   (*d*) *o*- and *p*-Nitrophenols.
  (*e*) Phenyl methyl ether (Anisole).   (*f*) *p*-Nitrosophenol.

2. (*a*) Benzophenone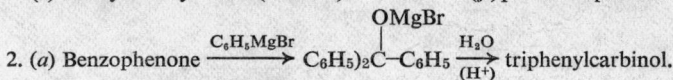

  (*b*) Ethyl benzoate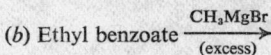

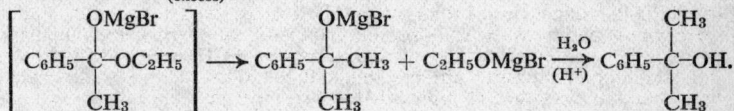

  (*c*) Benzene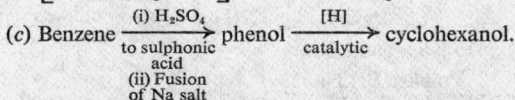

3. (*a*) Dissolve mixture in ether, extract successively with dilute sodium bicarbonate solution and dilute sodium hydroxide solution. After acidification the bicarbonate extract will give benzoic acid and the sodium hydroxide extract will give phenol. Benzene can be recovered from the ether layer.
  (*b*) Dissolve mixture in ether and extract with sodium hydroxide solution. Only phenol will be extracted and can be recovered by acidification of the alkaline extract. Removal of ether will furnish benzyl alcohol.
  (*c*) Extraction procedure as in (*a*). The bicarbonate extract will furnish picric acid while the sodium hydroxide extract will furnish phenol. Cyclohexanol remains in the ether layer and can be recovered.

Problem Set No. 12

1. (*a*) Diethylacetal of benzaldehyde: C$_6$H$_5$CH(OC$_2$H$_5$)$_2$ (cf. reactions of aliphatic aldehydes).
  (*b*) C$_6$H$_5$—CH$_2$OH + C$_6$H$_5$—COONa.
  (*c*) Diethyl benzalmalonate C$_6$H$_5$CH=C(CO$_2$C$_2$H$_5$)$_2$.

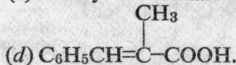

  (*d*) C$_6$H$_5$CH=C—COOH.   (*e*) C$_6$H$_5$CH=CH—CHO.
  (*f*) C$_6$H$_5$—CH(OH)COC$_6$H$_5$.

2. (*a*) Toluene 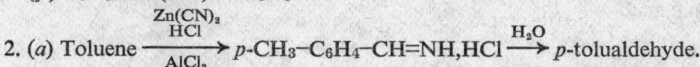 *p*-toluadehyde.

  (*b*) C$_6$H$_5$CHO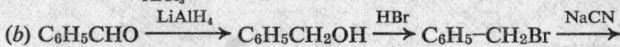

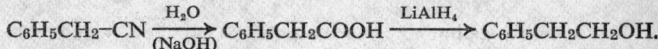

  (*c*) C$_6$H$_6$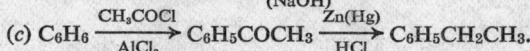

  (*d*) C$_6$H$_5$CHO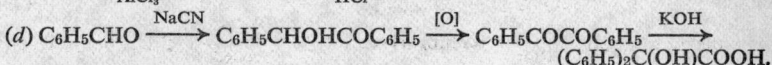

  (*e*) C$_6$H$_5$CHO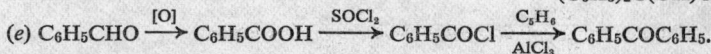

3. H_3C<benzene>$+ H_3C$<benzene>$-COCl$ $\xrightarrow[\text{Crafts Reaction}]{\text{Friedel-}}$
(A)

H_3C<benzene>$-CO$<benzene>$-CH_3$ $\xrightarrow{\text{NH}_2\text{OH}}$
(B)

H_3C<benzene>$-\underset{\underset{\text{N}-\text{OH}}{\|}}{C}$<benzene>$-CH_3$ $\xrightarrow[\text{[Beckmann Rearrangement]}]{\text{PCl}_5}$
(C)

H_3C<benzene>$-CONH$<benzene>$-CH_3$ $\xrightarrow{\text{H}_2\text{O}}$
(D)

H_3C<benzene>$-NH_2 + H_3C$<benzene>$-COOH.$

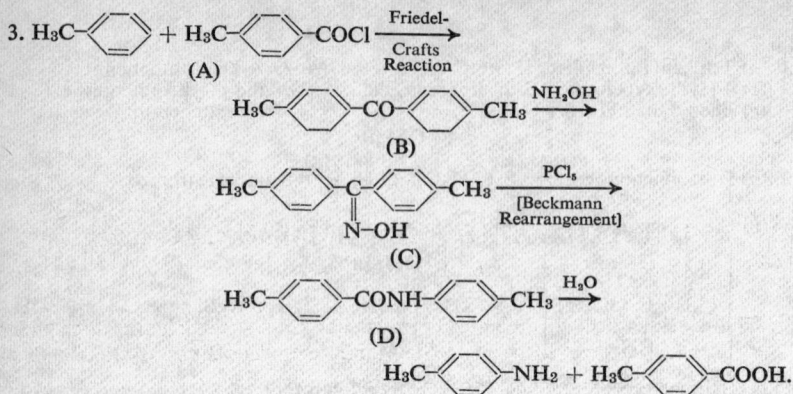

Problem Set No. 13

1. (a) m-Bromobenzoic acid.

(b) $C_6H_5CO-\underset{\underset{\text{C}_6\text{H}_5}{|}}{\overset{\overset{\text{CH}_3}{|}}{N}}$

(c) o-Benzoylbenzoic acid
$$\left[\begin{array}{c} \text{CO} \\ \text{HOOC} \end{array}\right]$$

(d) Aspirin

(e) Anthranilic acid
$$\left[\begin{array}{c} \text{NH}_2 \\ \text{COOH} \end{array}\right]$$

2. (a) $CH_3C_6H_5 \xrightarrow[\text{AlCl}_3]{\text{CH}_3\text{COCl}} p\text{-CH}_3C_6H_4COCH_3 \xrightarrow{\text{[O]}} p\text{-HO}_2C-C_6H_4-CO_2H.$

(b) $p\text{-CH}_3-C_6H_4CHO \xrightarrow[\underset{\text{Heat}}{\text{CH}_3\text{COONa}}]{(\text{CH}_3\text{CO})_2\text{O}} p\text{-CH}_3-C_6H_4-CH=CHCOOH.$

(c) $C_6H_5Br \xrightarrow{\text{Mg}} C_6H_5MgBr \xrightarrow{\text{CO}_2} C_6H_5COOH \xrightarrow[(\text{H}^+)]{\text{CH}_3\text{OH}} C_6H_5COOCH_3 \xrightarrow[\text{H}_2\text{SO}_4]{\text{HNO}_3}$
$m\text{-NO}_2\text{-C}_6H_4\text{-COOCH}_3.$

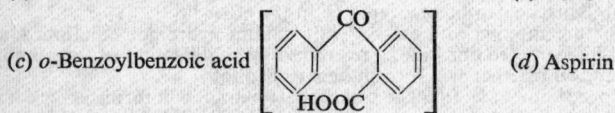

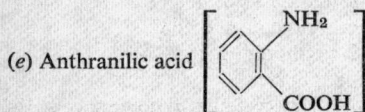

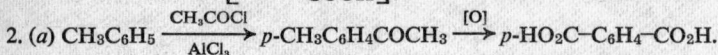

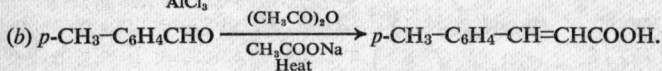

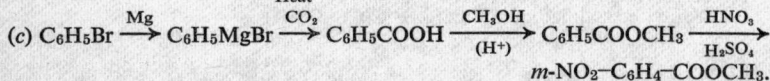

3. A is o-ethyltoluene. B can be either n-propylbenzene or isopropylbenzene, and C is phthalic acid.

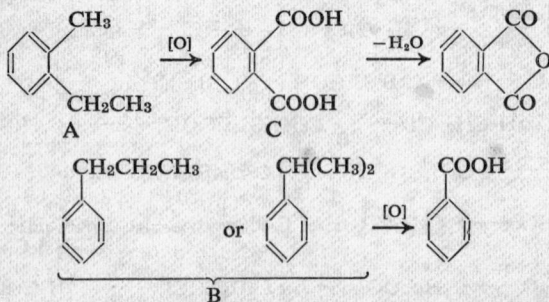

$$\underset{\text{A}}{\overset{\overset{\text{CH}_3}{\underset{\text{CH}_2\text{CH}_3}{\bigcirc}}}{}} \xrightarrow{\text{[O]}} \underset{\text{C}}{\overset{\overset{\text{COOH}}{\underset{\text{COOH}}{\bigcirc}}}{}} \xrightarrow{-\text{H}_2\text{O}} \overset{\overset{\text{CO}}{\underset{\text{CO}}{\bigcirc}}}{\text{O}}$$

$$\underbrace{\underset{}{\overset{\overset{\text{CH}_2\text{CH}_2\text{CH}_3}{\bigcirc}}{}} \quad \text{or} \quad \underset{}{\overset{\overset{\text{CH}(\text{CH}_3)_2}{\bigcirc}}{}}}_{\text{B}} \xrightarrow{\text{[O]}} \overset{\overset{\text{COOH}}{\bigcirc}}{}$$

GLOSSARY OF SOME COMMON ORGANIC COMPOUNDS

This glossary lists, alphabetically, some common organic compounds with an I.U.P.A.C. approved name. Compounds with trivial names approved by I.U.P.A.C. are indicated with an asterisk.

Common or trivial names	Systematic names (I.U.P.A.C.)
acetaldehyde	ethanal
acetaldehyde cyanohydrin	2-hydroxypropanonitrile
acetaldehyde diethyl acetal (acetal)	1,1-diethoxyethane
acetaldehyde hemiacetal	1-ethoxyethanol
acetamide	ethanamide
acetic acid	ethanoic acid
acetic anhydride	ethanoic anhydride
acetoacetic acid	3-oxobutanoic acid
acetone (dimethyl ketone*)	propanone
acetone cyanohydrin	2-hydroxy-2-methylpropanonitrile
acetophenone	methyl phenyl ketone or phenyl-ethanal or ethanoylbenzene
acetylacetone	pentan-2,4-dione
acetyl chloride	ethanoyl chloride or 1-chloroethanal
acetylene	ethyne
acetylene tetrachloride	tetrachloroethane
acrylonitrile	propenenitrile or propenonitrile
adipic acid	hexandioic acid
acrylic acid	propenoic acid
alanine (α-aminopropionic acid)	2-aminopropanoic acid
aldol (β-hydroxybutyraldehyde)	3-hydroxybutanal
alkyl halide *	halogenoalkane
aniline	phenylamine or aminobenzene
anisole (methyl phenyl ether) *	methoxybenzene
anthranilic acid	2-aminobenzenecarboxylic acid
benzaldehyde	benzenecarbaldehyde
benzamide	benzenecarboxamide
benzal dichloride	(dichloromethyl) benzene
benzidine	4,4-diaminobiphenyl
benzoic acid	benzenecarboxylic acid
benzonitrile	benzenecarbonitrile

Common or trivial names	Systematic names (I.U.P.A.C.)
benzoyl chloride	benzenecarbonyl chloride
benzyl alcohol	phenylmethanol
benzyl chloride	(chloromethyl) benzene
carbon tetrachloride	tetrachloromethane
choral	trichloroethanal
chloroform	trichloromethane
cinnamic acid	3-phenylpropenoic acid
crotonaldehyde	but-2-onal
diacetone alcohol	4-hydroxy-4-methylpentan-2-one
diethyl ether *	ethoxyethane
diethyl malonate (malonic ester)	diethyl propanedioate
ethanolamine	2-aminoethanol
ethyl acetate	ethyl ethanoate
ethyl acetoacetate (acetoacetic ester)	ethyl 3-oxobutanoate
ethyl alcohol	ethanol
ethyl chloride *	chloroethane
ethyl methyl ether *	methoxyethane
ethyl methyl ketone *	butanone or 2-oxobutane
ethyl phenyl ether *	ethoxybenzene
ethylene	ethene
ethylene dibromide	1,2-dibromoethane
ethylene glycol	ethane-1,2-diol
ethylene chlorohydrin	2-chloroethanol
formaldehyde	methanal
formic acid	methanoic acid
fumaric acid	trans-butenedioic acid
glycerol	propane-1,2,3-triol
glycerol tristearate	propane-1,2,3-triyl trioctadecanoate
glyceryl trinitrate (nitroglycerine)	propyl-1,2,3-trinitrate
glycine (aminoacetic acid)	aminoethanoic acid
isopropanol (isopropyl alcohol)	propan-2-ol
lactic acid	2-hydroxypropanoic acid
maleic acid	cis-butenedioic acid
malonic acid	propanedioic acid
mesityl oxide	2-methylpent-2-one-4-one
methyl alcohol	methanol
methyl methacrylate	methyl 2-methylpropenoate
methyl phenyl ether *	methoxybenzene
N-ethylacetamide	N-ethylethanamide
oxalic acid	ethanedioic acid
oxamide	ethanediamide
perbenzoic acid	benzeneperoxycarboxylic acid
phenetole	ethoxybenzene
phenol	hydroxybenzene
phthalic acid	benzene-1,2-dicarboxylic acid
phthalimide	benzene-1,2-dicarboximide
pinacol	2,3-dimethybutan-2,3-diol

Common or trivial names	*Systematic names (I.U.P.A.C.)*
pinacolone	2,2-dimethylbutan-3-one or 2,2-di-methy 3-oxo-butane
pyruvic acid	2-oxo propanoic acid
p-quinone	1,4-diketocyclohex-2,5-diene
semicarbazide	aminourea
stearic acid	octadecanoic acid
stilbene	1,2-diphenylethane
styrene	phenylethene
terephthalic acid	benzene-1,4-dicarboxylic acid
tertiary-butanol (tertiary-butyl alcohol)	2-methylpropan-2-ol
tertiary-butyl bromide	2-bromo-2-methylpropane
toluene	methylbenzene
vinyl alcohol (enol form of acet-aldehyde)	ethenol
vinyl chloride (chloroethylene)	chloroethene

Index

Chloramphenicol, 279
Chloranil, 248
Chloride ions, 11, 13
Chlorination, 33, 39
Chlorine, 10
  atoms, 10
  valence bond, 11, 14
Chloroform, 4
Cinnamic acid, 243, 303
Coal, source of aromatic compounds, 181
Compounds
  aromatic, 181
  diazo, 212, 216
  heterocyclic, 263, 272
  physiologically active, 274
  ring, 263
Condensation reactions, 81
Condensation polymerization, 82
  285
Coniine, 270
Conjugated systems, 173
Co-ordinate covalence, 15
Copolymerization, 285
Coupling, 221
Covalence, 14
Covalent bond, 12
Cracking, 39
Crystallization, 301
  fractional, 303
Cycloparaffins, 152

Dative covalence, 15
Deamination, 221
Dehydrohalogenation, 41, 42
Delocalization, 171
Detergents, 98
Diazo compounds, 212, 216
Diazotization, 218, 225
Dibasic acids, 99, 126, 256, 258
Dibenzalacetone, 243
Dichlorobenzene, 247
Diels–Alder reaction, 248
Diethanolamine, 64
Diethyl ether, 69
Digonal orbitals, 169
Dimethyl ether, 26
Diols, 64
Dioxane, 65
Dipole, 18
Disaccharides, 146, 151
Distillation, 304
  steam, 306
Drug action, mechanism of, 279

Drugs, 274, 283
  antibiotics, 277
    chloramphenicol, 279
    isoniazide, 278
    penicillin, 277
    streptomycin, 278
    tetracyclines, 278
  antihistamine, 280
    benadryl, 280
  antimalarials, 275
    atabrine, 275
    chloroquine, 275
paludrine, 275
    pentaquine, 275
    quinine, 275
  arsenicals, 274
  arsphenamine, 274
  atoxyl, 274
  sulphonamides, 276
    prontosil, 276
    sulphanilamide, 276
Dyes, 212, 223, 226
Dypnone, 246

Electromeric effect, 194
Electrons, 8
Electrophilic reagents, 170
Electrovalence, 13
Electrovalent bond, 12, 13
Elements, 7
  definitions of, 7
  periodic table of, 16
  structure of, 8
    beryllium, 9
    carbon, 9
    chlorine, 9
    helium, 9
    hydrogen, 9
    lithium, 9
    oxygen, 9
    sodium, 9
  valence bonds and, 10
Empirical formulae, determination of, 21
Enantiomorphs, 112
Enzymes, 59, 280
Ephedrine, 280
Esterification of alcohol, 63
Esters, 95, 228
Ethane, 31
Ethanol, 59
Ethanolamine, 64
Ethene, 40
Ethers, 4, 68, 72, 228, 301
Ethyl acetoacetate, 114